'How can we protect the extraordinary diversity of life on Earth if we cannot protect environmental defenders? This powerful, timely and insightful book tackles two converging global crises; humanity's war on nature and the systemic violence against the courageous communities and individuals struggling to defend life, nature and culture. Enriched by a diversity of voices from across the planet, *Environmental Defenders* identifies the root causes of exploitation and prescribes pathways towards peace, reconciliation and a sustainable future.'

– **David R. Boyd**, *UN Special Rapporteur on Human Rights and Environment, Associate Professor of Law, Policy, and Sustainability, School of Public Policy and Global Affairs, The University of British Columbia*

'A very important collection of narratives, essential to understand what motivates environmental defenders to become locked in a constant fight for communal territories when they also must struggle to keep themselves alive! To understand this, we need to immerse ourselves in this book, which brings touching testimonials of people who experience firsthand this long history of resistance.'

– **Angela Mendes**, *Coordinator of the Chico Mendes Committee and daughter of Chico Mendes, who was assassinated in 1988 for his struggle to defend the environment and people of the Amazon*

'*Environmental Defenders: Deadly Struggles for Life and Territory* is essential reading to understand the frontlines of the war for land, shaping the future of humanity and the planet. It will be a crucial handbook for organisations like mine to help us learn about the experiences of environmental defenders in places we have never yet been.'

– **Patrick Alley**, *co-founder of Global Witness and author of* Shadow Network: Chasing Down the Thieves and Crooks Who Secretly Run Our World *(2022)*

'This is an important collection of carefully selected research on nature and dynamics of violence over conservation and development projects in different parts of the world. Highly recommended for all scholars and students of the social movements in general and environmental protests in particular.'

– **Ashok Swain**, *Professor of Peace and Conflict Research, Uppsala University, Sweden*

Environmental Defenders

This book is about environmental defenders and the violence they face while seeking to protect their land and the environment.

Between 2002 and 2019, at least two thousand people were killed in 57 countries for defending their lands and the environment. Recent policy initiatives and media coverage have provided much needed attention to the protection and support of defenders, but there has so far been little scholarly work. This edited volume explains who these defenders are, what threats they face, and what can be done to help support and protect them. Delving deep into the complex relations between and within communities, corporations, and government authorities, the book highlights the diversity of defenders, the collective character of their struggles, and the many drivers and forms of violence they are facing, as well as the importance of emotions and gendered dimensions in protests and repression. Drawing on global case studies, it examines the violence taking place around different types of development projects, including fossil fuels, agro-industrial, renewable energy, and infrastructure. The volume also examines the violence surrounding conservation projects, including through militarized wildlife protection and surveillance technologies. The book concludes with a reflection on the perspectives of defenders about the best ways to support and protect them. It contrasts these with the lagging efforts of an international community often promoting economic growth over the lives of defenders.

This volume is essential reading for all interested in understanding the challenges faced by environmental defenders and how to help and support them. It will also appeal to students, scholars, and practitioners involved in environmental protection, environmental activism, human rights, social movements, and development studies.

Mary Menton is Research Fellow in Environmental Justice with the Sussex Sustainability Research Programme at the University of Sussex, UK. She is a part of the core team of Not1More, a collective that works to support at-risk environmental defenders.

Philippe Le Billon is Professor in the Department of Geography and the School of Public Policy and Global Affairs at the University of British Columbia, Canada. He is the author of multiple publications, including *Fuelling War: Natural Resources and Armed Conflicts* (Routledge, 2013), and collaborates with human rights and environmental investigation organizations.

Routledge Explorations in Environmental Studies

An Environmental History of Australian Rainforests Until 1939
Fire, Rain, Settlers and Conservation
Warwick Frost

Daoism and Environmental Philosophy
Nourishing Life
Eric S. Nelson

Ecological Law and the Planetary Crisis
A Legal Guide for Harmony on Earth
Geoffrey Garver

From Environmental to Ecological Law
Kirsten Anker, Peter D Burdon, Geoffrey Garver, Michelle Maloney and Carla Sbert

Climate Change Temporalities
Explorations in Vernacular, Popular, and Scientific Discourse
Edited by Kyrre Kverndokk, Marit Ruge Bjærke, and Anne Eriksen

Rights of Nature
A Re-examination
Edited by Daniel P. Corrigan and Markku Oksanen

Ecocritical Geopolitics
Popular Culture and Environmental Discourse
Elena dell'Agnese

Environmental Defenders
Deadly Struggles for Life and Territory
Edited by Mary Menton and Philippe Le Billon

For more information about this series, please visit: www.routledge.com/Routledge-Explorations-in-Environmental-Studies/book-series/REES

Environmental Defenders

Deadly Struggles for Life and Territory

Edited by Mary Menton and Philippe Le Billon

Routledge
Taylor & Francis Group
LONDON AND NEW YORK

First published 2021
by Routledge
2 Park Square, Milton Park, Abingdon, Oxon OX14 4RN

and by Routledge
605 Third Avenue, New York, NY 10158

Routledge is an imprint of the Taylor & Francis Group, an informa business

British Library Cataloguing-in-Publication Data
A catalogue record for this book is available from the British Library

Library of Congress Cataloging-in-Publication Data
Names: Menton, Mary, editor. | Le Billon, Philippe, editor.
Title: Environmental defenders : deadly struggles for life and territory / edited by Mary Menton and Philippe Le Billon.
Description: Abingdon, Oxon ; New York, NY : Routledge, 2021. | Series: Routledge explorations in environmental studies | Includes bibliographical references.
Identifiers: LCCN 2020057570 (print) | LCCN 2020057571 (ebook) | ISBN 9780367649647 (paperback) | ISBN 9780367649708 (hardback) | ISBN 9781003127222 (ebook)
Subjects: LCSH: Environmentalism—Social aspects—Case studies. | Environmentalists—Violence against—Case studies. | Environmental degradation—Prevention—Case studies. | Environmentalists—Biography.
Classification: LCC GE195 .E5827 2021 (print) | LCC GE195 (ebook) | DDC 304.2/8—dc23
LC record available at https://lccn.loc.gov/2020057570
LC ebook record available at https://lccn.loc.gov/2020057571

ISBN: 978-0-367-64970-8 (hbk)
ISBN: 978-0-367-64964-7 (pbk)
ISBN: 978-1-003-12722-2 (ebk)

Typeset in Bembo
by Apex CoVantage, LLC

Contents

Contributors

Raquel da Silva Alves is a member of the Jenipapo-Kanindé Indigenous people and a student at UFRB-BR.

Daniela Alves de Araújo is a member of the Jenipapo-Kanindé Indigenous people and a student at UFRB-BR.

Peter Bille Larsen is Senior Lecturer and researcher with GEDT at the University of Geneva. His work addresses the intersection between environmental conservation and social equity concerns, including work in the fields of transboundary conservation, environmental governance, international human rights standards, and sustainable development.

Martín Correa Arce is an independent scholar with long-term participation in societies-in-movements and practices of *Comunalidad.* Martín has published in *Sociedad y Ambiente.*

Alexander Dunlap is a postdoctoral research fellow at the Centre for Development and the Environment, University of Oslo. His work has critically examined police–military transformations, market-based conservation, wind energy development, and extractive projects more generally in both Latin America and Europe. He has published the books *Renewing Destruction: Wind Energy Development, Conflict & Resistance in a Latin American Context* (Rowman & Littlefield, 2019) and *The Violent Technologies of Extraction: Political Ecology, Critical Agrarian Studies and the Capitalist Worldeater* (Palgrave, 2020).

Paul R. Gilbert is Senior Lecturer at the Department of International Development, University of Sussex, UK.

Chitra Karunakaran Prasanna is Assistant Professor at the Department of Social Work, School of Social Sciences & Humanities, Central University of Tamil Nadu, Kangalancherry, Tamil Nadu, India.

Mohammad Tanzimuddin Khan is Professor at the Department of International Relations, University of Dhaka, Bangladesh.

Yves Lador is the permanent representative of Earthjustice to the United Nations in Geneva, focusing on the links between human rights and the environment.

Fran Lambrick is Co-Founder and Director of Not1More (N1M) and the producer and director of the documentary *I am Chut Wutty*.

Philippe Le Billon is Professor at the Department of Geography and the School of Public Policy and Global Affairs at the University of British Columbia, Vancouver, Canada.

Päivi Lujala is Professor and Academy of Finland Research Fellow at the Geography Research Unit, University of Oulu, Finland, with a background in both human geography and economics, as well as Senior Researcher at the Chr. Michelsen Institute (CMI), in Bergen, Norway.

San Mala is a young environmental activist who works with Not1More in Cambodia. In 2015, he was imprisoned for his activism, fighting a sand mine.

Ansumana Manneh is a core member of Our Resources, a grassroots organization that works to protect the environment and community rights to natural resources in Guinea Bissau and West Africa in general.

Mary Menton is Research Fellow in Environmental Justice, Sussex Sustainability Research Programme, University of Sussex.

Felipe Milanez is a journalist and Professor at the Institute of Humanities, Arts and Sciences and the multidisciplinary Culture and Society graduate program at the Federal University of Bahia. He is a columnist for *CartaCapital* and *Mídia Ninja*. He was editor of *National Geographic Brazil*.

Sarah Milne is Senior Lecturer at the College of Asia and the Pacific, Australian National University. She studies natural resource struggles and environmental intervention, particularly when it comes to community-based conservation; resource rights initiatives; and market mechanisms for conservation. Most of her research is focused on Cambodia, where she has been active as a conservationist, ethnographer, and advocate since 2002.

Melissa Moreano Venegas is Assistant Professor at the Department of Environment and Sustainability of the Simón Bolívar Andean University in Quito, and an environmental activist, working with diverse organizations since 2003. Member of the Collective of Critical Geography of Ecuador (Colectivo de Geografía Crítica del Ecuador) and of the CLACSO's working group on political ecologies of Abya-Yala.

Grettel Navas is a doctoral student and researcher at the Institut de Ciència i Tecnologia Ambientals (ICTA) Universitat Autònoma de Barcelona, Spain.

Yannick Ndoinyo was born and raised a Maasai. He has lived with his community all his life, set up the Loliondo community radio, and managed a local livelihoods support organization (RAMAT). He is Founder and Executive Director of Traditional Ecosystems Survival Tanzania (TEST), the only Indigenous conservation organization in Tanzania.

Rob Nixon is the Currie C. and Thomas A. Barron Family Professor in the Humanities and the Environment. He is affiliated with the Princeton Environmental Institute's initiative in the environmental humanities. Before joining Princeton in 2015, Nixon held the Rachel Carson Professorship in English at the University of Wisconsin-Madison, where he was active in the Center for Culture, History and Environment.

Jeanne Perrier is a research engineer at CNRS and did her doctorate at Paul Valéry University, Montpellier 3, France.

Louisa Prause is a post-doctoral researcher at Humboldt-Universität zu Berlin and part of the research group Biomaterialities, Albrecht Daniel Thaer-Institute of Agricultural and Horticultural Sciences, Humboldt-Universität zu Berlin, Berlin, Germany.

Antônia Silva Santos is a member of the Kanindé Indigenous people and a student at UFRB-BR.

Claudelice Santos is a Brazilian environmentalist and human rights defender. She became an activist following the murder of her brother and sister-in-law, who were killed for their efforts to combat illegal logging and deforestation in the Brazilian Amazon. She is the founder and director of the Zé Claudio and Maria Institute and studies law at UNIFESSPA (Federal University of South and Southeastern Pará).

Cora Shaw is a master's student in human geography at the University of British Colombia. Her research explores decolonization of conservation policy and practice with specific regard to Indigenous and Community Conserved Areas.

Jurema Machado de A. Souza is an anthropologist and Professor at the Center for Arts, Humanities and Letters, Federal University of Recôncavo da Bahia (UFRB-BR).

Justine Taylor is a campaigner with Not1More (N1M) and previously worked with Quaker United Nations Office as programme assistant on Human Impacts of Climate Change.

Karolien van Teijlingen is a human geographer with a PhD in social sciences. She currently works as a postdoc at the Radboud University Nijmegen (The Netherlands) and as an associate researcher at the Universidad Andina Simón Bolivar (Ecuador). Karolien is also part of the Critical Geography Collective of Ecuador (Colectivo de Geografía Crítica del Ecuador).

Judith Verweijen is Lecturer at the Department of Politics and International Relations of the University of Sheffield. She works on the micro-dynamics of militarization, including of conflicts around natural resources, in zones of protracted violent conflict, with a focus on the eastern Democratic Republic of the Congo.

Figures and tables

Figures

Tables

Preface

In defense of life in the broadest sense and against terror

During the III Latin American Congress of Political Ecology in 2019, different people – intellectuals, activists, community members, Indigenous people, *quilombolas*, and supporters engaged in denouncing violence against grassroots environmentalists – gathered in Salvador, on the Ondina campus of the Federal University of Bahia, in a First Encounter in Defense of Life in the Broadest Sense: joining hands, connections, and resistance against violence.

Brazil is going through a time of growing violence and fear. This atmosphere of terror is bolstered by a pedagogy of cruelty, materialized through violent attacks by gunmen, torture, and massacres. This physical violence is accompanied by symbolic violence, i.e. public discourse by parliamentarians, judges, members of the Bolsonaro government, or their allies in different spheres of public administration. For example, the Special Secretary for Land Affairs of the Ministry of Agriculture, the ruralist Luiz Antônio Nabhan Garcia: despite occupying a public office that must be managed by the principles of administrative law, he declared that he will neither receive nor negotiate with the Landless Workers Movement (MST) or its militants.

Bolsonaro's threats against activists were already publicly known, in a campaign based on hatred. But the politics of death in the current government and the preaching of fascist ideology in power must immediately be fought by society.

The present moment is characterized by setbacks through reversal of rights in government and Congress, by the promotion of hate speech, while gunmen and armed groups are active on the ground, carrying out assassinations, torture, and intimidation, setting fire to houses, and spreading panic and terror.

Federal agents and representatives, as well as some state agents that follow the same ideologies of intolerance, have been working with militias, paramilitaries, and state forces to promote violence. Acts of criminalization, invisibilization, and silencing, as well as threats, massacres, and assassinations, have increased exponentially along with the impunity and public justifications of these brutal crimes against humanity. This sphere of forces outside and inside the State structures collusion for control over life itself and needs to be urgently denounced to prevent further irreparable crimes against humanity. In this meeting, advocates, people of the land, of the forests, of the water, and academics and researchers, Brazilian and international, indicate ethical and humanitarian principles that can contribute to the defense of life in the broadest sense:

Principles in defense of life, collective and ecological

- **Struggles are collective and class-based:** those people that are publicly called "defenders" of the land and of the environment represent in reality the collective culture of entire communities that are fighting for life, for the defense of territories, and against the alienation of work. This fight goes beyond a theoretical choice to defend, to be a "defender". It is not a cause; instead, it is the fight to defend the web of life that is at the heart of the struggle.
- **The rights of nature:** the life of rivers, oceans, forests, air, soil, animals, and every "living being" are indisputable!
- **Racism structures violence:** combating violence and overcoming racism, whether against Indigenous peoples or against black people, is a necessary step towards building a new society. Removing the structural racism of invisibility and securing the conditions of historical reparation are the only means to establish justice and inclusive participation.
- **Patriarchy structures violence:** overcoming patriarchal logic and structures, eradicating gender-based violence, and ensuring women's broad participation and protagonism in their diverse forms of organization are fundamental aspects of collective struggles for a new society across territories.
- **Respect the protagonism of the people:** the words that emerge from these stories and the trajectories of these territories must go far and echo. But the history of each people and each culture, in these territories under attack, must be the basis of the fight.
- **Strengthening the base:** the institutional strengthening of grassroots organizations is fundamental for the survival and defense of life, as violence takes place in the territories, and it is therein that collective physical, cultural, and spiritual integrity must be guaranteed.
- **Media ethics:** to save lives in this dark moment and in this environment of terror, it takes a humanist commitment by public opinion-makers. The violence that contaminates the territories is also legitimized by media loudspeakers. Therefore, saving lives requires us to combat sensationalism, romanticism, and depoliticization. Above all, defend the most vulnerable in conflicts, exercise control over power, and combat injustice.
- **Ethics of alliances and consent:** dialogue with the communities involved should always precede any action or campaign that exposes its members to risk and vulnerability, regardless of whether the entity or the person sees themselves as an ally, respecting internal forms of decision-making and consultation. We have been witnessing growing campaigns

and sensationalist defamatory news against grassroots leaderships, which need to be responded to in the public opinion. Care is an urgent principle and must be followed, accompanied by permanent dialogue.
- **The right to be listened to:** faced with parastatal terror, it is communities that must have the power to decide who can enter their territories. It should be up to Indigenous peoples to decide which journalists and researchers to receive on their lands, in order to prevent the State from using its prerogative of control for the purpose of censorship and tutelage, as well as to grant that the right of consultation and communication be respected. The same goes for traditional communities.
- **Respect for territories:** unconditional respect for the intangibility and integrity of territories, whether urban, rural, forests, waters, or fields.

Letter approved at the General Assembly of the III Latin American Congress of Political Ecology, Salvador, March 20, 2019

The following are members or representatives of the following organizations and movements: Articulação dos Povos Indígenas do Brasil (APIB); Conselho Nacional das Populações Extrativistas (CNS); Comissão Pastoral da Terra (CPT); Conselho Pastoral dos Pescadores (CPP); Movimento Pela Soberania Popular na Mineração (MAM); Movimento de Pescadores e Pescadoras Artesanais (MPP); Acción Ecológica; Mídia Ninja; Escola de Ativismo; Global Witness; Not1More; Centro Terra Viva, Moçambique; Teia dos Povos, Maranhão; Associação Indígena Pariri; Lideranças de reservas extrativistas e lideranças indígenas; Associação de Advogados de Trabalhadores Rurais da Bahia (AATR); Sociedade Paraense de Defesa dos Direitos Humanos (SDDH); Movimento Xingu Vivo Para Sempre; Coordenadoria Ecumênica de Serviço (CESE); Fundação Zé Cláudio e Maria; Comitê Chico Mendes; Colectivo de Geografia Crítica; Pesquisadoras, pesquisadores, professoras e professores das seguintes universidades; Universidade Federal da Bahia; Universidade Federal do Recôncavo da Bahia; Universidade Federal do Rio de Janeiro; Universidade Federal do Ceará; Universidade do Estado do Pará; Universidade Federal do Pará; Universidade Federal do Oeste do Pará; Universidade Federal de São Paulo; Universidade Nacional da Colômbia; Universidade de La Plata; Universidade de British Columbia; Universidade de Oxford; Universidade de Sussex; Royal Institute of Technology in Stockholm (KTH); Universidade de Coimbra; Grupo de Trabalho Ecologia Política desde América Latina Abya Yala, do Conselho Latino-Americano de Ciências Sociais (CLACSO).

1 Introduction

Philippe Le Billon and Mary Menton

In March 2016, thousands of people took to the streets of Honduras' capital city to protest the killing of Berta Cáceres, a prominent environmentalist and Indigenous social movement leader opposing a hydroelectric dam project. Cáceres' international profile as a Goldman prize recipient had not deterred her killers, militaries and hired gunmen working on behalf of the dam company (Lakhani 2020). Between 2002 and 2019, close to two thousand people were killed in 57 countries for defending their lands and the environment (Global Witness 2020). Many of these 'environmental and land defenders' were, like Berta Cáceres, Indigenous and local people opposing large-scale resource projects, but the term is applied to a broad range of people defending their lands and environments and those seeking to protect defenders or support their cause, such as lawyers, journalists, and staff from environmental or human rights organizations. Beyond the reported number of defenders killed, countless others were stigmatized, criminalized, and violently repressed by resource-based companies and government authorities (Forst 2014; Rasch 2017; Navas et al. 2018; Le Billon & Lujala 2020).

This collective volume is about the violence faced by people seeking to protect their land and the environment from mostly large-scale, land-based projects. The volume was in large part born out of the intensity of exchanges and experiences that occurred during the Latin American Congress of Political Ecology in 2019 in Salvador, Brazil. This was no classic academic conference. Scholars were the minority and the Congress bristled with activities, harrowing testimonies from defenders, and public calls for mobilization (see Preamble, this volume). The event ended with much of the audience on stage singing a liberation hymn. Having enrolled more colleagues and defenders into this volume, including through a high-level event held in Geneva with defenders and UN Special Rapporteurs David R. Boyd and Michel Forst (see Bille Larsen & Lador this volume; Bille Larsen et al. 2020), we reconvened in September 2020 through sessions of the 2020 Political Ecology Network (POLLEN) conference. Bringing together academics and defenders, our collective explores the drivers and wider contexts of these murders, but also the different forms of violence experienced by environmental defenders, by those who fight for land rights, and other groups who fight against the powerful actors who perpetuate violence against them. We thank the Social Sciences and Humanities Research Council (Canada) and the British Academy (UK) for funding.

The book has three objectives. The first is to understand who these 'defenders' are, the various forms of violence that they face, and the reasons why such violence takes place. The second is to examine these processes within different sectors (e.g. coal mining, renewable energy projects, conservation projects) to understand their specificities and commonalities. The third objective is to understand how defenders seek to respond to this violence, and what can be done both locally and internationally to protect and support them. We thus engage with the violence surrounding both 'development' and 'conservation' projects, as well as with the range of responses taken by defenders, and their allies, to this violence.

The persecution of environmental and land defenders not only constitutes abuses against traditional human rights, such as rights to life, peaceful assembly, or freedom from arbitrary arrest, but also against more specific rights, such as the right to a healthy environment and Indigenous rights to free, prior, and informed consent (Knox 2017). Persecution also represents an attack against efforts to establish or protect the rights of nature (Boyd 2017), as well as against Indigenous or agrarian community struggles for more sustainable forms of livelihoods and traditional forms of environmental conservation (Martinez-Alier et al. 2010; Porto-Gonçalves 2016). In many cases, defenders resist environmentally destructive projects that would drastically undermine biodiversity and ecosystem services, thereby potentially contributing to slowing down the rate of environmental degradation, often at multiple scales, as local extractive projects resonate through broader scales, including water pollution, waste production, and greenhouse gas emissions (Gleason & Mitchell 2009; Pigrau & Borràs 2015; Temper et al. 2015; Glazebrook & Opoku 2018).

Not only do defenders seek to stop environmentally harmful projects by 'putting their bodies on the line', but their message and mobilization can also contribute to environmentally progressive shifts in policy and public opinion (Agnone 2007; Piggot 2018; Lambrick 2019). As such, the UN Environment Program (2018) defines *environmental* defenders – or environmental human rights defenders (EHRD) – as 'anyone . . . who is defending environmental rights, including constitutional rights to a clean and healthy environment, when the exercise of those rights is being threatened' (see also Verweijen et al., this volume; Knox 2017). Among the most highly reported killings, those of Brazilian rubber-tapper Chico Mendes in Brazil in 1988 and the execution of Ogoni activist Ken Saro-Wiwa in Nigeria in 1995 contributed to the internationalization of environmental movements, as well as to the defence of agrarian and Indigenous community rights (Martinez-Alier et al. 2016). The persecution of environmental defenders is thus not only a failure of rights duty borne by governments and corporations, but also an important and grievous dimension of environmental politics.

Several UN Special Rapporteurs have been vocal about the importance of protecting environmental human rights defenders (Forst 2014; Knox 2017). In

2018, the UN Environment agency launched an environmental governance policy dedicated to the support and protection of environmental defenders (UNEP 2018), and in 2019 the UN General Assembly recognized 'the contribution of environmental human rights defenders to the enjoyment of human rights, environmental protection and sustainable development', condemned the violence against defenders, and called upon states and business enterprises to respect their rights (UNGA 2019; see Annex this volume; Khanna & Le Billon 2019). Such high-level initiatives reflect a growing concern about the rise of anti-environmental movements and their backing by not only many conservative governments but also left-wing 'neo-extractivist' regimes (Rowell 2017; Tilzey 2019). Putting them into practice, however, is often highly challenging, especially when the 'rule of law' is either pitted against or easily instrumentalized to criminalize defenders (see Gilbert & Khan, this volume; Middeldorp & Le Billon 2019). Many NGOs and foreign donors also appear unwilling to address, and incapable of addressing, the roots of violence against defenders, being themselves caught within pragmatic logics of cooperation with political regimes deepening power inequalities (Grant & Le Billon 2020).

While finalizing the chapters for this book, stories from the frontlines continued to arrive via WhatsApp. Colleagues in Cambodia were jailed for speaking out against the government's environmentally destructive development policies; others have gone into hiding. A defender in Brazil just discovered he is on a hit-list, a $20,000 price on his head. Another was driven off the road by a pick-up truck; the driver yelled. 'You need to die, wretched bitch'. A woman from the Philippines told of rape threats via social media. As these stories and the Letter from Salvador (see Preamble) highlight, this book comes at a time of crisis for many environmental and land defenders. It draws on quantitative studies, social theory, and political analyses, but also, perhaps more importantly, on the lived experiences of defenders, on their stories and their struggles. It is at the intersection of these many different forms of violence that the defenders often face 'atmospheres of violence' – multi-dimensional, temporally and spatially dispersed, a mixture of direct attacks and continual fear from threats against themselves, their families, their communities, and their territories.

This volume has three main parts. The first – *On Defenders* – is about defenders, their perspectives, and the concept of the term itself, as well as the various forms, determinants, and gendered dimensions of violence faced by defenders. The second is about struggles against so-called *Dirty Projects* that are often more readily associated with resistance, such as mining, large agro-industrial, or road projects. The third is about struggles against supposedly more benign projects, or so-called *Green Projects*, such as renewable energy projects and biodiversity conservation areas. We conclude the volume with not only a reflection but also a call for action.

On defenders

Following this Introduction, three defenders provide first-hand accounts and perspectives about what can be done to protect their voices, lives, and struggles.

Drawing from their own experiences as defenders, social leaders, and global advocates, **Claudelice Santos** from Brazil, **San Mala** from Cambodia, and **Yannick Ndoinyo** from Tanzania share their experiences of violence. **Santos** shares the story of what led to the murder of her brother and sister-in-law, environmental defenders who fought to protect their land and forests. **San** speaks to his experience being jailed for his activism against a sand mine. **Ndoinyo** tells of experiences of violence in the name of conservation, of forced displacement and criminalization of Maasai people.

One of the issues faced by defenders is the way they are portrayed by governments and project proponents, including as anti-development radicals, foreign-funded activists, land squatters, or poachers. Often racialized when involving Indigenous peoples and historically marginalized populations, these representations form the basis of symbolic violence that characterizes processes of exclusion and criminalization, as well as of direct forms of violence. While terms such as 'defenders' or 'protectors' seek to counter this symbolic violence, they are not neutral terms, either. **Verweijen**, **Lambrick**, **Le Billon**, **Milanez**, **Manneh**, and **Moreano Venegas** explore the advantages and drawbacks of the term 'environmental defenders', looking at its analytical merits and deployment as a mobilizing tool, the extent to which such a term corresponds to the ways people involved in land and environmental struggles see and identify themselves, and whether the notion of 'defenders' individualizes these struggles and obscures long-standing histories of collective resistance.

As just mentioned, violence does not only take direct physical forms. **Menton**, **Navas**, and **Le Billon** conceptualize the various processes and forms of violence involved in these struggles and the ways they feed into each other to create 'atmospheres of violence'. This advances a multidimensional understanding of violence and, in particular, the cumulative impacts of multiple, intersecting violences. Given the growing importance of social media in creating virtual spaces in which defenders experience violence (through threats, smear campaigns, criminalization), and the spatially and temporally dispersed nature of their lived experiences, atmospheres of violence speak also to the creation of climates of fear and oppression, to violences against the bodies, minds, and territories of defenders.

Building on available databases, **Le Billon** and **Lujala** identify patterns of repression and potential determinants of killings. Globally, about a third of socio-environmental conflicts involve mass mobilization, arrests, and direct forms of violence. These 'high intensity' conflicts are more frequent in Asia and Latin America. Most of the killings of defenders over the past two decades have occurred in Brazil, the Philippines, Colombia, Honduras, Mexico, and Peru. Killings are more likely in countries with high levels of foreign direct investment, a dependence on mineral extraction, large Indigenous populations, or frequent protests. There are also more killings in countries that are neither strong democracies nor autocracies. More systematic reporting and analysis of repression – including at sub-national level – can help protect and support defenders, notably through conflict-sensitive investment policies and greater accountability for abuses.

Violence is also highly gendered. In their insightful chapter on the gendered dimensions of repression against defenders, **Moreano Venegas** and **van Teijlingen** reflect upon two features of the violence against land and environmental defenders, anti-extraction activists, and communities that oppose extractive activities in Ecuador. The first aspect is the gendered character of this violence and the second is the perils of individualization of the struggles in relation to this violence, and the benefits of its collectivization. Using a critical feminist geography perspective and interviews from long-term fieldwork in the Amazon region with communities affected by extractive activities, they argue that violence is gendered in multiple ways, including through the frequently patriarchal character of leadership in defender organizations, the feminization of the territory that notably results from the criminalization, and the subsequent flight of male defenders from community areas under threats, as well as the emancipatory political action of the women collectives, such as the *Mujeres Amazónicas*.

We finish this first part of the book with a chapter on Indigenous counter-conquest struggles. The struggles of defenders are often intimately linked to the defence of their 'territory', often indissociably connecting people, land, livelihoods, cultures, ways of life, and aspirations. In this chapter, **Milanez** reflects on epistemological contestation of colonial society by shamans and Indigenous political leaders. Indigenous intellectuals, drawing on their life experiences, have helped build creative counter-conquest movements in Brazil and, as put by Ailton Krenak, propose 'Ideas to Postpone the End of the World'.

'Dirty' projects

Opening this second part, **Nixon**'s chapter considers a variety of circumstances across the global South where external development agendas result in the impoverishment of affected communities. Through a mixture of intimidation, coercion, and deception, developers often, under the guise of lifting communities out of poverty, instead plunge them into destitution. The resulting immiseration easily eludes official metrics, as the affected community loses any semblance of subsistence security. Official data also often miss the contributions of agroforestry to a community's nutritional and spiritual wellbeing. As forest commonage is privatized and converted into plantations and ranches, communities frequently suffer from development-induced poverty. For many Indigenous communities, the measure of community health and wealth is inseparable from the long durée of ancestral time and the generations that are yet to be. In this regard, the extractivist timeframe that is prevalent among developers is frequently incompatible with a community's internal understandings of relative poverty and relative wealth.

Mega-projects in the Amazon have rightly focused a lot of attention on the struggles of Indigenous defenders. In their chapter, **Souza**, **Menton**, **Santos**, **Araújo**, and **Alves** look at the long and unresolved processes of demarcation and regularization of Indigenous lands in the northeast of Brazil that are the target of large-scale speculation by agribusiness and private commercial exploitation. They focus on forms of violence and resistance in three emblematic cases that point to intense and distinct forms of exploitation, contradicting Indigenous rights, harming the autonomy of Indigenous peoples, and impeding their free self-determination. Yet, they also see intense processes of resistance and mobilization aimed at reversing these violations and ending exploitation of their lands.

Bringing development through coal mining and power plants in one of the country's most at risk of climate change may seem like a short-sighted idea, yet according to its current energy master plan, Bangladesh's reliance on natural gas is to be reduced while coal-fired power generation is set to expand drastically. Previous efforts to expand domestic coal extraction have, however, been met with significant opposition. **Gilbert** and **Khan** delve into the case of the Phulbari open-pit coal mine, where six protestors were killed by security forces in 2006. Stalled since then, the project is being restarted through a joint venture agreement with PowerChina to develop four 1000MW mine-mouth power plants, with resettlement consequences for up to 220,000 people. Against this backdrop, they examine the role that land defenders have played in resisting mining and demanding compensation, and explore the domestic and international actors implicated in ongoing plans to expand open-pit coal mining and resettle residents of Phulbari.

Transportation infrastructure is often tied to 'dirty' projects feeding international trade, and **Karunakaran Prasanna** looks at struggles against forced eviction in the context of an eviction resulting from a road and port project in Kerala, India. Documenting the chronology of the struggle, she investigates some of the more subtle forms of violent manifestations of State power, including biased interpretations of legal frameworks for land acquisition, flawed public hearing processes, media misrepresentations, and bureaucratic apathy.

Investments in large-scale, land-based projects have increased over the past two decades, with a concomitant rise in resistance by community-based land defenders. Drawing from data on resistance movements, literature findings, and two case studies in Senegal, **Prause** and **Le Billon** compare the actions of land defenders in different types of land transformations for either agro-industrial or mining projects. They find that outcomes of resistances seem largely case-dependent and determined by political opportunities, while the motives, narratives, and more confrontational practices of land defenders differ across both sectors. They suggest that this can be explained through sector-specific material, discursive, and institutional factors which pose challenges for potential cross-sectoral alliances between different groups of environmental and land defenders.

Pointing out that violence against defenders is not only taking place in 'developing' or 'authoritarian' countries, **Taylor** shows that the level of harm done to peaceful protestors in the UK is increasingly at odds with the notion of democracy and supposed values sustaining its international reputation. A member of Not1More, an UK-based NGO seeking to end both violence against defenders and the impunity of perpetrators, Taylor and her colleagues recorded four hundred incidents against peaceful protestors at just one anti-fracking protest site in the UK. Her collection of testimonies not only shows the extent of the repression, including beatings and psychological intimidation, but also the toxic normalization of this violence, including among a general public that often thinks defenders deserved being harmed for standing in the way of economic growth.

'Green' projects

Large-scale infrastructural and energy projects such as dams are a major cause of socio-environmental conflicts. Intense hydro-power development has taken place since 2007 in Cambodia's Cardamom Mountains, with multiple Chinese-backed dams now operational while others remain proposed or suspended. In this chapter, **Milne** considers local experiences of dam development, especially for Indigenous communities whose lands and livelihoods have been directly affected. She compares the Stung Atay dam, which faced no overt local opposition, and the proposed Stung Areng dam, which became the subject of a remarkable anti-dam campaign linking transnational advocacy networks and urban Cambodian interests. She explores the factors that enabled and constrained resistance, the role state-backed violence played in these contrasting outcomes, and the consequences of overt resistance – including uncertainty and fear akin to slow violence – given shrinking political space in Cambodia.

Dunlap and **Correa Arce** examine the struggle against the new Électricité de France (EDF) wind park, Gunaa Sicarú, in Unión Hidalgo (UH), Mexico. Foregrounding Indigenous land defence, they refer to wind energy as 'wind factories' to discuss agrarian change in the region. Revealing the *counterinsurgency colonial model* as a foundational approach to extractive development, they argue that the distribution of money, *sicarios* (hitmen), and NGOs are instrumental to engineering 'social acceptance'. Moreover, the liberalism underlining NGOs, if not careful, advances processes of infrastructural colonization and, consequently, wider trajectories of (neo)colonialism.

Like renewable energy, wastewater treatment plants are often considered a 'green' form of development, improving sanitation and a more sustainable development of water resources. Yet, they can also form an object of infrastructural violence and coloniality. **Perrier** investigates the opposition of people from three villages to a treatment plant within the West Bank, from its beginning in the early 2000s to the present. The Palestinian Authority and international donors have invested massively to build such infrastructures, with sustainable development rationales and technical criteria supporting the inevitability of such projects with the aim of silencing any contestation. Many local villagers, however, perceived these land expropriations as similar to land confiscations by the Israeli army and as a form of infrastructural racism, leading to weekly demonstrations and legal complaints that have led in turn to various forms of violence by the Palestinian Authority.

Pointing at the increasing interactions between 'dirty' and 'green' projects, **Le Billon** discusses complicit relations between extractive violence and crisis conservation creating spaces of 'double exception' and a politics of enmity against local communities opposing various combinations of combining extraction and conservation, such as 'biodiversity offsets' for mining projects. As the deadly consequences of militarized conservation and resistance to extraction receive increased attention, the chapter points to the challenges posed to different types of defenders by exclusionary forms of 'sustainable' extraction and 'extractive' conservation brought about by the logics and praxis commonly adopted by authoritarian political formations, extractive corporations, and conservation organizations.

Menton and **Gilbert** address how some international environmental organizations (WWF, CI, TNC, and IUCN) who frame themselves as 'supporters' or 'protectors' of environmental and land defenders have been complicit in violence perpetrated by park guards and oil companies. They argue for the need for NGOs to work towards more radical decolonial forms of solidarity with environmental and land defenders.

Finally, **Shaw** provides a brief but deep dive into the struggles of Indigenous people seeking to establish and control conservation projects on their own terms. ICCAs provide a promising way of advancing conservation objectives while respecting the rights and will of local Indigenous populations. Yet, Indigenous conservation frequently faces various forms of opposition and repression, including from national authorities.

To conclude this volume, our collective returns to the growing number of initiatives by academic, civil society, and multilateral organizations seeking to protect environmental rights, enable civic spaces, and support defenders. This chapter by **Bille Larsen** and **Lador** sums up the findings and recommendations of a 'Geneva Roadmap' event organized to inform human rights and conservation policy processes. Bringing together leading organizations, the chapter takes stock of the situation of environmental defenders around the world, reviews existing support initiatives, debates possible response modalities, and advances more systematic collaboration in the long term. In response and as a follow-up to UN Human Rights Council Resolution 40/11, the Geneva Roadmap outlines policy recommendations to promote free and safe spaces for information and discussion on environmental matters (see Annex). The chapter is tied to an online portal – environment-rights.org – providing resources for environmental human rights defenders, which include initiatives and commitments by States, civil society, research and academia, as well as private actors to support the effective implementation of the right to act for the protection of the environment and environmental rights.

References

Agnone, J. (2007). Amplifying public opinion: The policy impact of the US environmental movement. *Social Forces*, *85*(4), 1593–1620.

Bille Larsen, P. B., Le Billon, P., Menton, M., Aylwin, J., Balsiger, J., Boyd, D., . . . Wilding, S. (2020). Understanding and responding to the environmental human rights defenders crisis: The case for conservation action. *Conservation Letters*.

Boyd, D. R. (2017). *The rights of nature: A legal revolution that could save the world*. Toronto: ECW Press.

Forst, M. (2014). *Report of the special Rapporteur on the situation of human rights defenders*. UN Human Rights Council. A/HRC/28/63.

Glazebrook, T., & Opoku, E. (2018). Defending the defenders: Environmental protectors, climate change and human rights. *Ethics and the Environment*, *23*(2), 83–109.

Gleason, J. M., & Mitchell, E. (2009). Will the confluence between human rights and the environment continue to flow? Threats to the rights of environmental defenders to collaborate and speak out. *Oregon Review of International Law*, *11*, 267.

Global Witness. (2020). *Defending tomorrow: The climate crisis and threats against land and environmental defenders*. London.

Grant, H., & Le Billon, P. (2020). Unrooted responses: Addressing violence against environmental and land defenders. *Environment and Planning C: Politics and Space*, 239965 4420941518.

Khanna, S., & Le Billon, P. (2019). *Protecting defenders: A review of policies for environmental and land defenders*. University of British Columbia. www.researchgate.net/publication/342671903_Protecting_and_Supporting_Defenders_A_Review_of_Policies_for_Environmental_and_Land_Defenders

Knox, J. H. (2017). *Environmental human rights defenders: A global crisis*. Policy Brief. Geneva: Universal Rights Group.

Lakhani, N. (2020). *Who killed Berta Caceres*. New York: Verso Books.

Lambrick, F. (2019). Environmental defenders: Courage, territory and power. In R. Gee & J. Pettit (Eds.), *Power, empowerment and social change*. Abingdon, UK: Routledge.

Le Billon, P., & Lujala, P. (2020). Environmental and land defenders: Global patterns and determinants of repression. *Global Environmental Change*, *65*, 102163.

Martinez-Alier, J., Kallis, G., Veuthey, S., Walter, M., & Temper, L. (2010). Social metabolism, ecological distribution conflicts, and valuation languages. *Ecological Economics*, *70*(2), 153–158.

Martinez-Alier, J., Temper, L., Del Bene, D., & Scheidel, A. (2016). Is there a global environmental justice movement? *The Journal of Peasant Studies*, *43*(3), 731–755.

Middeldorp, N., & Le Billon, P. (2019). Deadly environmental governance: Authoritarianism, eco-populism, and the repression of environmental and land defenders. *Annals of the American Association of Geographers*, *109*(2), 324–337.

Navas, G., Mingorria, S., & Aguilar-González, B. (2018). Violence in environmental conflicts: The need for a multidimensional approach. *Sustainability Science*, *13*(3), 649–660.

Piggot, G. (2018). The influence of social movements on policies that constrain fossil fuel supply. *Climate Policy*, *18*(7), 942–954.

Pigrau, A., & Borràs, S. (2015). 21 Environmental defenders: The green peaceful resistance. In L. Westra, J. Gray, & V. Karageorgou (Eds.), *Ecological systems integrity: Governance, law and human rights* (pp. 239–271). Abingdon, UK: Routledge.

Porto-Gonçalves, C. W. (2016). Lucha por la Tierra. Ruptura metabólica y reapropiación social de la naturaleza. *Polis, Revista Latinoamericana*, (45), 291–316.

Rasch, E. D. (2017). Citizens, criminalization and violence in natural resource conflicts in Latin America. *European Review of Latin American and Caribbean Studies*, *103*, 131–142.

Rowell, A. (2017). *Green Backlash: Global subversion of the environment movement*. Abingdon, UK: Routledge.

Temper, L., del Bene, D., & Martinez-Alier, J. (2015). Mapping the frontiers and front lines of global environmental justice: The EJAtlas. *Journal of Political Ecology*, *22*, 255–278.

Tilzey, M. (2019). Authoritarian populism and neo-extractivism in Bolivia and Ecuador: The unresolved agrarian question and the prospects for food sovereignty as counter-hegemony. *The Journal of Peasant Studies*, *46*(3), 626–652.

UN Environment Program. (2018). *Promoting greater protection for environmental defenders*. Policy. https://wedocs.unep.org/bitstream/handle/20.500.11822/22769/UN%20Environment%20Policy%20on%20Environmental%20Defenders_08.02.18Clean.pdf?sequence=1&isAllowed=y

UN General Assembly. (2019). *Recognizing the contribution of environmental human rights defenders to the enjoyment of human rights, environmental protection and sustainable development*. A/HRC/40/L.22/Rev.1.

[Part of this chapter first appeared as: Middeldorp, N., & Le Billon, P. (2019). Deadly environmental governance: Authoritarianism, eco-populism, and the repression of environmental and land defenders. *Annals of the American Association of Geographers*, *109*(2), 324–337.]

Part 1

On defenders

2 Conflicts in the Amazon

The assassination of Zé Claudio and Maria

Claudelice Santos

Brazil has been in the news recently, with Global Witness pointing out that the country had the third highest number of environmental and land defenders killed in the world in 2019. Within the country, the Amazonian state of Pará was singled out as one of the most dangerous regions for defenders. This reality, however, is nothing new for us in the Amazon. Scenes of brutal violence have played out for many years, increasingly in regions where wealth is abundant; vast expanses of fertile land attract the capitalist expansion of agrobusinesses; and the implementation of large public and private projects that mobilize seductive discourses on 'development' but act to the detriment of local peoples and communities.

The world recognizes two things about the Amazon region: its potential for global climate equilibrium and its potential for natural wealth from minerals, wood, water, land, and diverse other resources. Like the ocean, the Amazon was seen for a very long time as vast, plentiful, and inexhaustible. However, we have now reached a point where we cannot deny the historical levels of its exploitation and the damages that this has caused. The whole planet is now worried, but despite many decades in which this exploitation has caused pain, desperate suffering, and the eviction of entire peoples and communities, almost nothing has been done in practice by the global community to change the situation, which promotes the slow death of humanity. In this chapter, I will show how a sequence of actions and omissions by the state, with or without the additional power of private capital, determines who lives and who dies in the name of development – in the name of money.

For the peoples of the Amazon who live on and in the land, forests, and waters, land/territory is much more than just a physical space from which survival is taken. It is part of life itself: collective and individual trajectories, affectivity and spirituality. The land/territory is like a family member. But the conflicts and violence committed against the environment and those people who today are called defenders of human rights are central and constant elements, associated with the inefficiency of the State in its entire structure as well as attacks by private interests.

I will talk in this chapter about the region I live in, the south and southeast of Pará in the Brazilian Amazon. This region has a long history of human rights violations against squatters, union members, activists, and lawyers allied with the defense of agrarian reform and human rights. In return, the State has failed to respond with justice. The sequence of serious state errors – and, in fact, serious crimes – against the Amazon began in the colonial period. However, it was during the widespread violence between 1964 and 1985, under the military dictatorship and successive authoritarian governments, that "integrate it to not lose it" was known as the main policy for the Amazon. This narrative argued that the forest had to be colonized and was anchored in discourse of integrating the last green frontier into the nation's process of development. It was furthered with the slogan "Amazonia: a land without men, for men without land." The military was committed to describing the Amazon as "empty" and as a place of hope for "development," in addition to being seen as strategic for national interests – as if the Amazon had no peoples and communities that had been living there for centuries, even millennia.

One of the most violent periods of struggle for land/territory in southeastern Pará occurred during the 1980s, occurring under networks of land grabbers that still exist today who promote cartels of fear or implement large projects that steamroll peoples and communities – all amid total failure of the Brazilian State to promote justice.

According to a 2019 report entitled *Rainforest Mafias* (Human Rights Watch 2019), of 300 cases of defenders murdered in the Amazon, only 14 cases reached the courts (Gortázar 2019). This figure reveals a State that clearly intentionally aims to maintain the current state of tension, while also subliminally encouraging and legitimizing a widely denounced series of violations of rights. CPT (Pastoral Land Commission) has organized and published cases of conflicts in the countryside since 1985; however, we know that this story is much older and deeper, and more violent, than the numbers or timeframes reveal. In addition, the cases that are published are only those that, in one way or another, have reached the spheres of justice – there are countless cases that are never even investigated.

It is against this historical context of violence, in but one region of the Amazon, that I will mention three emblematic cases of massacres perpetrated against the working class, before turning to the violence that directly affected my own family.

In 1987, under the command of the then-governor of Pará, Hélio Gueiros, the military police cornered artisanal goldminers from the extinct "Garimpo da Serra Pelada" who claimed their rights, and opened fire.[1] It is important to note that the practice of panning for gold in the Serra Pelada was not illegal at that time; regardless, nothing justified the brutal violence to which the goldminers were subjected – not to mention that it was already part of the military-era plan for the area to later be mined by transnationals. The second case is the Eldorado dos Carajás massacre of 1996. Once again, the military police opened fire, summarily executing 20 landless rural workers and leaving many others with injuries they carry to this today. The police falsely claimed they had been attacked first. Most recently, in 2017, the Pau D'arco massacre occurred, with ten landless workers again harassed, tortured, and executed by the military, civilians, and DECA (police specialized in agrarian conflicts). These massacres are cases of direct violent action by the State through excessive use of force.

The aforementioned cases provide examples of how violent the southeastern region of Pará can be. I cannot fail to mention cases with tragic outcomes that result from the State's inefficiency when it does not fulfill its role, when it does not promote justice in investigating violent cases carried out by hired gunmen who act as a parallel law in the service of private power. Behind every act of violence against human rights defenders are organized interests with capital and political strength. Whoever crosses their path, whoever is "getting in the way" of these groups' goals, will be "excluded from society."[2]

The cases of murders and violence against defenders of human rights and the environment, those that have national and international repercussions, carry in their essence a history of violations and violence in the interests of public and private powers. Imagine what has been swept "under the carpet" of oblivion and neglect of the State, of the media that are not able (or, most of the time, do not want) to expose these cases of violence against defenders in an isonomic and fair way. I could make a gigantic list of cases of serious violations, against life and the dignity of people, mainly committed against workers and leaders of socio-environmental movements, but I will stick, instead, to the case of environmentalists Zé Claudio and Maria: an emblematic case to understand how these absurd series of events, which I will call serious errors, can be lethal to a defender of the forest, land, and human rights.

José ('Zé') Claudio Ribeiro da Silva and Maria do Espírito Santo, my brother and sister-in-law, were defenders of the forest and human rights. They acted on the front lines against land grabbers, loggers, ranchers, and businessmen in the region where they lived. They were also active in the struggle for the legal recognition of the community's territory. As a consequence of their fight, they were brutally murdered in May 2011.

To understand how the public and private powers promote and perpetuate violence against defenders, let's go back to Zé Claudio and Maria's history in the early 1990s, when they first started to participate in meetings that debated land tenure regularization in the area, land reform, and defense of the community's way of life and human rights. Community members had lived in the area for many years, but without any legal security that would guarantee permanence or regularization of the territory. Zé Claudio and Maria, together with the land rights movements, learned of the importance and the real need to collectively defend fundamental rights and guarantees. They were also very inspired by Chico Mendes' fight for the preservation of the forest; Mendes argued that only through a fight against the forces that destroyed could they keep the forest standing and thus guarantee the existence of forest peoples' way of life.

In 1992, they participated in the initial fight to create a union of small rural workers in the municipality of Nova Ipixuna, and in 1997 the local association

APAEP – Association of Small Rural Workers of the Praia Alta Piranheira Agroextractive Settlement Project – was created. Regularization of the area was approved on 22 August 1997.[3]

The Praia Alta Piranheira Agroextractive Settlement Project (PAE) is located is located on the banks of the Tocantins River,[4] in the municipality of Nova Ipixuna. It was created in an area of Brazil nut groves with 22 thousand hectares of land that is part of the Lake Tucuruí region (UFPA-NAEA 2018).[5] Although it is the duty of the State to promote agrarian reform, the creation of the PAE was not an initiative of the Brazilian State, but rather the result of many mobilizations by the local community. They wanted not a standard settlement project, but instead a model in which the community's way of life would be respected, in such a way that the community could work in forestry and agriculture. At that time, many debates had arisen about the creation of agrarian reform settlements with a conservationist perspective, a settlement model that took into account the historical way of life and care for the land by traditional communities and, as they called it, an "alliance of the peoples of the forest."

The local leaders endeavored to follow the still recent process of regularization of the area as a PAE. This modality has a Contract for Concession of User Rights (CCDRU) signed by residents of environmentally differentiated settlements (INCRA 2020). In these cases, no domain title is granted, but the document has the same value as other titling instruments granted by the National Institute of Colonization and Agrarian Reform (INCRA).

The first serious mistake was made after the creation of the PAE: failing to carry out complete and total land tenure regularization. Within the 22 thousand hectares of the PAE, there were six ranches. Procedures should have been taken by INCRA to avoid future conflicts. However, sequences of omissions and inertia by the State would continue over time.

In a letter written to the director of INCRA on 24 November 1999, the so-called owners of the ranches claimed that a year and a half had passed since that the PAE had been created, and yet they had not received any official communication informing them of: a) the conditions for those who live in the PAE area; and b) whether those who did not agree with the settlement project could be indemnified by INCRA for the improvements carried out on the lands. They also complained that no inspection had been carried out on the farms and that the lengthy inspection process caused them losses. They set a deadline for INCRA to carry out the procedure: 1 January 2000.

In the midst of these tensions with farmers and loggers over the creation of the settlement, Zé Claudio and Maria began to receive death threats, starting in 2000. On 14 July 2001, Zé Claudio received a very serious death threat and in the same year was added to the Pastoral Land Commission's (CPT) list of people facing death threats. CPT's work has been very important in monitoring violence in the countryside. At that point, a long period of tension, aggression, violations, arrests, and deaths began.

In 2003, due to INCRA's delay in resolving the situation of the farms within the PAE area, workers decided to occupy the Cupú Farm, one of the ranches located within the area demarcated for the PAE. The occupation was a way to pressure the State to regularize the area and settle families. At the time, Maria was the president of the local association and was in charge of the occupation. The farmer registered a complaint at the police station and Maria was arrested. She refused to give the delegate the names of the workers who were occupying the area.

The threats against Zé Claudio and Maria intensified due to pressure against the State regarding territorial irregularities within the PAE territory, as well as to the intensification of illegal logging; this left the couple once again at the center of a conflict, one that would drag on for years without the State taking any action to resolve conflicts or investigate complaints. The illegal exploitation of wood, especially of Brazil nut trees, intensified in the late 1990s and continued for several years with total inertia on the part of IBAMA (Brazilian Institute of the Environment and Natural Resources), even under constant condemnation voiced by the couple and community members who blocked roads and photographed timber trucks.

On 8 May 2004, Maria also joined the list of those threatened in CPT's report, three years after Zé Claudio. In the report, the name of Maria was marked with two asterisks (**) to indicate that she had been threatened with death more than once. Meanwhile, her partner's name remained on the list.

Zé Claudio and Maria identified themselves as farmers and harvesters, that is, they worked both the land and the forest. They knew the importance of the standing and preservation of the forest for the maintenance of not only the family but also the planet. They held the ancestral knowledge of our ancestors of Indigenous origin about the use of the forest and its diversity and passion for sustainable agriculture.

Zé and Maria led a simple but dignified life and were active together in social movements from the 1990s on. They led the movement in the municipality, and stood side by side with comrades in struggles; they promoted partnerships with universities, non-governmental organizations, and other institutions, always in order to promote social justice and to improve local organization and sustainable production through agriculture and extraction.

A short chapter on the couple's trajectory, intertwined with the regional context in the midst of so many actions, cannot account for all the details of years of fights for land, environmental, social, political, and economic issues. Zé and Maria not only denounced the violations but also sought solutions for the issues that affected their community. A brief look at the couple's history, emphasizing issues such as denunciations and organizing for possession of the land, can sometimes decontextualize or simplify the situation, disregarding the broader aspects of the struggle for dignity and fundamental rights.

Zé Claudio and Maria fought against deforestation, looking for partnerships that potentiated sustainable forest production by and for the community and that still reverberate today. In 2004, as a result of pressure from loggers, and in the search for alternatives for making a counterproposal, they entered into a partnership with the federal government through the Ministry of the

Environment and Federal University of the South and Southeast of Pará (UNIFESSPA) and carried out several projects that boosted the commercialization and production of forest products from the PAE. They also developed a Community Forest Management Plan that, in addition to promoting training in forest management for farmers/extractivists in the region, also carried out a survey of PAE's timber and non-timber forest potential. This work led to the creation of a women's group, Group of Artisanal Extractive Workers (GTAE), who continue today to resist the production of phytocosmetics and herbal medicines, by-products of forest oils.

In November 2010, at the TEDx Amazônia event, Zé Claudio said, "I live in the forest, I protect it at all costs, that's why I live with a bullet to my head" (TedX Talks 2011). Zé Claudio was already predicting a near future, the result of neglect, inoperability, and inefficiency in investigating the death threats reported by the couple.

On 24 May 2011, at seven-thirty in the morning, Zé and Maria were brutally murdered. They were traveling on a motorcycle to the city when they were caught in an ambush. The killers were strategically positioned on a streambank along the road where the couple would pass. The stream is at the foot of a small hill, where there is a kind of improvised, very worn bridge made of wooden logs. To pass, it was necessary to slow down. The hill provided a good vantage point for the assassins, who were camouflaged in the forest. Zé Claudio and Maria always rode a motorcycle, a common practice, especially in rainy times when the roads are in very poor condition. However, when they slowed to pass the bridge, they were surprised by shots and fell off the bike. The killers left the forest and continued shooting, removing them from the middle of the road. Zé Claudio was still alive when one of the gunmen cut off his right ear, revealing both the cruelty of the act of torture and a *modus operandi* typical of hired gunmen.

Between the first threats suffered by Zé Claudio and Maria and the murders, 11 years passed. According to reports, the main cause was land conflicts. But what is it that occurred over that time that led to the murder of two leaders of an agrarian reform settlement that was considered to have completed the land titling process? There is legislation on land regularization that guides the State authorities in procedures for the creation of a settlement (whether traditional or for collective use). However, it is at this point we see the most serious actions and omissions of the State in the case of PAE Praia Alta Piranheira. Previously I mentioned that the settlement was created with six ranches within the geographical space of the PAE. Yet, since the establishment of the settlement, the State has never acted effectively to resolve the conflicts that have been generated over the 13 years between its creation and 2011, when two people lost their lives. During these 13 years, Zé Claudio and Maria waged a war against the exploitation of the forest and against land grabbing, fighting against loggers of all sizes as well as farmers, large and small, who were also businessmen. How did a couple of smallholder farmers dare to "fight," facing people with the power of money, politics, and networks in the region? For many, these types of people must "be excluded from society," as evidenced by the narratives created to criminalize their struggles and to justify absurdities.

The regional context cannot be disconnected from the story. At that moment in time, Marabá was the center of the arc of deforestation, as the gateway to the Amazon. At the center of the debates was pressure for agricultural expansion and its dairy basin, expansion of the iron production industry, and implantation and expansion of large projects; once again, the narratives of development expand and gain space.

The State had never resolved PAE's environmental and agrarian conflicts, and in 2009 a medium-sized rancher, José Rodrigues, emerged who was willing to buy one of these conflict areas, and thus "buy the fight," as he put it when he decided to buy the land. In 2011, in an organized and planned manner, he set into motion his plan to murder Zé Claudio and Maria, who were considered protectors of the workers occupying the land under conflict. José Rodrigues and his cronies (other farmers and businessmen) knew that as long as Zé and Maria were alive, they would protect workers from the various forms of violence they suffered. For instance, in the episode in which the farmers, gunmen, henchmen, and the municipality's own police arbitrarily intimidated, psychologically tortured, burned the workers' homes, and coercively took one of the workers to the police station with the aim of getting him to sign a document to give up the land and pass it on to José Rodrigues; this is just one of the episodes.

The ground ceased to exist when we received the news of the murders. The day before, my siblings and I had met and decided that we would talk to them and ask them to leave for a while, because the threats were very strong. But we never had time; it was over that day. The pain of "what if" as I write these words is very strong, but it is necessary to tell the story; it is necessary to continue. From that day forward, we did not have a single day of rest as we struggled to overcome their deaths, transforming all the pain and revolt at their loss into a struggle for justice: justice for Zé Claudio and Maria, who throughout their lives fought for the lives of others, who faced rotten powers with honor and love for those who needed support.

So, in the midst of this avalanche of events, we followed every moment of the investigations, including some proceedings and some testimonies. Unfortunately, the evidence was not sufficient to accuse everyone we are convinced participated in killing my brother and his wife. But, in the list of the accused, three were recognized as the masterminds and executors of the double homicide: (landowner) José Rodrigues Moreira as principal and Lindonjonson Silva Rocha and Alberto Lopes do Nascimento as the executors. We are convinced that three minds alone did not produce all the strategies used to kill the couple; we know that there was a broader network involved, and at least a year and a half of discussions and plotting for action.

The first trial was held at the Marabá Criminal Court, two years after Zé Claudio and Maria's murders. It received national and international media coverage for the ridiculous absolution of mastermind José Rodrigues and the condemnation of the executors, Lindonjonson Silva Rocha and Alberto Lopes. The jury was chaired by Judge Murilo Lemos, who, in my opinion, did not show impartiality in the conduct of the trial, an indispensable condition for a judge. In the judgment of the first trial, Zé Claudio and Maria's performance in defense of human rights was attacked:

> *The victim's behavior contributed, in a certain way, to the crime,* because, as declared in plenary by the witness José Maria, the victim faced the brother of the accused Lindonjonson, the co-defendant José Rodrigues Moreira, trying to do justice by his own hands, using third parties (squatters/landless) to prevent the co-defendant from having possession of a rural property, thus worsening the land conflict, when the victim could have sought the support of the constituted authorities to bring the action of the co-defendant before the court.
>
> (TJ-PA 2011, emphasis added)

Such a statement reveals the wide spreading of narratives of hatred against defenders of human rights and the environment, which in turn comes to "justify" violence and even murder. In attributing responsibility to the couple for their own deaths, we see how strong and emblematic ideologies can influence the justice system. The judge expressed these ideologies clearly in placing responsibility on the victims for their own deaths. It represents a return to ideas of the dark ages, causing dehumanization and the normalization of absurdities against life.

In 2016, a second trial took place for the appeal, this time held in the state's capital city, Belém do Pará. This time, the jurors decided differently than the first trial in 2013. As a result, José Rodrigues' acquittal was annulled, and he was condemned to 60 years in jail.

However, at the time of writing, nine years after the murders, José Rodrigues remains a fugitive. In August 2020, an operation was carried out that resulted in the recapture of Lindonjonson Silva Rocha, who had escaped from the Marabá prison in November 2015. The details of this story can be analyzed in a way that aims not simply to find those responsible or hold institutions accountable, but instead to reveal the lack of institutional understanding and commitment in the realization and consolidation of the national and international standards to which Brazil is a signatory: consolidation of human rights above all in the construction of a society based on citizenship, and the dignity of the human person as a corollary to a democratic state of law.

Thus, almost 20 years have passed, yet the state of conflict that has generated a series of violations and murders, as well as the effective destruction of the Brazilian government due to the inoperability of INCRA, IBAMA, and justice agencies, remains – and we resist.

The case of the murders of Zé Claudio and Maria is considered emblematic. They were threatened with death for many years. Their complaints were never investigated. The assassination of Zé Claudio and Maria followed a plot that occurs frequently in the south and southeast of Pará and in many other parts of Brazil, wherever there is a struggle for the forest. It is a story of the struggle for land/territory and for the survival of defenders of human rights and the environment, and a story that is marked by the role of the State in promoting mega-projects and flexibilization of environmental laws, as well as in supporting the expansion of agribusiness that favors elites, whether local, regional, national, or even international. It is many years later, but it is as if we were still in the years of the dictatorship, as the State criminalizes those who defend the collective public interest, the right to land, to work, to nature, to life.

Audacity is what moves us, as Maria said. It is what we have today in the transformation of pain into struggles for justice, memory, and continuation of life. This comes no longer from Zé Claudio and Maria – lamentably, they are gone – but from the defenders who are still alive, fighting for the rights of their community as well as for their lives.

My family's struggle for justice for Zé Claudio and Maria has led to a lot of confrontations, including death threats, but none of this is able to intimidate us. During these years, we have built and organized networks in defense of defenders threatened with death. Through support and solidarity, we denounce in various spheres of power all the aggressions that the environment and its defenders suffer. We fight so that other defenders do not experience the same fate as Zé Claudio and Maria. Working with other women, we are creating a network for the protection of women human rights defenders. Partnerships with international organizations provide us with urgent support for frontline defenders who are at high risk of death. We are working to create a space, a "house of sighs," for defenders who are psychologically shaken due to death threats and are seeking relief. We also lead actions that promote maximum visibility of our struggles and follow individual cases so that these defenders do not feel alone, that they know they have support to continue fighting – everything that Zé Claudio and Maria did not have and that led to the dire consequences we have seen.

What's left? It is the pain and suffering for the losses that pass through us every day like knives. What is missing? It is the absence of the warmth and strength of Zé Claudio and Maria. What will remain? It will be the inspiration through the courage and strength they had in defending others, in loving nature to the point of putting themselves on the frontlines to protect them. And so, we continue to honor, feeling fear, feeling pride, and missing them. It hurts me just to imagine that, as I write these words, other defenders – of land, water, forests – are suffering, being murdered or begging for help without having anywhere to turn. We need to fill ourselves with strength and strategies to fight effectively with these defenders against destructive forces.

In order for us to have more justice and peace, the dreamed-of *bem vivir* (good-living), global powers must stop consuming products that cause pain and suffering to humans and non-humans, must cease their capitalistic view of nature and people and the anthropocentric perspective that disregards the environment and sees the rivers and the land as bargaining chips and without rights.

This must be overcome. We must have ethical limits on our consumption, from those who produce and finance to those who consume. The forest and its diversity must be granted rights. Countries need to make their industries and their negotiators responsible for the destructive exploitation of natural resources. Indigenous forest peoples, community members, family farmers, fishermen – all forest and water peoples need to be respected and to have their value recognized by humanity. We are ALL passengers here; when we die, the land will still be here: what mark do we want to leave? We are already leaving destruction, but we can and must change that, for ourselves and for future generations.

Notes

1 Located in the Serra dos Carajás, in Pará, it was the largest open pit in the country. Currently, there is only one crater, 70 to 80 meters deep, with mercury-polluted waters.
2 Giovane Queiroz, ex-governor and rancher, stated this in an interview for the film *Toxic Amazon*, a documentary produced by Bernardo Loyola and Felipe Milanez (2011) for *Vice* magazine, shortly after the murders of José Claudio and Maria.
3 The land is registered in the name of the Union under No. 01385, pages 001, book n 2-F of the Property Registry Office of the Marabá District in the State of Pará (Ordinance No. 42 of 21 August 1997).
4 Studies on the Tocantins River hydroelectric power plant started around 1957 and gained strength in the 1960s as part of the federal government's policies for the development and integration of the Amazon, and to serve the aluminum industry generated by the region's bauxite deposits.
5 Lago de Tucuruí Integration Region is one of the 12 regions adopted for planning purposes by the Government of the State of Pará, one of the regions with the smallest number of municipalities. It is located in the southeast of the State of Pará and comprises the municipalities of Breu Branco, Goianésia do Pará, Itupiranga, Jacundá, Nova Ipixuna, Novo Repartimento, and Tucuruí (UFPA-NAEA 2018).

References

Gortázar, N. G. (2019, September 17). Brazil has only tried 14 of the 300 murders of environmentalists in the past decade. *El País*. https://brasil.elpais.com/brasil/2019/09/16/politica/1568661819_648829.html

Human Rights Watch. (2019). *Rainforest mafias: How violence and impunity fuel deforestation in Brazil's Amazon*. www.hrw.org/report/2019/09/17/rainforest-mafias/how-violence-and-impunity-fuel-deforestation-brazils-amazon

INCRA. (2020). *Titulation*. Brasilia: Instituo Nacional de Colonização e Reforma Agrária.

Loyola, B., & Milanez, F. (2011). *Toxic Amazon* [Film]. Vice. YouTube. https://youtu.be/Mxj-_JKiDq8

TEDx Talks. (2011, February 24). *Killing trees is murder: Zé Cláudio Ribeiro at TEDxAmazonia* [Video]. YouTube. https://youtu.be/XO2pwnrji8I and www.youtube.com/watch?v=OSS2ALiU1ss&feature=youtu.be

TJ-PA (Justice Tribunal of Pará). (2011). Marabá, branch on domestic and family violence against women, homicide. (vara de Violência Doméstica e Familiar contra mulher; Homicídio) processo n° 0005851-94.2011.814.0028. Partes: Justiça pública e J.R.M, L.S.R. e A.L.N, TJ-PA 2011.

UFPA-NAEA. (2018). *Plano de Desenvolvimento Regional Sustentável (PDRS) e a Implantação de Usinas Hidrelétricas Estruturantes*. Universidade Federal do Pará. www.mme.gov.br/documents/36144/471801/Produto+2.pdf/ae7cae03-c9c3-5183-7e8e-c29eccf8889a

3 How young Cambodian environmental activists work under dictatorship

San Mala

The fall of the Khmer Rouge genocidal regime in 1979 opened a new era of resource exploitation in Cambodia, a small and poor country in Southeast Asia. After the death of an estimated 2 million innocent people, damages to infrastructure were extensive and the economy was down to zero. There was a need to rebuild, and natural resources such as timber, minerals, gold, and sand became the go-to resources for trade and economic reconstruction. Yet, most of those resources were confiscated for the benefit of individuals and political factions that came to assert their power over the population.

From the mid-1990s on, a London-based NGO, Global Witness, published a series of reports, such as "Cambodia, where money grows on trees", exposing how remaining Khmer Rouge groups that still ruled parts of the country were able to earn as much as $6.5 million per month in timber revenue (Global Witness 1996). These reports also exposed the deep level of corruption within the government that made this possible. After the Khmer Rouge regime ended, a new breed of dictatorship, led by Prime Minister Hun Sen and the Cambodian People's Party (CPP), came to power. Forcing its way into government after losing UN-sponsored elections in 1992, and then gaining full power after a coup d'état in 1997, Hun Sen's rule has since continued to undermine democratic rights and strip natural resources with impunity and without transparency, accountability, or openness to public participation, despite repeated criticism and condemnation from the international community.

The general election in 2013 brought a sense of hope, as the largest opposition party in Cambodia, the CNRP, won 55 of the 123 seats in parliament. Yet, while the CNRP believed it had won the most votes and could thus lead the government, the process was rigged by the National Election Commission, a puppet of the CPP ruling party, and Hun Sen stayed in power. Since then, a series of demonstrations and protests focusing on elections, human rights, democracy, and natural resources have taken place, mostly mobilizing young people and university students, including me.

> *Against the government, you will be offered three types of options:*
> *1. Imprisonment, 2. Fleeing, 3. Death*
>
> Dr Kem Ley, social development analyst shot dead in 2016

Environmental work is a global responsibility, and people all over the world should pay attention to, participate in, and care about it. Yet, for the Cambodian government, defending the environment puts you squarely into the category of political opponent.

In early 2014, I joined the Mother Nature Movement, a grassroots activist group, and worked closely with communities to combat 'development projects' that have negative impacts on the environment and local residents. Before deciding to join this movement, and thereby publicly becoming an 'environmental activist', I was already mentally prepared to accept the obstacles, threats, and problems that would arise.

I have been arrested three times by the police. The first time, in September 2014, I was with community members and other youths from the Mother Nature Movement, blocking a road to prevent local authorities from conducting a biased 'environmental impact assessment' for the Areng Valley Hydropower dam project, which we knew would simply 'rubber stamp' the project. Blocking the Environmental Impact Assessment (EIA) was ironically one of the few ways to block the project. In the end, many of us were arrested, questioned by the police, and then released by night-time.

The second time was in February 2015, when the Spanish founder of the Mother Nature Movement, Alejandro Gonzales-Davidson, was denied a visa and deported from Cambodia. At that time, he and I were working together. I accompanied him pretty much everywhere, just in case something went wrong, so there would be someone who could alert the rest of the movement and spread the word. This time, however, the immigration police from the General Department of Immigration decided not only to arrest Alex, a foreigner, but also me, a Cambodian citizen. Despite being detained, I was able to alert allies, and after the intervention of the UN High Commissioner for Human Rights, I was released.

Finally, in August 2015, I and two other members of the team were again arrested on charges of 'threats to destroy' during a protest against sand mining companies. I was imprisoned for 10 months and 14 days, despite the lack of evidence and witnesses to charge me. The arrest of the three of us, we believe, was an act of repression and retaliation against environmentalists by the Koh Kong provincial authorities after we acted against their development projects, which affected the citizens and the environment. Here is the story of my struggles and my bitter history as an environmental activist.

Demonstrating against the sand dredging company

In April 2015, just before the Khmer New Year, I, along with four other young people, agreed to go to a small island called Koh Sralao in Koh Kapi commune, Koh Kong district, in the southwest of Cambodia. We pretended to be young tourists and asked for a villager's house to stay in. As tourists, we had the task of building relationships with the villagers and chatting with them, in order to investigate and gather information from them regarding dredging. We did not ask the questions we wanted to ask, not directly, but we did ask about their livelihoods, whether they were doing well, or what challenges they were facing. They all complained about the hardships of life due to the dredging. They did not know the name of the company or whether they had a license or not. They only knew that the government had given the company rights to pump there, and the authorities had told them that it was unimportant whether the people agreed or not. The only thing that mattered to them was that the company had gotten permission from the central government.

The dredging took place in the area where the community fishes, and the extraction of sand from the creek resulted in the loss and destruction of the habitats of crabs, fish, and other marine animals, which are the main source of income for the island's people. In addition, the boat traffic and the company's machinery were floating on the water day and night, causing noise, and the boat's propellers cut off and destroyed some of their fishing gear.

Seeing that most of the villagers were unhappy and angry about such dredging activities, we decided to invite the affected villagers to a meeting. The idea was to gather opinions and concerns. As young people and students, we promised to help solve their problems as much as possible. The meeting was held on the tenth and eleventh nights of April 2015. At the second meeting, the people asked us to join them in protesting to drive those companies out of their villages.

We were very much surprised to hear that they wanted to protest against the dredging, as we did not expect that the people living on this island would have such courage. But maybe it was precisely because they lived so miserably that they could no longer stand it. Having the courage to openly protest is not so widespread in Cambodia, and for good reason, as most people live under the tight control of local authorities and do not know their basic rights or what to expect from a government in a democratic society.

The majority of Cambodians were traumatized by the Khmer Rouge genocidal regime, which massacred 2 million innocent people – especially those Cambodians living during the armed conflicts that afflicted the country between 1970 and 1997.

The people's request for us to join them in a demonstration occurred so quickly that we could not get ready. We initially thought that it might take at least two months to provide training on basic legal rights and to make them understand their obligations as citizens before exercising those rights in self-defence against the violation of various benefits.

However, to show our support and hope, we agreed to join them; however, we informed them how to demonstrate peacefully and non-violently, forbidding the destruction of company property or insulting the workers there.

On 12 April 2015, at 12:30 pm, the first sea-borne demonstration erupted, and it received a lot of attention from the local media (Hul 2015). The demonstration gathered about 70 families aboard ten of their fishing boats and travelled to Prek Kabong, about a kilometre from their home island. People raised banners and shouted for the company to immediately suspend its activities and show their licenses to the people. Since the company's dredging boat was taller than the fishing boats, we all decided to get on their boat to meet the company boss. The company manager told us that they were a legitimate company and had a valid license from the government, but they did not keep the license at the site; if we wanted to see them, we should visit the Department of Mines and Energy of Koh Kong or the company's headquarters in Phnom Penh.

Residents responded that if the company did not have a license, they would ask it to suspend its activities and leave immediately. People used fishing boats to chase and ride alongside the company's dredging boats, even if chasing was very difficult. At around 4 pm, the Koh Kapi commune police chief appeared, dressed in civilian clothes and using obscene language against young activists, accusing us of inciting people to protest against the company. The police chief showed us the company's license over his phone, and we copied it and read it aloud to the people. In it, we found that the license of this company, International Rainbow Co. Ltd, had expired two months earlier, and the location obtained in the license was not at that location; the company had violated the technical principles set by the Ministry. The Ministry of Mines and Energy, the licensing authority, confirmed the zone where the company could and could not extract sand, but apparently the company had been pumping it regardless. Because the company had nothing to do with the people, they decided to leave, but this departure occurred in a dissatisfying manner; they tried to cut out their engine when our boat had gone some distance. Seeing this, we all decided to get onto the company boat, stand next to the crew, and ask them to leave immediately because they did not have a license.

The demonstrations, which were reported in the local media and sustained by the villagers' solidarity, prevented the Ministry from renewing the company's license. Turning to the local authorities, they were very angry with us because now the people were not afraid of the authorities anymore. Before this, they had never dared to speak out or protest at all, and now they knew how to protest. A few weeks later, the local authorities prepared a document to sue and thumbprint the people in order to evict us from the island as well (Khuon 2015a).

Our success with this Koh Sralao community has been highly praised by the public. Therefore, we aimed to spread this success to another community located in Andong Teuk commune, Botum Sakor district, which is only 193 kilometres away from Phnom Penh.

We did the same thing on 26 July 2015, this time protesting against the sand dredging company Direct Access Ltd., demanding that they stop and leave the canal where they were dredging. This second area was different and more difficult than Koh Sralao due to geographical factors: the canal has a bridge in the middle, and it was the rainy season; the water level had risen, making it impossible for some dredging boats to cross (Aun 2015a). We held not only a demonstration, but also a press conference, to present the irregularities we discovered to the media as well as the impact of the dredging in Phnom Penh.

The demonstration lasted for four days and still could not stop the company's activities. We then learned that the authorities had arranged to hire people to frame us; they organized a group of people to cause trouble in order to then place the blame on us (Aun 2015b). So, we changed our direction to protest in front of Botum Sakor District Hall, demanding that the district governor stop the dredging company and solve this problem with the people. The district hall asked us to arrange for community and youth representatives to meet with them, but we refused (Sen 2015) – first, because this is an issue that affects all citizens, so it must be resolved publicly in front of everyone; and second, arranging for a representative would make it easier for the authorities to identify the leader, and that representative would then be the target of threats and arrests.

In the end, the district hall promised to stop the dredging and wait for a solution from the Ministry before we dissolved the meeting. However, on 11 July, the company resumed operations because they heard that on 14 July, a secretary of state from the Ministry of Mines and Energy would come to do an inspection, so they dared to resume their activities.

On 12 July, we held another demonstration to continue our demands, and on 14 July, three environmental activists, including me, were sent a letter by the Botum Sakor District Police for three urgent summonses. They required us to appear to respond to the company's complaint (Khuon 2015b). We declined to comment (Pech 2015) because we were busy monitoring the Ministry's inspection activities; we were afraid they would not reach the site and accurately measure the depth of the canal, which was getting deeper because of the sand dredging.

That time, due to our strong and active protests, and having already won twice – the campaign to stop the construction of the Stung Chhay Areng dam and the anti-dredging demonstration on Koh Sralao – the authorities decided to arrest me, Try Sovika, and Sim Samnang in order to stop our protest activities. It was the company's revenge on us for disrupting their business.

On 17 August 2015, the police surrounded our house to arrest us. We were prepared and had had hints that we would be arrested soon. So, we were not afraid at all. We told the police that if they had a court order, they could arrest us and we would not object.

The police took the three of us to the Koh Kong Provincial Court, and the prosecutor charged us under Article 424 of the Penal Code 'Threats to Commit Destruction Followed by an Order'. The allegation was based on a complaint by Direct Access Ltd. alleging that we threatened to burn the company's boat if they did not leave. The court decided to send the three of us to pre-trial detention at Koh Kong prison on the night of the 17th (Aun 2015c).

Our arrest sparked protests in the provincial town of Koh Kong by the community and environmental activists in the same group, demanding that the court release us immediately and unconditionally (Aun 2015d). The protest also created further resentment from provincial authorities; on 20 August 2015, police and prosecutors raided the offices of Mother Nature Movement and interrogated the activists and communities living there (Pech & Cuddy 2015).

Finally, the police used 'hot measures' by starting to subdue (Titthara & Cuddy 2015) and arrest 17 people (Aun 2015e), including activists, protesters demanding the release of the three of us, journalists, and human rights organizations who participated in the watch protest of 2 September. The crackdown and interrogation by the Koh Kong provincial authorities was used to silence voices against the exploitation of natural resources that served the interests of those in power and lacked transparency, accountability, and consideration of impact on the environment and livelihoods of local communities.

During the trial, the court found no witnesses or evidence that we had threatened the company and said we would burn their boats if they did not leave. Yet, the court sentenced us to 10 months and 15 days. All three of us were released on 1 July 2016.

Despite being unjustly imprisoned, I have continued to work as an environmental activist. The story of the sand dredging came to an end after the Mother Nature Movement changed its strategy by stopping public demonstrations. By investigating the data on Cambodia's sand exports, we also found corruption and smuggling sand from Cambodia to other countries, such as Singapore and Taiwan.

According to Singapore Customs, from 2007 to 2015, 73 million tons of sand were imported from Cambodia, valued at about $750 million. Cambodia, on the other hand, recorded only about 3 million tons of exports for a value of $5 million, so 70 million tons of sand worth about US$700 million went missing from the export list – and, presumably, so did the revenues (Seers 2016).

Separately, for the export of silica sand from Cambodia to Taiwan, we found that the Taiwan Customs Department showed that Taiwan imported 1.5 million tons of sand from Cambodia, at a cost of $32 million. According to Cambodian Customs and Taxes figures, only 28,900 tons were exported to Taiwan, equivalent to only $275,605 in the same period. Again, tens of millions of dollars had gone missing (Mother Nature Cambodia 2020).

In the end, the government decided to temporarily suspend dredging and to stop issuing new licenses, shutting down sand exports, which is a new achievement and a gain for the Cambodian people from the struggles of the entire nature movement. However, instead of the arrest of corrupt officials and levying of fines against defrauding companies, the revelations of the sand smuggling corruption resulted in the arrest of two more activists. Hun Vannak and Doem Kundy were arrested in September 2017, after they went to investigate and take photos of the export of silica sand from Cambodia to foreign countries.

Although limited, all of the successes and achievements from these protests and campaigns are due to the contribution of local communities, national and international civil society organizations, and the Cambodian people, who in the past have fully supported and cared for environmental issues. Together, they put pressure on the government to take action against corruption related to natural resource extraction.

References

Aun, P. (2015a, July 28). Activists board dredging barges in Koh Kong. *The Cambodian Daily*. https://english.cambodiadaily.com/news/activists-board-dredging-barges-in-koh-kong-89643/

Aun, P. (2015b, July 30). Anti-dredging protests in Koh Kong called off. *The Cambodian Daily*. https://english.cambodiadaily.com/news/anti-dredging-protests-in-koh-kong-called-off-90020/

Aun, P. (2015c, August 18). Anti-dredging activist jailed in Koh Kong. *The Cambodian Daily*. www.cambodiadaily.com/news/anti-dredging-activists-jailed-in-koh-kong-91880/

Aun, P. (2015d, August 20). Dozens of protesters demand release of activists in Koh Kong. *The Cambodian Daily*. https://english.cambodiadaily.com/news/dozens-of-protesters-demand-release-of-activists-in-koh-kong-92113/

Aun, P. (2015e, September 3). Police arrest 17 at protest in Koh Kong. *The Cambodian Daily*. www.cambodiadaily.com/news/police-arrest-17-at-protest-in-koh-kong-93241/

Global Witness. (1996, October 21). *Civil war goes on hold as the Cambodian government and Khmer Rouge cooperate over timber deals*. www.globalwitness.org/en/archive/civil-war-goes-hold-cambodian-government-and-khmer-rouge-cooperate-over-timber-deals/

Hul, R. (2015, April 13). Fishermen claim victory against sand dredgers. *The Cambodian Daily*. https://english.cambodiadaily.com/news/fishermen-claim-victory-against-sand-dredgers-81942/

Khuon, N. (2015a, April 22). Koh Kong activists threatened with eviction over dredging row. *The Cambodian Daily*. https://english.cambodiadaily.com/news/koh-kong-activists-threatened-with-eviction-over-dredging-row-82341/

Khuon, N. (2015b, August 15). Police summon three Koh Kong anti-dredging activists. *The Cambodian Daily*. www.cambodiadaily.com/archives/police-summon-three-koh-kong-anti-sand-dredging-activists-91746/

Mother Nature Cambodia. (2020, June 5). *Why are most exports of silica sands to Taiwan 'missing'?* [Video]. YouTube. https://youtu.be/BLmK6KG6iTg

Pech, S. (2015, August 17). Activists ignore summonses. *The Phnom Penh Post*. www.phnompenhpost.com/national/activists-ignore-summonses

Pech, S., & Cuddy, A. (2015, August 21). Police 'check' NGO's Koh Kong premises. *The Phnom Penh Post*. www.phnompenhpost.com/national/police-check-ngos-koh-kong-premises

Seers, P. (2016, December 16). Cambodia, Singapore sand trade figures prompt $750m question as corruption suspicions mount. *MLex*. https://nebula.wsimg.com/6268608f4862ed8aca9b124cfbecb0c3?AccessKeyId=37A067AFCC99EDE100B6&disposition=0&alloworigin=1

Sen, D. (2015, July 30). Anti-dredging activists refuse to meet with gov. *The Phnom Penh Post*. www.phnompenhpost.com/national/anti-dredging-activists-refuse-meet-gov

Titthara, M., & Cuddy, A. (2015, September 3). Koh Kong crackdown. *The Phnom Penh Post*. www.phnompenhpost.com/national/koh-kong-crackdown

4 Human rights violations in the name of conservation

Case of Ngorongoro district

Yannick Ndoinyo

This chapter examines human rights abuses and the use of violence in the name of conservation in Ngorongoro district, Tanzania. Various categories and levels of authorities use conservation rhetoric and sweeping statements around biodiversity conservation to legitimize the use of force, grab community lands, and impose external governance mechanisms. These approaches have proved to be wrong and ineffective, and they belittle community and Indigenous practices and their institutions.

Background of national parks in Tanzania

Historically, the national parks were created through either forceful evictions or bogus treaties between Indigenous peoples and colonial administrators for promises that were never meant to be fulfilled (Minja 2020; Shivji & Kapinga 1998). Through the nature of their creation and function, national parks became fortresses by locking Indigenous people out, or by establishing near-carceral centres within them where drastic rules, abuses, and brutal torture were applied to resident populations. National parks created after independence followed the same pattern of evictions and violations of human rights. They have always been created within the ancestral lands of Indigenous peoples. These lands were not just for grazing; they provided haven, identity, and religious attachment to these people. So, when they were alienated, Indigenous peoples and local communities were deprived of more than just 'geographical' territories.

Indigenous peoples who are the original owners of these lands faced serious negative stereotyping, such as trespassers, poachers, and negligent persons. This is the origin of the eventual criminalization that followed. To deter the people from accessing resources and cultural sites, heavy fines were devised and imposed on people who breached the many rules newly imposed by conservation.

Such brutal imposition of conservation took place near my home. In the late 1950s, after pressure from international conservationists for the Maasai to leave the Serengeti grasslands and Moru kopjes (small hills), a sham negotiation was initiated; only the colonial administrators were aware of the intended outcome.

Twelve illiterate traditional leaders signed an agreement in 1958 to renounce their ownership of the Serengeti plains, although the majority of the people, including the traditional leaders themselves, did not approve of the agreement. The Maasai people were forcefully evicted in 1959 from 'core conservation areas' but were granted access in perpetuity to Loliondo and Ngorongoro Conservation Area (NCA) as a multiple land-use area administered by the government, in which natural resources would be conserved primarily for their interest, but with due regard for wildlife. This promise was not kept. Tourism revenue has become the paramount interest in maintaining and managing these areas (Bluwstein et al. 2018), while the human rights situation has deteriorated and even been worsened by the designation of the NCA as a UNESCO World Heritage Site (Melubo & Lovelock 2019). For example, the statistics from Tanzania National Parks Authority show a record number of tourists for a period of ten years (2009/2010–2018/2018) – totalling 9.5 million tourists (TANAPA 2019). Although this is a success for the parks, it ignores the realities of the context, especially the discontentment and acute poverty of the people neighbouring these parks.

The governance and management of these NPs

Immediately after independence, the government of Tanzania enacted laws and acts to provide for the governance of newly established national parks (Nelson & Makko 2005; URT 1959). Based on this legal framework, the most common approach adopted for conservation throughout the country has been the deployment of military personnel to 'guard' the parks against the re-entry of local people to either graze their animals or access items as simple as fuelwood and herbs. This framework has also, in effect, legalized the torture and maiming of the local and Indigenous communities, who were, and continue to be, branded as harmful to conservation, wildlife, and nature. The militarization of the National Parks and Protected Areas evolved from game scouts and park rangers in early years to para-military units present today, popularly known as 'conservation commissioners'. Instead of portraying conserved areas as friendly, soothing natural and harmonious landscapes offering hope and inspiration, the military is used to scare people and legalize any force to be applied. It paints a grim picture of hate, force, and danger all the time. The successful 'governance' of these areas is measured through forceful evictions – burning of homes, shooting, and maiming: acts that are interpreted as evidence that the military is achieving the desired results. Most boundaries of conserved areas in Tanzania are permanent war zones where the only enemy is the local community.

In 1975, the remaining Maasai living inside Ngorongoro Crater were violently evicted, and that same year, cultivation of food crops was banned in NCA. This cultivation ban was lifted in 1992 but re-introduced in 2009 after threats from UNESCO to declare the area unfit. The people of the NCA live under the colonial-style rule of the Ngorongoro Conservation Area Authority (NCAA). They are not allowed to grow crops or build modern houses.

People are suffering from poverty and children from high levels of malnutrition (Kipuri & Sørensen 2008; Melubo & Lovelock 2019). In 2017, Maasai were excluded from important grazing areas as a result of the NCAA general management plan and have been living under threats of further evictions for many years. In September 2019, a plan was developed and presented to the Minister for Natural Resources and Tourism for adoption without consultation of the local community in the area. If this plan is adopted, it will completely go against the basic principles of free, prior, and informed consent (FPIC) as well as the United Nations Declaration of the Rights of Indigenous Peoples, signed by Tanzania. The people of Ngorongoro have stood firm in opposition to the plan. Following decentralization measures, the newly formed states formulated laws to govern these lands, which completely disown Indigenous populations and alter their traditional and cultural institutions.

Human rights violations and criminalization of Indigenous human rights defenders

The practice of conservation in Tanzania is characterized by rampant violations of human rights that have affected the Indigenous peoples and their livelihoods for decades. The local people are criminalized, charged, and arraigned in courts for crimes such as crossing (trespassing) the boundaries of protected areas, grazing, or simply affirming their ancestral land rights by demanding justice and access to fundamental rights (LHRC 2014; Mayengo et al. 2017). In other extreme cases, which have become more frequent, they are shot and killed.

The case of Pormoson Ololoso

This is the story of a young man who found his fate on 8 August 2017, in the Oloosek area of Ololosokwan village, Tanzania. He was grazing his livestock while keeping quite a safe distance from the Serengeti National Park. However, without provocation, a ranger from the Park opened fire and shot the herder, Pormoson Ololoso, hitting him with three live bullets, in both thighs and his left arm. It is said that there was a slight disagreement as the Ranger was accusing Pormoson of grazing his cows in the Park in the past. Now the rangers were extorting him for money, which Pormoson refused. Pormoson was taken to several hospitals in Kenya and Tanzania for treatment that cost his family huge sums of money. Luckily, he has fully recovered and joined his family again.

In Loliondo division, there is a law prescribing the governance and ownership of land through local institutions. Here land is governed by Tanzania's rather progressive land laws, under which village land belonging to the village assembly (all adult villagers) is managed by the village council under Village Land Act No. 5 of 1999 (URT 1999). Still, international conservation, tourism investors, and central government have never ceased looking to further expand fortress conservation practices onto village lands (Gardner 2012). This has far too often led to cases of extreme violence. In 2009 and 2017, there were major operations with illegal invasions of village land ordered by the district commissioner, who is the highest representative of the central government in the district, appointed by the president. Hundreds of homesteads were burned while there were beatings, arrests, seizing of cattle, blocking of water sources, and alleged cases of rape.

All these began in 1992, when the government leased the community land to Otterlo Business Corporation (OBC) without consent or even consultation with the local community, completely ignoring the UN'S FPIC principle about Indigenous peoples. OBC is a hunting company that organizes luxury safari hunting for Sheikh Mohammed of Dubai and other members of the royal family from UAE.

In 1992, OBC obtained the right to hunt in the Loliondo Game Controlled Area, 4000 km^2 of pristine lands owned by the Maasai pastoralists. Increasingly in the past two decades, OBC has lobbied the Tanzanian government to create a protected area for exclusive hunting tourism out of the 1,500 km^2 of dry season grazing refuge for the Maasai pastoralists. In 2009, this led to the brutal invasion of village land, with mass arson and other human rights crimes committed by the police and rangers. In an extreme drought year, hundreds of *bomas* were burned to the ground, people were beaten, and cattle were pushed into even drier areas. During the stampede a young girl, 7-year-old Nashipai Gume, was lost and has never been found.

In 2010, a land use plan was drafted without the consultation of the people and presented to the councillors for approval. The plan proposed turning 1,500 km^2 of land into a conservation area, totally against the Village Land Act No. 5 of 1999 (URT 1999) or the wish of the people (NDC & NLUPC 2010). The councillors, exercising their statutory powers, annulled the land use plan, and in collaboration with the community in 2011, the Maasai managed to stop that draft district land use plan.

However, the government – refusing to believe that it had been defeated by a handful of pastoralists – devised a plan worthy of their colonial predecessors: divide and rule. This situation worsened week by week, and by 2016 the police had become more repressive to please both the investors (OBC and the American Thomson Safaris, which claim ownership of 12,617 acres) and the government. The police continued to intimidate and silence many people. While the Ministry of Natural Resources and Tourism tirelessly worked for the full land alienation wanted by OBC, intimidation, threats, persecution, and illegal arrests increased, weakening local leaders more than in 2009 and 2013. For example, I was arrested along with four people and detained for 11 days in August 2016, maliciously prosecuted for 'espionage and sabotage'. All these abuses were done and implemented in the interests of conservation – and, of course, tourism revenue.

The day of my arrest: personal recollection

On 19 July 2016, I was attending the ordinary meeting of the Full Council for Ngorongoro District. It was a normal day for myself and 37 other councillors.

As the session ended for the day and I joined other colleagues to leave the building, a plainclothes officer charged towards where I was seated and beckoned for me to follow him. In Loliondo, everybody knows everybody, so I knew him immediately when I spotted him. I did not know his reason for being there!

When I stepped outside, he told me that I was required to report to the police station for a small matter. He pushed me in the police vehicle, and as we were leaving, he saw another female councillor, murmuring to other police officers that they must bring her as well. We arrived at the police station and I was locked up in a cell. I found two other people there, school teachers. Later in the evening, the lady councillor's husband (a former member of Parliament) was brought in and pushed into my cell. We were not questioned, nor did we have our statements taken that day, not until very late the next day. I was not told the reason for being detained. I heard from friends who visited me in the cell that my charges involved deals with international terrorists and government overthrow.

In the afternoon of the following day, I was called for questioning. In the room, there were at least eight people, all from the government, representing immigration, chief of police, army, intelligence/secret service, and local administrators. They asked me many unnecessary questions, all related to the land. However, one of the statements that has stuck with me to this day was 'as an elected councillor, did you advise or tell your people to relocate elsewhere in Tanzania to free the land for the Arab?' The Maasai NGOs should have done research to identify empty lands for the Maasai to go instead of staying on the land that belongs to someone else and cause troubles! I was released the evening of 20 July 2016, with orders to return when summoned in future. I was instructed to inform my people that the land has been given to the Arab investor and the people should go to Dodoma, Mbeya, or anywhere else in Tanzania! An impossible order to obey!

In late 2016, the Prime Minister visited Loliondo and was confronted by the people. He promised to address their plea, and then he formed a famous commission, popularly known as the Gambo Commission. After almost a year of work, the committee presented the report detailing the opinion and irreducible demands of the people with regards to their land rights.

An unexpected event happened in August 2017, at a time when everyone was waiting for a decision by the Prime Minister. There was a repeat of the 2009 invasion of village land by security forces and conservation agents, with mass arson, beatings, and seizing, even shooting of cattle in the last days of the operation, which went on well into October 2017 (Weldemichel 2020). There were alleged cases of rape committed by Serengeti rangers assisted by NCA rangers, OBC rangers, anti-poaching, local police, and others. After months of extreme brutality, the Minister for Natural Resources and Tourism was sacked in a Cabinet reshuffle, and his successor immediately stopped the operation and made some big promises, including that OBC would leave come January 2018, a promise which at time of writing (December 2020) is yet to be fulfilled.

To respond to the violations and brutality, authorities from four villages in Loliondo, representing the rest of the villages, filed a court case against the government in the East African Court of Justice (EACJ) during the invasion of village land (EACJ 2017; Oakland Institute 2018). In all the testimonies, cross-examinations and court arguments, the government claimed that the operation was undertaken to evict illegal occupants who have settled and built settlements in Serengeti National Park. This, of course, is not true, and we are still waiting for a decision from the Court to settle this matter. In no way could the decision to burn homes, beat people, and confiscate their property be justified in the name of conservation.

In 2018, a military camp was set up in Loliondo, an act most-termed as a way to provide reinforcements to the existing hostile mechanisms. As expected, the presence of the camp and regular activities of the soldiers – especially on the disputed land – constituted serious forms of intimidation and harassment that instilled fear. Towards the end of the year, the soldiers – in total violation of temporary court orders by the EACJ (Oakland Institute 2018) – continued to burn down *bomas* in wide areas around hunting camps, seriously intimidating activists and leaders.

The UN has passed famous declarations that could protect Indigenous rights to land and life, like UNDRIP, principles like FPIC, and the Convention on Biological Diversity (CBD), but sadly, these are regarded as toothless instruments (Cooper & Noonan-Mooney 2013; Fitzmaurice 2015). Tanzania is a signatory to UNDRIP, which was passed 17 years ago – but it does not, in the least, recognize any group or tribe to be Indigenous in the country (IWGIA 2011, 2012).

Conclusion

Entitlement to life is a universal human right that should be observed regardless of the context. Although conservation rhetoric is applauded for ensuring sustainability of the planet (and mitigating climate change), it should never compromise the human rights of people, least of all the Indigenous and local communities. Conservation authorities and government agencies should revisit their mantra of protected areas to make them humane. These authorities should adhere to and respect human rights by all means. If tourists come and visit these areas by the thousands to enjoy, relax, and relieve fatigue, why should the Indigenous and local communities suffer insurmountable costs in order to live close to these areas? At this point in time, the losses that these people have to bear, that are incurred almost daily, should be compensated, including their lost lands. At the very minimum, conservation should be inclusive to accommodate all interests proportionally with regards to the rights of local communities, especially those that have risked so much for the existence of conservation and protected areas. The Maasai people admire wildlife and practice their own forms of nature conservation. We should be recognised and given an opportunity to contribute as much.

References

Bluwstein, J., Homewood, K., Lund, J. F., Nielsen, M. R., Burgess, N., Msuha, M., Olila, J., Sankeni, S. S., Millia, S. K., Lazier, H., Elisante, F., & Keane, A. (2018). A quasi-experimental

study of impacts of Tanzania's wildlife management areas on rural livelihoods and wealth. *Scientific Data, 5*, 180087.
Cooper, D. H., & Noonan-Mooney, K. (2013). Convention on biological diversity. *Encyclopedia of Biodiversity*, 306–319.
EACJ. (2017). *Reference no. 10 of 2017 Ololosokwan village council & 3 others vs the Attorney General of the government of the United Republic of Tanzania*. East African Court of Justice. www.eacj.org/?cases=reference-no-10-of-2017-ololosokwan-village-council-3-others-vs-the-attorney-general-of-the-government-of-the-united-republic-of-tanzania
Fitzmaurice, M. (2015). The 2007 United Nations declaration on the rights of Indigenous peoples. *Austrian Review of International and European Law Online, 17*(1), 137–265.
Gardner, B. (2012). Tourism and the politics of the global land grab in Tanzania: Markets, appropriation and recognition. *Journal of Peasant Studies, 39*(2), 377–402.
IWGIA. (2011). *Indigenous peoples in Tanzania*. International Work Group for Indigenous Affairs. www.iwgia.org/en/tanzania/654-indigenous-peoples-in-tanzania
IWGIA. (2012). *Country technical note on Indigenous peoples' issues: United Republic of Tanzania*. International Work Group for Indigenous Affairs. www.ifad.org/documents/38714170/40224460/tanzania.pdf/59a6ddbc-fb50-4ae0-a4df-9277a89152d7
Kipuri, N., & Sørensen, C. (2008). *Poverty, pastoralism and policy in Ngorongoro: Lessons learned from the Ereto I Ngorongoro Pastoralist project with implications for pastoral development and the policy debate*. London: International Institute for Environment and Development.
LHRC. (2014). *Operesheni Tokomeza Ujangili report*. Dar es Salaam: Legal and Human Rights Centre.
Mayengo, G., Bwagalilo, F., & Kalumanga, V. E. (2017). Human wildlife conflicts to communities surrounding Mikumi National Parks in Tanzania: A case of selected villages. *International Journal of Environment, Agriculture and Biotechnology, 2*(4), 1777–1784.
Melubo, K., & Lovelock, B. (2019). Living inside a UNESCO World Heritage Site: The perspective of the Maasai community in Tanzania. *Tourism Planning & Development, 16*(2), 197–216. doi:10.1080/21568316.2018.1561505
Minja, G. (2020). *Exploration and evaluation of the alternative wildlife management options for the Loliondo game controlled area in Tanzania: Multi-criteria analysis*. Doctoral dissertation. University of Adelaide.
NDC, & NLUPC. (2010). *The United republic of Tanzania Ngorongoro district council draft district land use framework plan (2010–2030)*. Dar es Salaam: National Development Corporation/National Land Use Planning Commission.
Nelson, F., & Makko, S. O. (2005). Communities, conservation, and conflicts in the Tanzanian Serengeti. In B. Child & M. W. Lyman (Eds.), *Natural resources as community assets: Lessons from two continents* (pp. 123–145). Madison, WI: Sand Country Foundation/Aspen Institute.
Oakland Institute. (2018, September 27). *Maasai villagers win a major victory in the East African Court of Justice in case against Tanzanian government*. www.oaklandinstitute.org/maasai-victory-east-african-court-justice-tanzanian-government
Shivji, I., & Kapinga, W. (1998). *Maasai rights in Ngorongoro Tanzania*. London: International Institute for Environment and Development.
TANAPA. (2019). *Ten years arrival trends*. Dar es Salaam: Tanzania National Parks.
URT. (1959). *The National Parks Act, 282*. United Republic of Tanzania.
URT. (1999). *The Village Land Act, 5*. United Republic of Tanzania.
Weldemichel, T. G. (2020). Othering pastoralists, state violence, and the remaking of boundaries in Tanzania's militarised wildlife conservation sector. *Antipode, 52*(5), 1496–1518.

5 "Environmental defenders"

The power/disempowerment of a loaded term

Judith Verweijen, Fran Lambrick, Philippe Le Billon, Felipe Milanez, Ansumana Manneh, and Melissa Moreano Venegas

"Environmental defenders", also called "environmental and land defenders" or "environmental human rights defenders (EHRD)" are recent terms for an old phenomenon: people fighting to protect themselves, their community, their land, and their ecosystems against a range of threats, including dispossession, pollution, and unsustainable resource use. In many cases, these threats stem from extractive industries, agro-businesses, and large-scale infrastructure and energy projects. The term "environmental and land defenders" was initially used to describe environmentalists and lawyers fighting destructive projects through US courts in the 1970s (Anderson & Rosencranz 1975; Wandesforde-Smith 1974). In the 2000s, the terms gained currency within the United Nations (UN) human rights apparatus. In this context, environmental and land defenders were considered a subset of human rights defenders working on economic, social, and cultural rights (Forst 2014; Knox 2017). In parallel, the designation "environmental (human rights and/or land) defenders" and related subject matter have increasingly been taken up by non-governmental organizations (NGOs) (e.g., Global Witness 2012), journalists (e.g., Watts & Vidal 2017) and a rapidly growing number of academics (Butt et al. 2019; Le Billon & Lujala 2020; Middeldorp & Le Billon 2019; Rasch 2017; Scheidel et al. 2020).

The institutionalization of the designation "environmental (human rights) and land defenders" has gone hand in hand with the development of an international infrastructure to support at-risk defenders. In 2018, UN Environment adopted a policy on "Promoting Greater Protection for Environmental Defenders" (UNEP 2018), and in 2019, the UN Human Rights Council unanimously passed a resolution on "Recognizing the Contribution of Environmental Human Rights Defenders".[1] In Latin America, the Escazú Agreement on Access to Information, Public Participation and Justice in Environmental Matters in Latin America and the Caribbean, adopted on 4 March 2018, sets out the requirement that states protect human rights defenders engaged in protecting the environment.[2] Furthermore, numerous NGOs, such as Not1More and those united in the Defending Land and Environmental Defenders Coalition, assist defenders, for instance, through judicial assistance, helping them stay under cover or building their protection skills. NGOs also engage in raising awareness about defenders at risk, including by keeping tallies of killed defenders (see e.g., Global Witness 2020) and reaching out to media outlets.

Many of these NGOs are based in the Global North, while the majority of the defenders they help are based in the Global South. This raises the question whether the term "environmental/land defenders" is not more a construct developed in the North to categorize people and actions in the South without this corresponding to their own views and discourses. What do those engaged in fighting socio-environmental injustices themselves think of the term "environmental defenders"? Is this a term they identify with and that accurately reflects their own sense of belonging and of what they do? Or is this more an outside categorization that does not correspond to how they designate themselves and their struggles? Relatedly, to what extent does the term, and the infrastructure that has emerged around it, help those labelled defenders and advance their struggles? Are there also certain dangers related to being identified or self-labelling as "environmental/land defenders"?

This contribution addresses these questions by drawing on our field research on those considered "defenders" in a number of different settings, including Brazil, Cambodia, Canada, Colombia, Guinea-Bissau, the Democratic Republic of the Congo (DRC), Ecuador, and the UK. Despite the fact that they often do not use the label themselves, we will simply call them "defenders" herein, given that it has proven difficult to find a single uncontested term that accurately captures this heterogeneous group. We do not aim to provide definite answers to the questions raised above – which would require more systematic research – but intend to share initial observations to provoke further discussion. Moreover, our observations are limited to our own research sites: we cannot speak for those considered defenders in other contexts.

The rest of this chapter proceeds as follows. The first part reflects on why the term "environmental defenders" is loaded and unevenly used. We then turn to the perspectives of defenders themselves. The next section looks at the extent to which the use of the term "environmental defenders" has helped or hindered struggles for socio-environmental justice. We conclude by reflecting on the practical implications of our findings.

The uneven geographies of a loaded term

Words are not neutral tools of speech: they always carry particular meanings and values. Through its second component, the term "environmental defenders" resembles "human rights defenders". As such, it is inscribed in the logic and language of human rights. This is even stronger the case with the term "environmental human rights defenders". The human rights project has its origins in the European Enlightenment. Among the many criticisms levied against this project are its universalizing tendencies, its Eurocentrism, and its origins in liberal individualism (Baxi 2007; Douzinas 2000). The notion "environmental defenders" carries some of this baggage, too. It directs attention to particular individuals, who are singled out for their activism, rather than seeing this activism as a product of collectives as a whole. In addition, it tries to place a heterogeneous group within a single category, assuming that the corresponding label is adequate across regions and cultures. Despite this baggage, the discourse of human rights has also been acknowledged to have worldwide emancipatory potential. In addition, it has been a source of hope and aspirations for a better future (Sikkink 2017). Moreover, there are growing efforts to "decolonize" the human rights project and construct a counter-hegemonic theory and practice of human rights (Barreto 2013). Environmental defenders' association with human rights is therefore a mixed blessing.

The first component of the term, "environmental", which is sometimes complemented by "land", also has particular connotations. It is grounded in the idea of the environment as a notion that is clearly separate from "society", which reflects a particular (western) worldview (Descola & Pálsson 1996). In addition, it suggests that those labelled "environmental or land defenders" fight primarily for the environment and/or their land. However, the "right to a clean and healthy environment" as established in international law has a relatively narrow and anthropocentric remit (Atapattu 2002). Indeed, those protecting particular rivers, mountains, rock formations, trees, or lands may be more motivated by the spiritual value these entities hold, or the inherent value they have, than by the desire to have a clean environment (for humans) per se. Furthermore, defenders may strive predominantly to preserve their lifeworld (of which the environment is only one part) or their livelihoods (which may depend on a clean environment, or "intact" nature, but also on "resources" such as wildlife). As stated by Martinez-Alier (2014: 240):

> The thesis of the "environmentalism of the poor" does not assert that as a rule poor people feel, think and behave as environmentalists. This is not so. The thesis is that in the many resource extraction and waste disposal conflicts in history and today, the poor are often on the side of the preservation of nature against business firms and the state. This behaviour is consistent with their interests and with their values.

Other defenders may aim more for the protection of nature, ecosystems, biodiversity, animal rights, or earth itself rather than "the environment". For many forest communities, for instance, the forest constitutes their entire world, implying they see no ontological difference between the forest and the world. Therefore, they fight to preserve the world at large, not their or "the" environment. For these reasons, the designation "environment" may not always accurately represent activists' objectives, motivations, and identification.

In sum, as with most categorizations, the term "environmental defenders" is loaded and contested, as it homogenizes a diverse set of actors and their projects. Yet, it is unclear to what extent and how this has affected its uptake among those considered defenders themselves as well as among other groups. There appear to be great variations in the use of the term "environmental defenders" per (language) area, as reflected in a simple Google search.[3] The term "environmental defenders" in English yields 205,000 results, compared to 13.8 million for the term "environmentalist". The French "*défenseurs de l'environnement*" – which signifies environmentalists and nature conservationists in general – has 469,000 results, while only 3.3% of internet users are primarily Francophone.[4] The Spanish "*defensores del ambiente*" and "*defensores del medioambiente*" yield 290,000 and 297,000 results, respectively, with Spanish speakers representing 7.9% of internet users. Finally, the Portuguese "*defensores ambientais*" gives just 13,700 results, with Portuguese speakers constituting 3.3% of global internet users. For the English and Portuguese terms, the first entries include UN Environment, Global Witness, and the International Union for Conservation of Nature, while for the Spanish "*defensores del medio ambiente*", Protection International and Human Rights Watch are among the first entries. This illustrates how the term is mostly used in UN and NGO circles.

Aside from being used unevenly in different languages, the term "environmental defenders" is not equally used across the Global North and the Global South. While the French term is commonly also employed to designate defenders, environmentalists, and conservationists in Francophone countries in the Global North, it is rarely used in Francophone countries in Africa. For instance, fieldwork in the DRC learned that most activists speaking French were either unfamiliar with the term or simply did not use it. Contrary to the French term, the English "environmental defenders" is rarely used to describe people in the Global North. A more common designation for this group is "environmental activist", which yields 2,240,000 results on Google. These disparities raise the question of where and by whom the term "environmental defenders" is taken up and when not.? To answer these questions, we must listen to those labelled defenders themselves.

Perspectives of defenders

In contexts where the term "environmental (human rights)/land defenders" and its various translations are not often used, we see a plethora of other terms in use. For instance, our fieldwork in Ecuador shows that other Spanish terms circulate among urban activists and the media, such as "life defenders", "nature defenders", or "ecologists". Ecuador was the first country to enshrine the rights of (mother) nature in its constitution, which were inaccurately translated as *Pachamama* in the Indigenous language Kichwa. This term has generated much debate, since it reduces a notion with a broad meaning in Andean–Amazonian philosophy to a narrow understanding of "nature". The designation of environmental defenders has similarly generated vivid debate. After a harsh confrontation with defenders, President Rafael Correa introduced the pejorative term "*ecologistas infantiles*" (childish ecologists) for activists struggling against extractive industries – a term that gained traction in pro-extractivist circles.

The activists we interviewed identified with the term defenders only partially. For example, one activist whose brother was killed in 2002 for organizing communities against oil extraction in the Amazon told us: "I prefer the term fighter (*luchador*)". He further explained: "We fight for life, because we do not want the planet to be stained as it is stained" (personal interview, August 2019). Another interviewee, who was a colleague of the murdered activist, described himself as a "*defensor de derechos humanos y ambientales*" (defender of human and environmental rights). He is a self-taught lawyer and coordinates the *Oficina de derecho ambiental* (Office of Environmental Law), where they give legal advice and representation to peasants and farmers affected by the oil industry. However, he also identifies as an "*ambientalista*" (environmentalist) (personal interview, August 2019).

Among Indigenous activists in Ecuador, the term "territory" often arises. As one of them explained:

> Here we are defending the territory, more than anything we are all conservationists of ecology (*conservadores de la ecología*). We are defending our territory for our children. Because later, when the mining company enters, it will totally destroy our territory. So we don't want that, we defend our territory for the good of our children and for the entire good of the Ecuadorian country.
>
> (personal interview, May 2019)

Some also affirm their ethnic identity associated with what can be seen as an "environmental vocation":

> We are Shuar. We are *conservadores y nativos de aquí* (conservationists and native from here), we do not cut the trees. . . . I was born here, my father was born here, may he rest in peace. That is why we do not want contamination, because if the mining company will be established, it is very polluting, all practices of the company.
>
> (personal interview, May 2019)

We may also consider that in this particular context, the community has been subject to military violence, hence the term "defender" may be associated with "defence" against a peril (of being subject to that kind of violence again).

In Cambodia, interviewees gave a wide variety of responses to the question of how they would describe themselves. Some would label themselves simply as "a member of the community", a "human rights defender", or "one of the people". In certain contexts, the term "land defender" resonated more than "environmental defender". In the DRC, the term "community" (*lisanga* in Lingala; *communauté* in French) also figured prominently in how defenders described themselves, especially when they were customary chiefs. When asked how he identified, a chief engaged in a struggle against an industrial logging company answered: "I am the leader of the community. Our aim is to protect tomorrow's life, to prepare the future of our children. The forest is something to look after" (personal interview, May 2018). Another local leader, fighting against an industrial palm oil company, said:

> I consider myself a community leader. I am afraid that they [palm oil company] will take away the little bit of land that remains. If they also take that

> away, I cannot but cry. If we had not resisted the company would have taken it all. And if I die, my children will continue to cry. It's better to die than to live with these stupidities. That's why I fight, so that it does not happen to me while I live.
>
> (personal interview, May 2018)

In the UK, research into violence against "environmental defenders" reveals that the term is not generally used. The more commonly used term is, rather, "environmental activists". Moreover, people engaged in environmental movements, especially in direct action, often call themselves "protectors" (Brock 2020). This is notably established in the anti-fracking movement and contrasts with the labels "protestors" and "activists". This last term has negative connotations, particularly in official statements and anti-terrorism policies set out by the State. Guidance on policing anti-fracking protests issued in 2015 by the National Police Chiefs Council (NPCC) contained a diagram on the "structure of protest" that defined "activism" as involving criminality (criminal damage) and saw it as the last stage before "extremism" (Jackson et al. 2019). In 2020, counter-terrorism police in the South-East of the UK listed Extinction Rebellion, a movement encouraging civil disobedience to put pressure on governments to take action on climate change, as an "extremist" group, stating that "an anti-establishment philosophy that seeks system change underlies its activism" (BBC 2020). By labelling themselves "protectors", those involved in protest are able to counter the negative narratives that accompany the idea of "environmental activists", and to emphasize the motivation and purpose of their actions, related to protecting land, ecosystems, and the environment. Moreover, in certain contexts the term "protector" may highlight how they protect one another against police violence (Jackson et al. 2016).

In Brazil, the term "*defensores ambientais*" gained much attention after its use in UN Environment and Global Witness reports, which pointed to Brazil as the most violent country in the world for defenders. These reports considered a wide range of disparate groups as "environmental defenders", including all rural workers, peasant union leaders, Indigenous leaders, rubber tappers, members of the *Movimento dos trabalhadores sem terra* (MST, Landless Workers' Movement), traditional populations, and community leaders in forest areas. These groups indeed constitute the largest share of the victims of violence from landlords and others, as tallied year by year by reports of the *Comissão pastoral da terra* (CPT, Pastoral Land Commission), which supports small farmers and the landless who, however, in many cases they do not consider themselves "environmental defenders".

Rural social movements rather use the terms "grassroots environmentalist" (*ambientalista popular*) or "communitarian environmentalist". These terms were introduced by Professor Moacir Gadotti, one of the founders of "ecopedagogy" and a follower of Paulo Freire, the author of the seminal *Pedagogy of the Oppressed* (1970 [1968]). Ecopedagogy aims to develop forms of critical ecoliteracy and knowledge grounded in sustainability, biophilia, and planetarity (Kahn 2010). For instance, Zé Claudio and Maria, collectors of nuts in the

Amazon who were assassinated by gunmen hired by ranchers, used to designate themselves as "*ambientalistas populares*". They would defend life together with the forest (Milanez 2020). Zé Claudio once said at a TEDx conference: "I live from the forest. I will protect her by any means. For this, I live with a bullet in my head at any time".[5] The United Nations honoured them in 2012 in memoriam as "Forest Heroes"[6] – a designation preceding that of Forest Defender.

Indigenous peoples and other "traditional populations" in Brazil often prefer to use the term "defenders of life", with life seen in a broader sense. Since the killing of the famous rubber tapper leader Chico Mendes, Indigenous peoples have incorporated the fight for the environment as a struggle against dispossession. This shows again the multifaceted nature of the struggles of those labelled "defenders", and how these are not adequately captured by the term "environmental defender".

Enabling or endangering struggles?

Just as terms are never neutral, they are also never "mere words"; they have – sometimes profound – social and political effects. Here we consider four ways in which the term "environmental (and land) defender" affects socio-environmental struggles, whether helping or hindering them: 1) legitimizing and delegitimizing particular struggles, movements, actors, and actions; 2) the martyrization of defenders, which may raise the visibility of their cause but also puts them at increased risk; 3) individualization, which on the one hand eases access to international support networks but on the other endangers defenders by creating individual responsibility and alienating them from communities that can provide them with crucial support; and 4) political ostracization, due to placing defenders in certain criminalized categories of activists and leaders.

Legitimation and delegitimation

The term "environmental and land defender" can be used both to legitimize and to delegitimize particular people and, by implication, their actions, causes, and the groups or movements they are part of. A good example of how the term can legitimate struggles is by asserting Indigenous sovereign authority over settler–colonial authority. In Canada, an Indigenous youth arrested for breaching a court injunction against a pipeline blockade asked the judge, "why do you keep calling us protesters when we're not? We're land defenders". When the judge failed to respond, the youngster replied,

> Why can't you respect me? I thank the land and the sacred water, and I protect all the medicines here. . . . I'd like to know why you won't give us respect for protecting our land. Why must we be treated as a criminal for defending mother earth – everyone's mother here? We've done nothing wrong here. You're on unceded territory here . . . we have the right to defend our land.
>
> (cited in Simpson & Le Billon 2021)

The challenged settler-state judge responded by threatening to remove the *defendant* from the courtroom – thereby re-inscribing the violent primacy of the colonial order in his narrative while denying the youth his Indigenous political status as a legitimate *defender* of his land, medicines, and environment. This rhetorical move was clearly aimed at delegitimizing his struggle (Simpson & Le Billon 2021).

Martyrization

As demonstrated by revolutionary political groups, the figure of the martyr can play an important role in the construction of social movement identity and mobilization (Guerra 2018; Krutzsch 2019). A key example in relation to environmental defenders is the "Forest Hero", which was institutionalized in the UN system in 2011, with the annual UN Forest Hero Award.[7] Another example is the Goldman Environmental Prize, which states on its website that it "honors grassroots environmental heroes".[8] By associating an individual or a community with a broad cause – environmental and land justice – rather than their personal grievances and the specific conditions of their struggle, "environmental defenders" become framed as martyrs. This shift is generally operated through narratives emphasizing the persecution of defenders and the righteousness of their cause. The resulting martyrization can broaden public support, consolidate alliances, increase the legitimacy of their struggle, and therefore enhance their chances of success (Rowell 2017; Scheidel et al. 2020). Yet, persecution and other legal harassment can also take a toll on movements and deter new members.

The martyrdom effects of singling certain people out as heroic "environmental defenders", as well as the effects on the outcomes of resistance, depend on a range of factors. These include: the political cultures at play, the intensity and duration of grievances, the prior profile of the defenders and their cause, the nature and strength of their solidarity network, effects on bodies and minds, and the processes of mobilization that follow (Conde & Le Billon 2017; Rahmouni Elidrissi & Courpasson 2019; Nixon 2016). In strong police states, extensive surveillance and systematic repression can methodically undermine socio-environmental movements. In such contexts, martyrdom may not give way to mobilization but may simply lead to attrition. As a Chinese environmental NGO leader explained, "if we all become martyrs, then who is left to do the work?" (Lu 2007: 3). Somewhat similarly, environmental defenders who are persecuted in countries with high levels of political violence and homicides may not see much domestic or international media attention to their cause, even in the case of murder. In Honduras, it took the death of a very high profile "defender" – Berta Cáceres – to bring about some concrete if very limited action by a government under which at least 100 "defenders" had been killed (Middeldorp & Le Billon 2019).

The martyrdom effects of the environmental defenders label also depend on the position, outlook, and strategies of the organization and individuals involved. Whereas some members of grassroots movements may be literally willing to die for their cause, many environmental organizations selectively engage in actions according to the relative degree of protection they can benefit from. In this regard, larger environmental NGOs sometimes expose grassroots defenders to risks that they themselves do not have to face. Northern NGOs, in particular, bear a measure of responsibility for the fate of defenders in the South. They promote challenging authorities and business interests without providing adequate protection for local defenders, who bear the brunt of retribution by companies and state authorities (Grant & Le Billon 2019, 2020).

Individualization

Martyrization points to another effect of the term "defender", namely, how it tends to emphasize the individual over the broader communities and organizations involved in environmental and land struggles. While such individualization can help raise public awareness by literally "giving a face" to socio-environmental struggles, it also puts individualized defenders at increased risk, as they become likely targets for opponents seeking to intimidate communities and deter leadership. Individualization can also create tensions within affected communities by singling out particular individuals or their families, who are then held responsible for acts of retribution or repression that affect communities as a whole. Tensions can also result where communities are divided on the struggles in question, and opponents take it out on individual defenders. As observed during fieldwork, in Guinea-Bissau, environmental defenders are seen as an obstacle by community members who exploit natural resources to make a living, and who perceive having the right to exploit these resources.

Individualization is exacerbated by the focus that many advocacy and media reports place on killings, rather than the broad range of pressures exercised on communities. However, some academic and policy reports on environmental and land defenders have been cautious to emphasize the many forms of violence to which defenders, movements, and communities are exposed and the collective character of the defence of land and the environment (e.g., Scheidel et al. 2020). For instance, in its definition of defenders, the UN Environment Programme (UNEP 2018) specifically mentions "groups of people". Yet, the logics and practices of advocacy and media reporting around the term "environmental defenders" and the focus on those killed can individualize "their" struggles, thereby obliterating the deeply rooted nature of resistance.

Political ostracization

Authoritarian regimes generally have repressive legislation and employ (extra) judicial mechanisms against human rights defenders, as they are seen as political opponents to ruling elites and a threat to single-party ideology. As such, the term "environmental defender", and especially EHRD, may cast defenders – and their communities – within the logics of broad political opposition, rather than the more limited (if related) issues associated with threats to land, livelihoods, and environmental protection (Middeldorp & Le Billon 2019). In

Columbia, the term "environmental leaders" (*líderes ambientales*) is associated with that of "social leaders" (*líderes sociales*), who are mostly community leaders or unionists that are cast as members of leftist movements. This framing renders them a frequent target of paramilitary forces, criminal organizations, and some government security forces, which accuse them of being enemies of "development" or even allies of leftist guerrillas (Pérez 2018).

During the 2014 electoral campaign in Guinea-Bissau, a time when illegal logging was at its peak, environmentalists were targeted by politicians who were fuelling their campaigns with revenues from illegal logging. Consequently, the term "environmental defenders" became a dangerous designation, and there was no protection for environmentalists during this period. Silence was the only tool for these activists to protect themselves against intimidation and harassment by timber barons with close connections to high-ranking government officials.

Political ostracization and associated criminalization, however, occur in not only authoritarian states but also (supposed) democracies (Brock & Dunlap 2018; Brock 2020). Canada is a case in point. In the words of the Canadian minister in charge of natural resources, environmental defenders "threaten to hijack our regulatory system to achieve their radical ideological agenda. . . . They use funding from foreign special interest groups to undermine Canada's national economic interest" (NRC 2012; see also Le Billon & Carter 2012; Matejova et al. 2018). The so called "War on Terror" intensified the stigmatization and criminalization of activism in both North America and the EU (Balfour 2014; Brock & Dunlap 2018). For instance, Europol qualifies various forms of protest and action against resource extraction companies as "single issue terrorism", which has led to increasing surveillance and criminalization (Monroy 2011). While the term "environmental defenders" is rarely used in these contexts, this situation does highlight how those engaged in socio-environmental struggles easily come to be seen as a broader threat to ruling elites and their vested interests.

Conclusion

Presenting the voices of those designated as "environmental (and land) defenders" and evaluating the effects of this designation on the struggles they are engaged in, this chapter has demonstrated that there are both drawbacks and advantages to this label. Within our research contexts in South and North America, Africa, Asia, and Europe, only a few groups and movements label and identify themselves as "environmental (and land) defenders" or "environmental human rights defenders". There appears to be a broad consensus that this term does not accurately reflect their identities and struggles and, in many cases, is seen as an "outside" designation used primarily by international media, NGOs and the UN. Despite this, there appear to be limited concerted efforts to actively contest the term.

Many of those labelled defenders collaborate with international media reporting on their cause, accept "Forest Hero" awards and other prizes, and make use of the resources and infrastructure aimed at "defending defenders". Clearly, this is often a strategic move to mobilize international media and policy attention, attract resources, and enhance their own and others' safety. However, as this chapter has shown, these different forms of support and recognition may have inadvertent consequences and lead to increased repression and risks. Another reason why there are no coordinated efforts to change the notion of environmental defenders is that there is no readily available alternative umbrella term that adequately captures all the different groups included, and their varying drivers and objectives. At the same time, these groups and individuals do have certain things in common: they fight unjust practices by multinationals and governments that inflict ecological and social damage. As such, despite their diversity, they often have many experiences and views in common.

Where do these observations lead us in respect of the continued use of the term? We suggest that media, NGO, and UN agencies reporting on defenders intensify their efforts to (also) present the labels these groups use themselves, to more accurately depict their self-identification, motives, and objectives. We also believe that organizations supporting defenders should do more to mitigate the potential counterproductive effects of their activities. For instance, to avoid martyrization and individuation, organizations could give awards and other prizes to collectives rather than individuals. This may help avoid a situation where certain highly visible defenders run most of the risks. Moreover, instead of focusing on a few prominent and vocal "heroes", they could present a broader array of participants in resistance. In addition, they should abandon the narrow focus on killings and other forms of spectacular violence and foreground the entire spectrum of violence and repression to which defenders are exposed. Furthermore, organizations supporting defenders should intensify monitoring the political climate in which defenders operate and conduct profound risk assessments before providing any assistance. Finally, similar to all organizations from or mostly financed by the Global North, they should be hyper-reflexive about their privileges and the profound socio-economic and often racial inequalities separating them from the defenders they are committed to support.

Notes

1 Human Rights Council 40th session, February 2019, RES/40/11, Recognizing the contribution of environmental human rights defenders to the enjoyment of human rights, environmental protection and sustainable development, www.right-docs.org/doc/a-hrc-res-40-11/ [accessed 11 December 2020].

2 Regional Agreement on Access to Information, Public Participation and Justice in Environmental Matters in Latin America and the Caribbean, https://repositorio.cepal.org/bitstream/handle/11362/43583/4/S1800428_en.pdf [accessed 11 December 2020].

3 The Google search for all terms listed was conducted on 11 December 2020.
4 Figures derived from Internet World Stats, "Internet World Users by Language", www.internetworldstats.com/stats7.htm [accessed 10 December 2020].
5 See "Killing trees is murder: Zé Cláudio Ribeiro at TEDxAmazonia", November 2010, www.youtube.com/watch?v=XO2pwnrji8I [last accessed 14 December 2020].
6 See "Forest Heroes – Jose Claudio Ribeiro and Maria do Espirito Santo, Special Award", www.youtube.com/watch?v=iJ5NH3-sFkY [last accessed 14 December 2020].
7 See www.un.org/esa/forests/outreach/forest-heroes/index.html.
8 See www.goldmanprize.org/about.

References

Anderson, F. R., & Rosencranz, A. (1975). The future of environmental defense. *American Bar Association Journal, 61*(3), 316–323.

Atapattu, S. (2002). The right to a healthy life or the right to die polluted?: The emergence of a human right to a healthy environment under international law. *Tulane Environmental Law Journal, 16*(1), 65–126.

Balfour, L. (2014). Framing redress after 9/11: Protest, reconciliation and Canada's war on terror against Indigenous peoples. *The Canadian Journal of Native Studies, 34*(1), 25–41.

Barreto, J. M. (Ed.). (2013). *Human rights from a third world perspective: Critique, history and international law.* Newcastle upon Tyne: Cambridge Scholars.

Baxi, U. (2007). *The future of human rights.* Oxford: Oxford University Press.

BBC. (2020, January 10). Extinction rebellion: Counter-terrorism police list group as "extremist" in guide. www.bbc.co.uk/news/uk-51071959 [last accessed 12 December 2020].

Brock, A. (2020). "Frack off": Towards an anarchist political ecology critique of corporate and state responses to anti-fracking resistance in the UK. *Political Geography, 82*, 102246.

Brock, A., & Dunlap, A. (2018). Normalising corporate counterinsurgency: Engineering consent, managing resistance and greening destruction around the Hambach coal mine and beyond. *Political Geography, 62*, 33–47.

Butt, N., Lambrick, F., Menton, M., & Renwick, A. (2019). The supply chain of violence. *Nature Sustainability, 2*(8), 742–747.

Conde, M., & Le Billon, P. (2017). Why do some communities resist mining projects while others do not? *The Extractive Industries and Society, 4*(3), 681–697.

Descola, P., & Pálsson, G. (Eds.). (1996). *Nature and society: Anthropological perspectives.* Abingdon, UK: Routledge.

Douzinas, C. (2000). *The end of human rights: Critical thought at the turn of the century.* London: Hart.

Forst, M. (2014). *Report of the Special Rapporteur on the situation of human rights defenders.* UN Human Rights Council. A/HRC/28/63.

Freire, P. (1970 [1968]). *Pedagogy of the oppressed.* London: Continuum.

Global Witness. (2012). *A hidden crisis: Increase in killings as tensions rise over land and forests.* London.

Global Witness. (2020). *Defender tomorrow: The climate crisis and threats against land and environmental defenders.* London.

Grant, H., & Le Billon, P. (2019). Growing political: Violence, community forestry, and environmental defender subjectivity. *Society & Natural Resources, 32*(7), 768–789.

Grant, H., & Le Billon, P. (2020). Unrooted responses: Addressing violence against environmental and land defenders. *Environment and Planning C: Politics and Space*, 2399654420941518.

Guerra, L. (2018). *Heroes, martyrs, and political messiahs in revolutionary Cuba 1946–1958.* New Haven, CT: Yale University Press.

Jackson, W., Gilmore, J., & Monk, H. (2019). Policing unacceptable protest in England and Wales: A case study of the policing of anti-fracking protests. *Critical Social Policy*, *39*(1), 23–43.
Jackson, W., Monk, H., & Gilmore, J. (2016). Pacifying disruptive subjects: Police violence and anti-fracking protests. *Contention: The Multidisciplinary Journal of Social Protest*, *3*(2), 81–93.
Kahn, R. (2010). *Critical pedagogy, ecoliteracy and planetary crisis: The ecopedagogy movement*. Bern: Peter Lang.
Knox, J. H. (2017). *Environmental human rights defenders: A global crisis*. Policy Brief. Geneva: Universal Rights Group.
Krutzsch, B. (2019). *Dying to be normal: Gay martyrs and the transformation of American sexual politics*. Oxford: Oxford University Press.
Le Billon, P., & Carter, A. (2012). Securing Alberta's tar sands: Resistance and criminalization on a new energy frontier. In M. A. Schnurr & L. A. Swatuk (Eds.), *Natural resources and social conflict* (pp. 170–192). London: Palgrave Macmillan.
Le Billon, P., & Lujala, P. (2020). Environmental and land defenders: Global patterns and determinants of repression. *Global Environmental Change*, *65*, 102163.
Lu, Y. (2007). Environmental civil society and governance in China. *International Journal of Environmental Studies*, *64*(1), 59–69.
Martinez-Alier, J. (2014). The environmentalism of the poor. *Geoforum*, *54*, 239–241.
Matejova, M., Parker, S., & Dauvergne, P. (2018). The politics of repressing environmentalists as agents of foreign influence. *Australian Journal of International Affairs*, *72*(2), 145–162.
Middeldorp, N., & Le Billon, P. (2019). Deadly environmental governance: Authoritarianism, eco-populism, and the repression of environmental and land defenders. *Annals of the American Association of Geographers*, *109*(2), 324–337.
Milanez, F. (2020, June 30). The defense of life within the forest: Documenting courage and audacity through film. *Revista: Harvard Review of Latin America*. https://revista.drclas.harvard.edu/the-defense-of-life-within-the-forest/
Monroy, M. (2011). Using false documents against "Euro-anarchists": The exchange of Anglo-German undercover police highlights controversial police operations, *Statewatch*, *21*, 1–8.
Nixon, R. (2016). Environmental martyrdom and defenders of the forest. *Lecture presented at The University of Texas at Austin*.
NRC. (2012, January 9). *Open letter from the Honourable Joe Oliver, Minister of Natural Resources*. Ottawa: Natural Resources Canada.
Pérez, C. E. (2018). Los enemigos del desarrollo: Sobre los asesinatos de líderes sociales en Colombia. *Iberoamérica Social: Revista-red de estudios sociales* (11), 84–103.
Rahmouni Elidrissi, Y., & Courpasson, D. (2019). Body breakdowns as politics: Identity regulation in a high-commitment activist organization. *Organization Studies*, 0170840619867729.
Rasch, E. D. (2017). Citizens, criminalization and violence in natural resource conflicts in Latin America. *European Review of Latin American and Caribbean Studies/Revista Europea de Estudios Latinoamericanos y del Caribe*, (103), 131–142.
Rowell, A. (2017). *Green backlash: Global subversion of the environment movement*. Abingdon, UK: Routledge.
Scheidel, A., Del Bene, D., Liu, J., Navas, G., Mingorría, S., Demaria, F., Avila, S., Roy, B., Ertör, I., Temper, L., & Martínez-Alier, J. (2020). Environmental conflicts and defenders: A global overview. *Global Environmental Change*, *63*, 102104.
Sikkink, K. (2017). *Evidence for hope: Making human rights work in the 21st century*. Princeton, NJ: Princeton University Press.
Simpson, M., & Le Billon, P. (2021). Reconciling violence: Policing the politics of recognition. *Geoforum*, Forthcoming.

UNEP (United Nations Environment Programme). (2018). *Policy on promoting greater protection for environmental defenders*. www.unenvironment.org/explore-topics/environmental-rights-and-governance/what-we-do/advancing-environmental-rights/uneps [last accessed 11 December 2020].

Wandesforde-Smith, G. (1974). The study of environmental public policy: A preliminary directory. *Human Ecology*, *2*(1), 45–62.

Watts, J., & Vidal, J. (2017, July 13). Environmental defenders being killed in record numbers globally, new research reveals. *The Guardian*. www.theguardian.com/environment/2017/jul/13/environmental-defenders-being-killed-in-record-numbers-globally-new-research-reveals [last accessed 14 December 2020].

6 Atmospheres of violence

On defenders' intersecting experiences of violence

Mary Menton, Grettel Navas, and Philippe Le Billon

Environmental and land defenders suffer violence in many forms. Most studies have so far focused on specific ones, with a particular focus on murder (Butt et al. 2019; Le Billon & Lujala 2020; Middeldorp & Le Billon 2019; Scheidel et al. 2020), forced displacement (Jeffery 2018; Tan 2020; Tripathi 2017), and gender-based violence (Deonandan et al. 2017; Sippola 2018; Moreano Venegas & van Teijlingen, this volume). As we argue here, violence takes many different overlapping and combining forms, resulting in what some defenders experience as a deep sense of insecurity and describe as 'a climate of fear', a sense of being constantly under attack. Such a climate, we suggest, results from what we call 'atmospheres of violence': assemblages of actors, institutions, logics, processes, and materialities characterized by pervasive and persistent forms of violence.

Such an understanding of violence *against* defenders requires a nuanced and multi-dimensional approach to an understanding of violence *around* environmental and land conflicts (Navas et al. 2018). In this chapter, we explore some of these different forms of violence, their manifestations, and their intersections. In this respect, we build upon the concept of 'violent environments' formulated two decades ago by Peluso and Watts (2001) to describe violent *processes* influenced by local histories as well as wider power relations and patterns of resource use and environmental transformation (see also Le Billon & Duffy 2018). We expand this concept, focused on temporality and spatialities through the idea of 'atmospheres of violence', a term seeking to emphasize: (i) a multi-dimensional understanding of violence, (ii) the assemblages that (re)produce violent processes, and (iii) the cumulative impacts of multiple and intersecting violences on defenders.

While we provide an overview of some of the most common forms of violence that defenders experience, we do not exclude the possibility that defenders experience others that we do not cover in this chapter. Given the growing importance of social media in creating virtual spaces in which defenders experience violence (through threats, smear campaigns, criminalization), and the spatially and temporally disperse nature of their lived experiences, 'atmospheres of violence' speaks also to the creation of climates of fear and oppression, to violences against the bodies, minds, and territories of defenders.

By environmental defenders, we mean 'any person (or group of people) who defends environmental and human rights, including constitutional rights to a clean and healthy environment, when the exercise of those rights is being threatened' (UNEP 2018), noting that environmental defenders can self-identify as such, but many engage in these activities through sheer necessity, because their lives and livelihoods are threatened with environmental damage (see also Verweijen et al., this volume). Perhaps even more than conscious 'environmental activists', many of the people or groups we consider here are also 'defenders', as they seek to protect, assert, or reclaim their 'land' – including in terms of ownership or user rights – under conditions of historical marginalization and ongoing repression (Sauer & de Castro 2019; Jay Chen 2020).

Galtung's triangle of violence

Violence is often framed as a triangle of different forms of violence: direct, structural, and cultural, as per Galtung (1969, 1990). In this case, direct violence represents a particular brutal event in a given space and time. It includes (but is not limited to) physical violence, sexual violence, psychological violence (e.g. death threats, intimidation), and repression of protests. In many cases, direct violence is more visible and apparent. At least one actor involved in the violence is visible, either those who suffer the violence or those who inflict it. Structural violence, on the other hand, is more opaque, as it stems from the social and societal structures that impact on individual and/or collective wellbeing. Rampant inequalities, systematic deprivation, discriminations, institutional racism, biased and corrupt criminal justice systems, and lack of basic services are all forms of structural violence that impact defenders. Cultural or symbolic violence, on the other hand, is the use of cultural elements like religion, ideology, or ethnicity to legitimize structural or direct violence. Cultural violence against defenders is framed by discursive criminalization and accusations of defenders being 'anti-development', 'violent radicals', or 'enemies of the state', 'anti-patriotic', and 'foreign funded activists' (Becerril 2019; Matejova et al. 2018; see Taylor, this volume). Beyond Galtung's typology, violence should be considered not simply as an 'act' or 'outcome' but rather as a relational process going far beyond the narrow frame of thinking about violence exclusively as location- and time-bound (Springer & Le Billon 2016).

Such a perspective allows for a fuller grasp of not only the multiple forms of violence, but also of their relations across time and places. Forced displacement of Indigenous peoples, for example, can thus be understood through its many facets – such as symbolic violence associated with accusations of 'backwardness' and supposed Indigenous desires to 'become modern like us' – and its extensive historical and geographical connections between Indigenous and settler societies, supply chain networks, and international political economies.

Repression of defenders

The repression of defenders often makes the news as a result of police brutality during peaceful protests, exposed threats, or targeted killings. State and corporate interests often respond through insidious rather than open forms of repression, but also through various forms of divide-and-rule and cooptation strategies not discussed in this chapter.[1] Police brutality against public demonstrations remains common, notably to end or deter lengthy and costly strikes and blockades (Franks et al. 2014; Dunlap 2019). Yet, everyday repression often takes a bureaucratic form – with a cumbersome, partial, costly, biased, and unfair 'rule of law', judicial system, and criminal code being used to constrain, discipline, and punish defenders. Such judicialization and criminalization of defenders can involve burdensome registration processes for civil society organizations, restrictions on foreign NGO funding, severe criminal charges based on minor or groundless offenses, or strict conditions for the expression of dissent (Deonandan & Dougherty 2016). The criminalization of dissent also takes discursive forms of delegitimization and a securitization of resource sectors, resulting in 'more investigation, infiltration, and disruption of radical environmental groups regardless of whether any law is actually violated, longer terms of incarceration for convicted activists, and the harassment of mainstream environmental groups' (Smith 2008: 564).

As Birss (2017) notes, insidious forms of repression – including judicialization, criminalization, defamation, intimidation, and execution through intermediaries – are often perceived as more effective by challenged authoritarian formations and corporations seeking to maintain a minimal degree of 'legitimacy' to sustain investments. Whereas the use of lethal force in response to (peaceful) demonstrations can result in direct condemnation and further mobilization of opponents, insidious forms of repression can deflect direct critique. In the case of judicialization and criminalization, states and corporations instrumentalize law to their benefit and can often take an open stance without fear of repercussion. Defamation, intimidation, and execution, in contrast, are generally conducted through intermediaries, including former military officers, private security contractors, and criminal entrepreneurs. Through 'subcontracting' repression, governments and corporations have the opportunity to publicly condemn abuses, deny complicity, and at least deny full responsibility. In turn, activists often fear insidious forms of repression the most, as these do not easily lend themselves to collective forms of protection, bring fear within everyday lives, and reduce accountability likelihood.

Deadly repression is understood as the result of impunity for perpetrators associated with the lack of independent and effective judiciary, the absence of (independent) media and other reporting, tight and unaccountable networks between political, economic, and military elites, social 'habituation' homicides on the part of authorities – notably as a result of recent civil wars – and state-tolerated/encouraged vigilante activity (Cruz 2011; Hill & Jones 2014). Deadly escalation also often results from high uncertainty in the capacity and behavioral norms among protagonists in a context of contentious politics, a situation characterizing intermediary political regimes falling between 'full' autocracies and democracies (Davenport 2007; Pierskalla 2010).[2] In such contexts, government authorities and corporations are frequently unwilling to follow the praxis of negotiated conflict settlement, while social movements refuse to back down, either based on principles or on the premise that sustained contestation will further erode authoritarian power, even if at the cost of deadly repression.

Slow violence

Contrary to what is commonly understood as violence in its direct and visible forms (Galtung 1969) or as a result of unequal power relations between social groups (Bourdieu 1979) and structures (Galtung 1990; Farmer 2004), Nixon's (2011) notion of 'slow violence' involves a long process characterized by little or invisible short-term impacts with devastating long-term and irreparable aftermaths. Slow violence seeks to move beyond these more visible forms of violence to include violence as a process and not simply as an act. For Nixon (2011), slow violence often escapes attention because it is dispersed across time and space; it occurs incrementally with cumulative effects. The slow-motion toxicity and its health outcomes, and the slow accumulation of toxins and consequent violence against bodies, is slow violence. Many defenders experience or are concerned with slow violence – including for inter-generational and environmental health issues – and their struggles often center around the extractive activities, processing factories, and agro-chemical processes that cause slow violence (see Souza et al., this volume).

Necropolitics

Defenders also suffer from necropolitics: the politics of determining who shall live and who shall die (Mbembe 2003). While some human (and non-human) lives are valued, others are seen, and treated, as dispensable or 'expendable'. Often the basis of colonial policies distinguishing between useful and useless land uses and populations, the concept has been reworked through ideas in the contemporary era of such terms as 'surplus population' for 'un-needed laborers' (Li 2017) or 'yield gap' for 'under-utilized lands' that could supposedly yield more calories per hectare under an agro-industrial model (Le Billon & Sommerville 2017). Indigenous land defenders in Brazil speak out against necropolitics and 'politics of extermination' of the government, manifest in both intentional violence and acts of omission (Menton et al. 2021; APIB 2019). Necropolitics thus refers not only to direct forms of killing and letting-die, but also to the systematic destruction of ecosystems and related livelihoods as a result of changes in modes of production and transformations of landscapes, including through polluting activities in the name of 'development' (Davies 2018) or protected areas in the name of 'conservation' (Cavanagh & Himmelfarb 2015).

Gendered violence

In addition to the other forms of violence, women environmental defenders also face gender-based violence (López & Bradley 2017; Castañeda Camey et al. 2020). Rape and threats of sexual violence are often used as weapons to control feminized bodies (Segato 2016) and to silence, stigmatize, and control women who dare to speak out in defense of the environment and their territories, as well as to discourage them from their activism (Barcia 2017; Castañeda Camey et al. 2020; Deonandan & Bell 2019). Woman defenders in particular also face discrimination and delegitimization in their role as defenders in their own organizations, which sometimes leads them to create their own collectives (Bolados & Sánchez 2017) and regional networks (EJAtlas 2020) to fight for nature. Furthermore, women are also disproportionately impacted by slow violence, in their role as caregivers for children and family members who have been affected, and more generally by micro-scale, intimate, and 'visceral' violence (see De Leeuw 2016).

Settler/colonial violence

Many defenders fight for land rights and/or to protect Indigenous territories of life (Milanez, this volume, Souza et al., this volume). The deadly environmental and land struggles of these peoples are inscribed in the long-lasting and still-present colonial violence, calling for what Malcolm Ferdinand (2019: 32) conceptualizes as '*écologie décoloniale*', an intersectional movement seeking to 'articulate the confrontation of contemporary ecological crisis with the emancipation from the historical colonial fracture'. This needs to go well beyond the traps of a 'colonial politics of recognition' (Coulthard 2014), whereby settler recognition of 'past harms' against and 'current rights' of Indigenous peoples are instrumented to continue the advancement of settler/colonial projects; including through the logics of private contracts such as Impact Benefits Agreements shutting down dissent *within* communities for the sake of unevenly distributed jobs and financial rewards (Peterson St-Laurent & Le Billon 2015). These demands, we suggest, advance a 'socio-ecological politics of recognition' sensitive to specific Indigenous ontologies, ways of life, and aspirations. In this respect, many environmental and land struggles by defenders are also *decolonizing* struggles, and as such as the violence that they face needs first to be recognized as colonial settler/colonial violence. Much of the attention to the ecological dimensions of settler/colonial violence has to do with the toxic legacies of extractivist and settler forms of colonialism, from the silver mines of Potosi (Machado 2014) to the nuclear tests in Polynesia (Keown 2018) – with people still suffering the consequences of pollution and defenders fighting back for reparations and compensations.

Epistemic violence

We understand here 'epistemic violence' as introduced by Gayatri Chakravorty Spivak in her essay, 'Can the Subaltern Speak?' (1988). Such violence constitutes not only a cultural or symbolic form of violence against defenders, but also a pervasive form of destruction and replacement of ways of understanding (and relating to) the world. As a form of colonization of the mind, epistemic violence is both the outcome and the process of dominant discourses about 'development', 'economic growth', and 'progress', to name a few associated with 'Euro-modern rational-scientific knowledge' (Quijano 2000); this includes dogmatic capitalist approaches that often reduce *live* to 'the economy' and *nature* to 'resources' (Escobar 2006), but also the regimes of truth advancing 'sustainable development' narratives as the efficient and environmentally sensitive capitalization of nature (Banerjee 2003). Epistemic violence can divide communities, pitting, for example, 'traditionalists' against 'modernists' who buy into the narratives of project proponents that seek to render land 'investable' through persuasive promises of future modernization, wellbeing, and prosperity (Le Billon & Sommerville 2017; Nixon, this volume). Often disregarding local traditional knowledge, this form of violence is generally driven by the hubris of material control, calculations of accumulation, and the racialized undercurrents of many ethos of 'progress' and 'empowerment' (McCarthy et al. 2018). In this regard, epistemological violence plays a crucial role in the 'violent technologies of extractions' that transform, degrade, and reconfigure communities, territories, and environments (Dunlap & Jakobsen 2020), thereby demonstrating its destructive ontological power.

Extractive violence

Despite a vast literature on the violence of extraction and its relevance to many defenders' struggles, very few studies have so far used the term 'extractive violence'. Among these, Sehlin McNeil (2017: 23) defines extractive violence as a 'type of direct violence against nature and/or people and animals that is caused by extractivism and that primarily affects peoples closely connected to land', emphasizing the importance of Indigenous perspectives and requiring people affected by extractive violence to have 'a deep spiritual connection to Country'. Extractive violence can be seen as a particularly category of 'environmental violence', occurring 'when development plans threaten the livelihoods of people and their possibilities of cultural reproduction by appropriating, transforming, and destroying natural resources and the environments in which these are embedded' (Narchi 2015: 9).

Extractive violence involves three main processes. First is the *violence of dispossession* that characterizes extraction, including the taking of land, the removal of resources, or the degradation of socio-environmental ecosystems. Often accompanied by some kinds of transactions, such as cash payments, temporary jobs, and 'alternative' livelihoods, dispossession remains generally marked by unfairness, uneven allocation, or the false 'equivalences' that such 'compensation' seeks to create between incomparable entities across incommensurable ontological and epistemic differences (Leifsen et al. 2017; Martinez-Alier 2001). Dispossession can be both material and ontological, through the delinking of communities from their territories, facilitated by the appeal or imposition of capitalist modernity and the environmental degradation that renders traditional livelihoods increasingly unviable. Here, Indigenous deterritorialization runs parallel with state/company-led (re)territorialization, as a mutually imbricated process (Di Giminiani 2015). Second is the *violence of coercion* often exercised on local communities to impose extractive activities, especially when resistance to projects takes a more organized shape (Middeldorp & Le Billon 2019; Navas et al. 2018), when clashes occur within or between communities over granting consent (Jaskoski 2020), when project are forcedly imposed despite refusal of consent (Steinberg 2016), or when there are tensions over compensation schemes. If coercion is often understood as direct threats or use of physical violence, we understand it as also taking many other forms, including deception, manipulation, and corruption consolidating the dominance of extractive regimes (Marston & Perreault 2017), as well as affecting decision-making within and by Indigenous communities (Cariño 2005; Nest 2017). Third are the physiological and psychological harms associated with the *violence of pollution and degradation of socio-environmental systems* resulting from extractive activities (Watts 2005), including through the temporally dispersed 'slow violence' against affected communities (Nixon 2011) and the 'ecological violence' perpetrated against the non-human within communal territories (Navas et al. 2018). Such a perspective on violence can help, for example, to reinterpret some forms of 'prior consultation' processes around extractive projects; Middeldorp and Le Billon (2021) suggest that they can constitute soft instruments of dispossession as they facilitate and legalize ownership and use transfers, frequently involving some forms of coercion – including among members of affected communities holding different perspectives and interests with regard to such projects – and frequently resulting in exposure to pollution and other socio-environmental impacts, as projects often end up being implemented despite a lack of consent.

Green violence

Many of the forms of violence experienced by defenders are driven by systems of accumulation and economic growth embedded within the dominant power structures of neoliberal capitalist systems and colonial histories of domination of spaces and bodies. It would, however, be an oversimplification to focus solely on these forms of violence when they relate to 'dirty' extractive projects such as mining or gas fracking. Violence also stems from industries and sectors that frame themselves as 'green' and 'sustainable'. As Dunlap and Correa Arce (this volume) show, renewable energies and 'green energy' are often accompanied by conflicts and death threats against local people; they manifest as 'murderous energy'. 'Green violence' (as per Büscher & Ramutsindela 2016) is violence carried out under the auspices of protecting nature: forced displacement of local communities and Indigenous peoples in order to create protected areas (Ndoinyo, this volume; Le Billon, this volume; Fanari 2019) and, indeed, direct violence against local communities via 'green wars' (Marijnen & Verweijen 2016), the militarization of conservation (Duffy 2016), and the necropolitics of some environmental NGOs (see Menton & Gilbert, this volume).

Psychological violence: on the cumulative impact of threats, intimidation, smear campaigns, and fear

> They're developing new strategies – they used to be more quick to kill us. Now, first they kill our character, attack the memory and reputations of our martyrs, attack our legitimacy, frame us as criminals, and try to shame us and scare us into silence.
>
> (Claudelice Santos 2018; also see Santos, this volume)

Psychological violence is a process that impairs another person's psychological integrity and potential. In the academic literature, it is most often studied in terms of domestic violence and verbal abuse in the workplace. It includes threats, intimidation, coercion, and harassment and is very relevant to the lived experiences of environmental and land defenders. Psychological violence against defenders comes in many forms: (i) threats and intimidation, (ii) hate speech, micro-aggressions and slurs, (iii) smear campaigns and delegitimization, and (iv) cumulative psychological impacts of physical attacks, fear, and repression. Indeed, physical attacks and assassination attempts are also forms of psychological violence, as they are intended to scare the defenders into backing down.

Threats and intimidation against leaders, their families, and/or their communities create an environment of fear and stress that undermines the mental health and physical wellbeing of threatened defenders, who sometimes leave their homes and territories, fearing for their lives.

In Brazil, environmental defenders reported that threats and intimidations came, for example, via threatening phone calls to elderly relatives, strangers shouting at their children as they walk home from school, or supposedly well-meaning informants telling them 'you better watch out' or 'there's a price on your head' (pers. com.). The threats and intimidation affect the whole family, the whole community, and often the entire social movement. The day after Bolsonaro was elected president of Brazil in 2018, his supporters drove into the Tuxá village of Rodelas-Bahia and shouted: 'Now you're gonna see. Now you're gonna get it' (Tuxá community member, pers. com.). In sharing this story, it was called a 'micro-aggression' – yet, the cumulative impact of such aggressions is more than micro. Hate speech has been shown to lead to both physical and mental health consequences for victims (Gelber & McNamara 2016).

At times, a defender's 'community' may itself be part of intense psychological violence, precisely because of the immediate proximity and even intimacy of village-level relations. As Grant and Le Billon (2019) explain in relation to NGO-sponsored community-forestry (CF) projects in Cambodia, the subjectivity of forest defenders can be re-affirmed by the constant threat of delayed retributive violence within their *own* village, which repeatedly re-inscribes their position as 'defenders' in relation to other villagers – sometimes the vast majority – who would prefer to log and clear land within the community forest. Such threat of retributive violence causes some defenders, especially those in the Management Committee, to live in constant fear. As one Committee member explained, 'every step, every minute of breathing, we are scared. We expect that someone will kill us and maybe our families too.' The children of the most active defenders indeed often experience intimidation, which makes them afraid to go to school, or return in tears after hearing classmates repeat the death threats they hear their parents make against CF members. Some interviewees reported being hesitant to join village activities or avoiding driving alone on quiet roads – in effect conducting themselves according to a 'defender' subjectivity performed not only 'in the forest' but also within a daily life suffused with fear.

With the popularity of social media, online hate speech against defenders has increased and is effectively written in indelible ink – the internet never forgets (Brown 2018). Smear campaigns, whether online or through word-of-mouth rumors or trumped-up charges, again seek to inflict emotional harm and induce fear: fear, in this case, of being imprisoned or of losing their legitimacy and reputations.

Atmospheres of violence

The intersection between these different forms of violence creates climates of fear and the multi-dimensional experience of violence vis-à-vis atmospheres of violence. The intersection of these violences is cumulative. The different forms of violence experienced simultaneously build a broader sense of risk and fear, deepening a sense of insecurity and precariousness. Many defenders suffer mental health consequences out of these constant traumas, whether insidious invisible traumas that come with being monitored or verbally aggressed, or surviving assassination attempts. These consequences can manifest in sleep disorders, anxiety and panic attacks, depression, loneliness, and even suicide. The lived experience of atmospheres of violence is not a linear sum of parts; it is not an $a + b = c$ equation. The intersection of multiple forms of violence, experienced over time and space, even intergenerationally and in different locations, creates a combination of perceptions and experiences of ever-present aggression. Their bodies, cultures, and territories are, indeed, often being assaulted from different directions, by both visible and invisible actors and processes.

Yet, despite this constant onslaught, defenders continue to fight to protect their forests, their waters, and their lands. They do so because their fight is about more than themselves as individuals. It is about the defense of life, in the broadest sense (see Preamble, this volume), and the hope that their fight can bring change and an end to these atmospheres of violence.

Notes

1 Responses also include cooptation, corruption, and the exacerbation of tensions and divisions within socio-environmental movements and communities, notably through

Corporate Social Responsibility contracts with NGOs (Baur & Schmitz 2012) and 'compensation' packages or outright payments dividing communities and buying off leaders (Schilling-Vacaflor & Eichler 2017).

2 Contentious politics refers to 'concerted, counter-hegemonic social and political action, in which differently positioned participants come together to challenge dominant systems of authority, in order to promote and enact alternative imaginaries' (Leitner et al. 2008).

References

APIB. (2019). *O governo Bolsonaro e a sua política genocida*. https://apiboficial.org/2019/03/24/governo-bolsonaro-e-sua-politica-genocida/

Banerjee, S. B. (2003). Who sustains whose development? Sustainable development and the reinvention of nature. *Organization Studies*, *24*(1), 143–180.

Barcia, I. (2017). Women human rights defenders confronting extractive industries: An overview of critical risks and human rights obligations. *Association for Women's Rights in Development (AWID) and Women Human Rights Defenders International Coalition (WHRDIC)*. www.awid.org/sites/default/files/atoms/files/whrds-confronting_extractive_industries_report-eng.pdf

Baur, D., & Schmitz, H. P. (2012). Corporations and NGOs: When accountability leads to co-optation. *Journal of Business Ethics*, *106*(1), 9–21.

Becerril, M. S. W. (2019). Frames in conflict: Discursive contestation and the transformation of resistance. In C. Mouli & E. Hernandez Degado (Eds.), *Civil resistance and violent conflict in Latin America* (pp. 175–204). Cham: Palgrave Macmillan.

Birss, M. (2017). Criminalizing environmental activism. *NACLA Report on the Americas*, *49*(3), 315–322.

Bolados, P., & Sánchez, A. (2017). Una ecología política feminista en construcción: El caso de las 'mujeres de zonas de sacrificio en resistencia', Región de Valparaíso, Chile. *Psicoperspectivas Individuo y Sociedad*, *16*(2), 33–42.

Bourdieu, P. (1979). Symbolic power. *Critique of Anthropology*, *4*(13–14), 77–85.

Brown, A. (2018). What is so special about online (as compared to offline) hate speech? *Ethnicities*, *18*(3), 297–326.

Büscher, B., & Ramutsindela, M. (2016). Green violence: Rhino poaching and the war to save Southern Africa's peace parks. *African Affairs*, *115*(458), 1–22.

Butt, N., Lambrick, F., Menton, M., & Renwick, A. (2019). The supply chain of violence. *Nature Sustainability*, *2*(8), 742–747.

Cariño, J. (2005). Indigenous peoples' right to free, prior, informed consent: Reflections on concepts and practice. *Arizona Journal of International and Comparative Law*, *22*, 19–39.

Castañeda Camey, I., Sabater, L., Owren, C., & Boyer, A. E. (2020). *Gender-based violence and environment linkages: The violence of inequality*. Gland: IUCN.

Cavanagh, C. J., & Himmelfarb, D. (2015). 'Much in blood and money': Necropolitical ecology on the margins of the Uganda protectorate. *Antipode*, *47*(1), 55–73.

Coulthard, G. S. (2014). *Red skin, white masks: Rejecting the colonial politics of recognition*. Minneapolis: University of Minnesota Press.

Cruz, J. M. (2011). Criminal violence and democratization in Central America: The survival of the violent state. *Latin American Politics and Society*, *53*(4), 1–33.

Davenport, C. (2007). State repression and political order. *Annual Review of Political Science*, *10*, 1–23.

Davies, T. (2018). Toxic space and time: Slow violence, necropolitics, and petrochemical pollution. *Annals of the American Association of Geographers*, *108*(6), 1537–1553.

De Leeuw, S. (2016). Tender grounds: Intimate visceral violence and British Columbia's colonial geographies. *Political Geography*, *52*, 14–23.

Deonandan, K., & Dougherty, M. L. (Eds.). (2016). *Mining in Latin America: Critical approaches to the new extraction*. Abingdon, UK: Routledge.

Deonandan, K., & Bell, C. (2019). Discipline and punish: Gendered dimensions of violence in extractive development. *Canadian Journal of Women and the Law*, *31*, 24–57.

Deonandan, K., Tatham, R., & Field, B. (2017). Indigenous women's anti-mining activism: A gendered analysis of the El Estor struggle in Guatemala. *Gender & Development*, *25*(3), 405–419.

Di Giminiani, P. (2015). The becoming of ancestral land: Place and property in Mapuche land claims. *American Ethnologist*, *42*(3), 490–503.

Duffy, R. (2016). War, by conservation. *Geoforum*, *69*, 238–248.

Dunlap, A. (2019). 'Agro sí, mina NO!' The Tía Maria copper mine, state terrorism and social war by every means in the Tambo Valley, Peru. *Political Geography*, *71*, 10–25.

Dunlap, A., & Jakobsen, J. (2020). *The violent technologies of extraction*. Palgrave Pivot.

EJAtlas. (2020). *Mujeres Latinoamericanas Tejiendo Territorios*. https://ejatlas.org/featured/mujeres [last accessed 14 December 2020].

Escobar, A. (2006). Difference and conflict in the struggle over natural resources: A political ecology framework. *Development*, *49*(3), 6–13.

Fanari, E. (2019). Relocation from protected areas as a violent process in the recent history of biodiversity conservation in India. *Ecology, Economy and Society/INSEE*, *2*(1), 43–76.

Farmer, P. (2004). An anthropology of structural violence. *Current Anthropology*, *45*(3), 305–325.

Franks, D. M., Davis, R., Bebbington, A. J., Ali, S. H., Kemp, D., & Scurrah, M. (2014). Conflict translates environmental and social risk into business costs. *Proceedings of the National Academy of Sciences*, *111*(21), 7576–7581.

Ferdinand, M. (2019). *Une écologie décoloniale-Penser l'écologie depuis le monde caribéen*. Paris: Le Seuil.

Galtung, J. (1969). Violence, peace and peace research. *Journal of Peace Research*, *6*, 167–191.

Galtung, J. (1990). Cultural violence. *Journal of Peace Research*, *27*, 291–305.

Gelber, K., & McNamara, L. J. (2016). Evidencing the harms of hate speech. *Social Identities*, *22*(3), 324–341.

Grant, H., & Le Billon, P. (2019). Growing political: Violence, community forestry, and environmental defender subjectivity. *Society & Natural Resources*, *32*(7), 768–789.

Hill, D. W., & Jones, Z. M. (2014). An empirical evaluation of explanations for state repression. *American Political Science Review*, *108*(3), 661–687.

Jaskoski, M. (2020). Participatory institutions as a focal point for mobilizing: Prior consultation and Indigenous conflict in Colombia's extractive industries. *Comparative Politics*, *52*(4), 537–556.

Jay Chen, C. J. (2020). Peasant protests over land seizures in rural China. *The Journal of Peasant Studies*, 1–21.

Jeffery, L. (2018). Development and forced displacement. *The International Encyclopedia of Anthropology*, 1–8. https://doi.org/10.1002/9781118924396.wbiea1698

Keown, M. (2018). Waves of destruction: Nuclear imperialism and anti-nuclear protest in the indigenous literatures of the Pacific. *Journal of Postcolonial Writing*, *54*(5), 585–600.

Le Billon, P., & Duffy, R. V. (2018). Conflict ecologies: Connecting political ecology and peace and conflict studies. *Journal of Political Ecology*, *25*(1), 239–260.

Le Billon, P., & Lujala, P. (2020). Environmental and land defenders: Global patterns and determinants of repression. *Global Environmental Change*, *65*, 102163. https://doi.org/10.1016/j.gloenvcha.2020.102163

Le Billon, P., & Sommerville, M. (2017). Landing capital and assembling 'investable land' in the extractive and agricultural sectors. *Geoforum, 82*, 212–224.

Leifsen, E., Gustafsson, M. T., Guzmán-Gallegos, M. A., & A. Schilling-Vacaflor. (2017). New mechanisms of participation in extractive governance: Between technologies of governance and resistance work. *Third World Quarterly, 38*(5), 1043–1057.

Leitner, H., Sheppard, E., & Sziarto, K. M. (2008). The spatialities of contentious politics. *Transactions of the Institute of British Geographers, 33*(2), 157–172.

Li, T. M. (2017). After development: Surplus population and the politics of entitlement. *Development and Change, 48*(6), 1247–1261.

López, M., & Bradley, A. (2017). Rethinking protection, power and movements: Lessons from women human rights defenders in Mesoamerica. Making Change Happen (No. 6). JASS (Just Associates). https://justassociates.org/sites/justassociates.org/files/jass_mch6._rethinking_protection_power_movements_4.pd

Machado, H. (2014). *Potosí, el origen: genealogía de la minería contemporánea*. Buenos Aires: Mardulce.

Marijnen, E., & Verweijen, J. (2016). Selling green militarization: The discursive (re) production of militarized conservation in the Virunga National Park, Democratic Republic of the Congo. *Geoforum, 75*, 274–285.

Marston, A., & Perreault, T. (2017). Consent, coercion and cooperativismo: Mining cooperatives and resource regimes in Bolivia. *Environment and Planning A: Economy and Space, 49*(2), 252–272.

Martinez-Alier, J. (2001). Mining conflicts, environmental justice, and valuation. *Journal of Hazardous Materials, 86*(1–3), 153–170.

Matejova, M., Parker, S., & Dauvergne, P. (2018). The politics of repressing environmentalists as agents of foreign influence. *Australian Journal of International Affairs, 72*(2), 145–162.

Mbembe, J. A. (2003). Necropolitics. *Public Culture, 15*(1), 11–40.

McCarthy, L., Touboulic, A., & Matthews, L. (2018). Voiceless but empowered farmers in corporate supply chains: Contradictory imagery and instrumental approach to empowerment. *Organization, 25*(5), 609–635.

Menton, M., Milanez, F., Souza, J. M. A., & Cruz, F. S. M. (2021). The COVID-19 pandemic intensified resource conflicts and indigenous resistance in Brazil. *World Development, 138*, 105222.

Middeldorp, N., & Le Billon, P. (2019). Deadly environmental governance: Authoritarianism, eco-populism, and the repression of environmental and land defenders. *Annals of the American Association of Geographers, 109*(2), 324–337.

Middeldorp, N., & Le Billon, P. (2021). Empowerment or imposition? Extractive violence, Indigenous peoples and the paradox of prior consultation. In J. A. McNeish & J. Shapiro (Eds.), *Our hyper-extractive age: Expressions of violence and resistance*. Abingdon, UK: Routledge.

Narchi, N. E. (2015). Environmental violence in Mexico: A conceptual introduction. *Latin American Perspectives, 42*(5), 5–18.

Navas, G., Mingorria, S., & Aguilar-González, B. (2018). Violence in environmental conflicts: The need for a multidimensional approach. *Sustainability Science, 13*(3), 649–660.

Nest, M. (2017). *Preventing corruption in community mineral beneficiation schemes*. U4/CMI. Bergen.

Nixon, R. (2011). *Slow violence and the environmentalism of the poor*. Cambridge, MA: Harvard University Press.

Peluso, N. L., & Watts, M. (Eds.). (2001). *Violent environments*. Ithaca, NY: Cornell University Press.

Peterson St-Laurent, G., & Le Billon, P. (2015). Staking claims and shaking hands: Impact and benefit agreements as a technology of government in the mining sector. *The Extractive Industries and Society*, *2*(3), 590–602.

Pierskalla, J. H. (2010). Protest, deterrence, and escalation: The strategic calculus of government repression. *Journal of Conflict Resolution*, *54*(1), 117–145.

Quijano, A. (2000). Coloniality of power, Eurocentrism and Latin America. *Nepantla*, *1*(3), 533–580.

Santos, C. (2018, April 12). Protecting the protectors. *Speech at Skoll World Forum*. Oxford.

Sauer, S., & de Castro, L. F. P. (2019). Struggles for land and territorial rights in Brazil. In O. De Schutter & B. Rajagopal (Eds.), *Property rights from below: Commodification of land and the counter-movement*. Abingdon, UK: Routledge.

Scheidel, A., Del Bene, D., Liu, J., Navas, G., Mingorría, S., Demaria, F., Avila, S., Roy, B., Ertor, I., Temper, L., & Martínez-Alier, J. (2020). Environmental conflicts and defenders: A global overview. *Global Environmental Change*, *63*.

Schilling-Vacaflor, A., & Eichler, J. (2017). The shady side of consultation and compensation: 'Divide-and-rule' tactics in Bolivia's extraction sector. *Development and Change*, *48*(6), 1439–1463.

Segato, R. L. (2016). *La guerra contra las mujeres*. Madrid: Traficantes de sueños.

Sehlin McNeil, K. (2017). *Extractive violence on Indigenous country: Sami and aboriginal views on conflicts and power relations with extractive industries*. Umeå: Umeå University.

Sippola, K. (2018). *Mapuche women's land rights activism and state-led gender based violence under neoliberal globalization in Chile*. Doctoral dissertation. The Ohio State University.

Smith, R. K. (2008). Ecoterrorism: A critical analysis of the vilification of radical environmental activists as terrorists. *Environmental Letters*, *38*, 537–576.

Spivak, G. (1988). Can the subaltern speak? In R. Morris (Ed.), *Can the subaltern speak? Reflections on the history of an idea* (pp. 21–78). New York: Columbia University Press.

Springer, S., & Le Billon, P. (2016). Violence and space: An introduction to the geographies of violence. *Political Geography*, *52*, 1–3.

Steinberg, J. (2016). Strategic sovereignty: A model of non-state goods provision and resistance in regions of natural resource extraction. *Journal of Conflict Resolution*, *60*(8), 1503–1528.

Tan, Y. (2020). Development-induced displacement and resettlement. In T. Bastia & R. Skeldon (Eds.), *Routledge handbook of migration and development* (pp. 373–381). Abingdon, UK: Routledge.

Tripathi, S. (2017). Development, displacement and human rights violations. *Indian Journal of Public Administration*, *63*(4), 567–578.

United Nations Environment Program. (2018). *UN Environment's policy on environmental defenders*. Nairobi: Environmental Rights Initiative, UNEP.

Watts, M. J. (2005). Righteous oil? Human rights, the oil complex, and corporate social responsibility. *Annual Review of Environment and Resources*, *30*, 373–407.

7 Environmental defenders

Killings, perpetrators, and drivers of violence

Philippe Le Billon and Päivi Lujala

A growing body of scholarly literature is studying socio-environmental conflicts and the persecution of environmental defenders to better understand risk factors (Butt et al. 2019; Clark 2009; Jeffords & Thompson 2016; Middeldorp & Le Billon 2019; Scheidel et al. 2020). The case study literature suggests that killings of environmental and land defenders are particularly prominent in countries experiencing high levels of inequality and corruption, historical marginalization of Indigenous and peasant communities, liberalization of foreign and private investment into land-based sectors, weak rule of law, and recent reversals in partial democratization processes, with many of these killings taking place within a broader context of high homicidal violence and high impunity rates for perpetrators (Middeldorp & Le Billon 2019).

In this chapter, we examine global patterns of repression across socio-environmental conflicts and determinants of killings of environmental defenders, using the Global Atlas of Environmental Justice on socio-environmental conflicts and the Global Witness dataset of environmental and land defender killings. Following this introduction, we briefly discuss the different categories of defenders and then outline patterns of socio-environmental conflicts, defenders killed, perpetrators, and determinants of killings. We conclude by stressing the need for greater protection of defenders and stricter investment vetting processes in countries with high risks of killings, including the respect for Indigenous consent rights, and outline further research to identify high risk sectors, areas, and authorities to prevent killings and increase accountability.

Environmental and land defenders

Environmental and land defenders are defined as 'people who take peaceful action to protect environmental or land rights, whether in their own personal capacity or professionally' (Global Witness 2017: 43). The term *environmental and land defenders* encompasses a broad range of people, including Indigenous people threatened by large-scale resource extractions, dams, agribusiness, and illegal logging, mining, or land settling (Lynch et al. 2018); landless peasants (re)claiming farmlands or long-established rural communities facing large-scale 'land grabs' by multinationals (Borras & Franco 2013); and grassroots and professional environmental advocates. The term covers people that may have very different understandings of their relationship with the environment and legitimate land entitlement. Many rural communities, for example, seek to defend their access to land for the sake of securing agrarian livelihoods. In doing so, they come to defend ways of life that can involve forest clearings to create farmland and assert de facto land rights in ways that may seem environmentally destructive and legally tenuous (Ghazoul & Kleinschroth 2018), even if such processes can (re)create forested and highly biodiverse anthropic rural landscapes (Hecht 2010). In contrast, park wardens seek to protect particular species and habitats, often through the militarized enforcement of human exclusion rules – 'fortress conservation' – that have evicted rural communities, undermined traditional livelihoods, and historically 're-wilded' the environment in the name of protecting game and 'natural' biodiversity (Duffy 2016).

The case study literature suggests that defenders fall within five main categories according to their social identities and main motivations (Table 7.1). These categories are not exclusive, with many defenders associated with several (e.g. Indigenous environmental activist serving as community forest patroller). Defenders in these categories often face similar threats, although there can be specific ones due to their profile and activities. Some defenders are well connected, are organized, and have high public profiles, including as recipients of international environmental prizes (e.g. Goldman Prize), while others are largely anonymous outside their area of residence and may be isolated within their own community due to their environmental activities (Grant & Le Billon 2019).

Table 7.1 Main categories of environmental and land defenders

Categories	*Main motivations*	*Main threats*
Indigenous people	Protect territory, culture, and ecology	Colonization and land encroachment, large-scale resource projects, logging and mining
Rural community members, farmworkers, landless peasants	Access and sustain land and livelihoods	Agribusinesses and large-scale resource projects
Environmental activists, social movement activists, artists, and public intellectuals	Prevent environmentally destructive activities, promote environmental and social justice	Large-scale resource projects, government crackdown on opposition
Lawyers, journalists, judges	Report and defend environmental and social rights, especially those of marginalized communities	Broad economic and political interests of ruling elites, business owners, and organized crime
Conservation, forestry, and police officers	Enforce environmental and forestry laws	Poaching, illegal logging, mining, land settling, and organized crime

The term 'defender' tends to emphasize the individual over the broader communities and organizations involved in environmental and land struggles. While such individualization can help raise public awareness by literally 'giving a face' to socio-environmental struggles,[1] it can also make individualized defenders more likely targets as perpetrators seek to intimidate communities and deter leadership, as well as create tensions within affected communities by singling out particular individuals or their families. The focus placed by many advocacy and media reports on killings, rather than on the broad range of pressure exercised on communities, can also exacerbate such individuation. As such, a focus on individual killings and individuals killed risks both misrepresenting and rendering less visible the communities to which the defenders belong. Some academic and policy reports on environmental and land defenders have been cautious in this respect, with for example UNEP (2018) specifically mentioning 'groups of people' in its definition of defenders, but the logics and practices of advocacy and media often individualize these struggles.

Overall, the concept of environmental and land defenders is best applied to people tying together community, territory, environmental protection, and livelihoods. These people are generally aspiring to maintain or (re)create land-use practices that help to sustain anthropogenic, yet often highly biodiverse ecosystems that tend to be more benign for the environment than agro-industrial practices, hydro-power or irrigation dams, and extractive activities such as industrial mining or clear-cut logging (Ghazoul & Kleinschroth 2018).[2]

Patterns of socio-environmental conflicts and killings of defenders

In many cases, defenders resist environmentally destructive projects that would drastically undermine biodiversity and ecosystem services. The Institute of Environmental Science and Technology (ICTA) at the Universitat Autonoma de Barcelona, together with 'activist partners', including socio-environmental activists and researchers, have systematically documented cases of socio-environmental conflict (Temper et al. 2015; Scheidel et al. 2020). The resulting EJAtlas defines socio-environmental conflicts as 'mobilizations by local communities, social movements, which might also include support of national or international networks against particular economic activities, infrastructure construction or waste disposal/pollution whereby environmental impacts are a key element of their grievances'.[3] Out of the 2,957 EJAtlas reported cases, 1,279 cases (43%) were categorized as 'medium intensity' conflicts, including visible mobilization and street protests, while 847 (29%) cases were categorized as 'high intensity' conflicts involving mass mobilization, arrests, and direct forms of violence. Different forms of repression occurred, with a criminalization or biased use of the law against defenders in 20% of all cases, violent targeting of activists in 18% of all cases, and killings in 12% of all cases. Killings were even more frequent when Indigenous people were involved (19% compared to 8%; see Scheidel et al. 2020).

A regional breakdown identifies greater rates of high intensity conflicts in Asia (39%) and Latin America (31%), with more frequent violent targeting of activists (24%) and deaths (19%) in Latin America than in Asia (22% and 13%, respectively). Sub-Saharan Africa and Western countries (Europe, US, Canada, Australia, New Zealand) have lower reported high-intensity conflicts, repression, violent targeting, and deaths. However, in contrast to Western countries, Sub-Saharan Africa saw the lowest level of 'success' for environmental and land defenders, along with the Middle East and North Africa region, while island states (Caribbean, West Indian Ocean, South Pacific), Western countries, and Latin America saw the highest. Looking at the main sectors involved, the percentage of high intensity conflicts was lowest in the fossil fuels and climate justice sector (25%) and highest in extraction (33%), while biomass and land conflicts (21%) were the deadliest.

Defenders killed

A second database, by Global Witness, collates information about killings of defenders from local media, national-level organizations such as that of the *Comissão Pastoral da Terra* (CPT) in Brazil, and global human rights databases such as HuriSearch. The objectives of such databases are to raise awareness of the killings, honor the memory of the defenders, support their struggles, enhance their protection, promote corporate and government policy change, and pursue accountability. Comprehensively identifying killings of environmental and land defenders across the world, however, is a major challenge, notably as a result of a lack of local monitoring, investigation, and reporting; suppression of information by authorities; difficulties in reaching local reporting organizations; and contexts of broader conflicts, making it challenging to identify specific cases of defender killings. Furthermore, killings only represent the 'tip of the iceberg' in terms of the harms of repression and other forms of violence associated with resource extraction and land dispossession (Butt et al. 2019). In 2018, former UN Special Rapporteur on Human Rights and the Environment John Knox estimated that 'for every 1 killed, there are 20 to 100 others harassed, unlawfully and lawfully arrested, and sued for defamation, amongst other intimidations' (UNEP 2018). A review of violence linked to 34 large-scale mining projects and activities by Canadian mining companies in 14 Latin America between 2000 and 2015 documented 44 deaths, 4 disappearances, 15 sexual assaults, 403 injuries, 537 arrests, detentions, and charges, and 195 warrants and legal complaints (Imai et al. 2016). Thus, for every death or disappearance, nine people were physically injured or sexually assaulted, and 17 people faced judicial measures or 'criminalization'. Looking at landless peasant struggles in Brazil between 1986 and 2006, there were three reported death threats and a murder attempt for every person assassinated (Girardi 2008). As Rasch (2017: 132) observed, growing resistance within natural resource conflicts 'goes hand in hand with an increased use of penal law and anti-terrorist legislation as a way of disqualifying social protest as well as an intensification of the use of violence and the surge of human rights violations'.

To sum up, killings represent only a small proportion of coercive actions against environmental and land defenders, and they should only be considered as a partial indicator of the level of repression. Furthermore, the physical and psychological effects of repression are only two forms of harms experienced by communities affected by resource-based projects; others include exposure to pollutants, loss of land and livelihoods, or socio-cultural conflicts (see Butt et al. 2019; Watts 2005).

The Global Witness dataset identifies 1,945 reported killings of environmental and land defenders that took place in a total of 56 countries between 2002 and 2019. This conservative estimate points to a sharp rise in annual killings between 2009 and 2015, coinciding with the primary commodity boom but possibly in part related to increased reporting. Six countries accounted for 77% of reported killings: Brazil (n=643), Colombia (n=260), Philippines (n=259), Honduras (n=156), Mexico (n=99), and Peru (n=80). The five most deadly countries in terms of killings per capita were Honduras, Nicaragua, Colombia, Guatemala, and Brazil. Honduras was an extreme outlier, with per capita killings four times higher than Nicaragua, the second deadliest country for defenders (see Middeldorp & Le Billon 2019). Regionally, 80% of reported killings were concentrated in Latin America, 20% in Southeast and South Asia, and 5% in Africa, suggesting even sharper regional differences than those reported through EJAtlas.

The number of women defenders killed over the 2002–2018 period represents 9% of the total killings (n=150), with a growing proportion of women defenders killed since 2010. Women defenders were mostly killed in Brazil (30%), the Philippines (19%), Colombia (15%), and Honduras (9%), and correlate closely with the number of males killed. During the 2014–2018 period, 276 Indigenous defenders were killed, representing 32% of total defenders killed in that period, with most of the killings of Indigenous defenders taking place in the Philippines (25%), Colombia (22%), Mexico (10%), Nicaragua (10%), and Brazil (9%). Based on reported killings in 2014 and 2015, the age of defenders killed varied from 11 to 75 years old, with the vast majority of killed defenders being in the 35–65 age bracket (65%) rather than youths (32%) or elders (3%), thus pointing at a likely targeting of active social leaders.

Perpetrators

While defenders killed are generally well identified once reported by the media or human rights organizations, people behind the killings frequently are not. Partly because of that, the perpetrators are rarely prosecuted. Global Witness (2014) documented only ten convictions for the murders of seven defenders, out of 908 cases identified between 2002 and 2013, with a further 34 perpetrators under arrest and facing charges. In Colombia, out of 122 defenders killed between July 2010 and June 2016, 102 killings were investigated, nine led to a verdict, and only eight in a conviction (Global Witness 2018).

The anonymity and impunity of perpetrators often result from the modus operandi of killings (e.g. hired gunmen); participation or complicity and cover-ups by authorities and local elites in the killings (e.g. direct responsibility in killings); corruption or pressure on the judicial system (e.g. bribes, clientelism); fears of reprisal against potential whistle-blowers; and a lack of investigations (e.g. social marginality of defenders, remoteness of location, limited means of police and prosecutors). Killings generally result from shootings or beatings by government armed forces, police, corporate security personnel, or mobs and thugs during public events such as mass protests, occupation, and blockades; or from targeted murders by 'hitmen', taking place at the home of the defender or in the street. Some targeted killings also take place during or shortly after protests, for example, to kill leaders who otherwise benefit from community-level protection. If individual killers, and the people who hire them, are rarely formally identified, there is generally more information on the category of perpetrators and sectors involved, including in the GW dataset.

Many studies point to patterns of repression associated with resource-based sectors (DeMeritt & Young 2013; Vadlamannati et al. 2019). Killings generally occur as part of escalating processes of disputed resource exploitation, social mobilization, and repression (Bebbington & Bury 2013; Dunlap 2019), with direct physical forms of violence being widely documented as part of land control strategies (Peluso & Lund 2011; Grajales 2011). Out of the 859 killings reported by GW during 2014–2018, 31% involved land conflicts, including those with landless peasants making claims on disputed lands controlled by established farmers and land speculators. Agribusinesses were associated with 14% of the killings, mining and extractive activities with 20%, logging with 9%, water and dams with 7%, poaching with 8% and other sectors (e.g. fishing, wind farms) with 5%.[4]

Based on 346 killings for which the GW dataset reported the category of perpetrators (identified for killings in 2015, 2017, and 2018), about a third were directly perpetrated by government authorities (police n=59, army n=58, government officials n=8), to which can often be added paramilitaries (n=31), armed militias (n=17), mobs supporting incumbent party (n=2), and private security guards (n=11). Militarized opposition forces – guerrillas – were associated with nine killings. Many killings were carried out by 'hitmen' (n=52) or criminal gangs/organized crime (n=46), as well as people directly involved in environmental or land exploitation (i.e. landowners/speculators (n=30), poachers (n=26), land settlers (n=11), loggers (n=4), and miners (n=1)). This suggests that environmental and land exploiters directly perpetrate killings relatively rarely. Rather, exploiters pursuing land-based projects in the face of local resistance mobilize not only state security forces and private security firms, but also criminal organizations, paramilitaries, and vigilante groups – especially in areas where the state lacks or outsources territorial control – in order to repress defenders (Cruz 2011). In Brazil, syndicates of local landowners and land speculators have hired assassins (e.g. murder of Sister Dorothy Stang; see Campbell 2015) and are part of larger political forces, such as the Ruralist Democratic

Union, opposing landless peasant movements through legal reforms and paramilitary groups (Hammond 2009; Mendes 2018).

Whereas some governments and corporations use their own security personnel to exert deadly repression, notably in the context of public street protests and blockades, more insidious forms of repression – including targeted killings – are generally subcontracted through middlemen to hired gunmen or criminal gangs (Global Witness 2014), making it more difficult to identify the full network and trace the chain of command. Criminal organizations and illegal business entrepreneurs also commit murders to advance their own interests, notably among poaching gangs, illegal loggers, and miners. Killings among (neighboring) communities occur in the context of local land conflicts, including between traditional residents – especially Indigenous populations – and newly settled populations (e.g. *colonos* or *mestizo* in Nicaragua; see Sylvander 2018). This can give way to complex situations involving 'cycles of violence' – including feuds and revenge killings – in which land settlers, Indigenous populations, and large-scale agribusinesses confront each other. These situations can be further complicated, and become deadlier, when narcotics production and trafficking and (counter)insurgency are also involved (e.g. in Colombia, Honduras, Myanmar, and the Philippines). Finally, there are many degrees of responsibility and forms of complicity involved, from carrying out the killing itself, to recruiting the killer(s), ordering and paying for the killing, knowingly promoting and/or investing in a resource project that could possibly result in a killing, and benefiting from the project without having taken part in the decision (e.g. pension fund holders; commodity consumers).

Determinants of killings

Contemporary killings of environmental and land defenders are part of a long history of colonialization and resource exploitation (Butt et al. 2019; Lynch et al. 2018; Totten et al. 2002;). Propelled by accumulative economic regimes (Moore 2015) and often underpinned by racial and socio-economic hierarchies (Virdee 2019), resource exploitation drastically accelerated after the onset of the Second World War (Krausmann et al. 2009). The globalization of mass consumption and economic emergence of China in the late 1990s further increased global commercial demand for land and natural resources, thereby pushing extraction frontiers and exacerbating conditions for socio-environmental conflicts, especially in resource-rich countries with populations resisting the burdens of pollution, displacement, cultural and livelihood loss, and social inequalities (Escobar 2006; Muradian et al. 2012). In this context, many local communities seek to assert and defend their rights in the face of powerful political and commercial alliances between government authorities, local economic elites, and primary commodity companies (Temper et al. 2015). The frequent absence of effective conflict prevention and resolution processes (e.g. free, prior, and informed consent by Indigenous groups) and the use of deceptive and coercive tactics by project proponents often lead to further resistance and conflict escalation (Conde & Le Billon 2017). As a communication revolution enabled many defenders to become more connected and their struggles more visible (Kirsch 2014), the perceived need of resource extraction proponents to 'silence' defenders and deter their supporters can increase, but so can the potential for backlash and even greater mobilization (Aytaç et al. 2018; Bob & Nepstad 2007).

Killings are in part facilitated by patterns of impunity for perpetrators, generally associated with the lack of independent and effective judiciary and media reporting; tight and unaccountable networks between political, economic, and military elites; social 'habituation' to homicides on the part of authorities – including as a result of recent war and state-tolerated/encouraged vigilante activity (Cruz 2011; Hill & Jones 2014). Deadly conflict escalation also often results from high uncertainty in the capacity and behavioral norms among protesters, corporate actors, and security forces in a context of contentious politics (Leitner et al. 2008), a situation characterizing intermediary political regimes falling between 'full' autocracies and democracies (Davenport 2007; Pierskalla 2010). In such contexts, government authorities and corporations are frequently unwilling to follow the praxis of negotiated conflict settlement, while some defenders and their movements refuse to back down on the premise that sustained contestation will further erode abuses of power, even if at the cost of deadly repression. The case study literature suggests that the likelihood of killings of environmental and land defenders thus seems particularly acute in middle-income countries with semi-authoritarian regimes, a recent history of armed conflicts and/or high homicides rates, and a high prevalence of conflicts around resource exploitation projects, as seen in Latin America (see Bebbington & Bury 2013; Jeffords & Thompson 2016; McNeish 2018; Middeldorp & Le Billon 2019; Temper et al. 2015). Butt et al. (2019) have shown that weak rule of law – based on the World Justice Project index – correlates with higher rates of environmental and land defender killings, echoing more general findings that the most significant factor increasing political killings besides civil war is a lack of judicial independence (Hill & Jones 2014).

In Le Billon and Lujala (2020), we studied the factors the previous literature has linked to environmental and land defender killings using multivariate negative binomial regressions. In our country-level analysis, we found that among economic factors, higher levels of incoming foreign direct investment (FDI) and extraction of minerals are associated with a higher number of killings. We also found strong evidence that the poorest countries have the lowest number of killings and the number of killings increases with per capita income levels, although this impact diminishes for the richest countries. When it comes to political factors, we found that there is a curvilinear relationship between regime type and the number of environmental and land defender killings: there are more killings in countries that are neither strong democracies or autocracies. We also found that defender killings are more likely in countries with many protests related to political processes. When it comes to demographic factors, we found that higher shares of Indigenous population in a country are positively related to a higher number of killings. There are also more killings in countries with larger populations, and possibly also in countries with high share of young males, population density, or homicide rates. Our results were inconclusive when it comes to the association between the killings of defenders and armed conflict or corruption.

Conclusion

Environmental and resource governance emphasizes the importance of local community and civil society participation to achieve social equity and environmental sustainability goals. Yet, repression often undermines such participation, including through the assassination of prominent defenders and members of their community and support network. This persecution is now more systematically reported, which allows for the identification of possible determinants of conflicts and killings. Among these, EJAtlas and Global Witness stand among the most comprehensive efforts, yet both suffer from limitations and possible biases in reporting. Any conclusions within this chapter should thus be treated with caution, as the overall numbers of conflicts and killings are not necessarily representative of the actual numbers in a particular country or region, while the association between a defender being killed and a specific sector is not always evidence of a direct connection.

With these caveats in mind, our findings suggest that conflicts are more frequently of 'high intensity' in Asia and Latin America than in the rest of the world, with the violent targeting of activists, including killings, being more frequent in this latter region. About a third of conflicts across the main sectors were of high intensity, especially mineral ore and building material extraction, with biomass and land conflicts being the most frequently deadly. According to the Global Witness database, Indigenous peoples constitute the group most at risk of killings, a result supported by a recent quantitative study (Le Billon & Lujala 2020). Threatened by large-scale resource extractions, dams, agribusiness, and land settling (Lynch et al. 2018), they represent nearly a third of reported environmental and land defenders killed. About a third of killings are perpetrated by government or corporate security personnel, while much of the rest is subcontracted to paramilitaries and hired hitmen. These findings call for tighter controls on investments taking place in countries with high-risk characteristics, as well as respect for the consent rights of Indigenous communities and stronger protection measures for defenders.

Further research is needed in a number of areas. In terms of documenting repression, potential 'blind spots' with under-reporting should be further investigated, through increased communication with local human rights and civil society organizations in countries with suspected low reporting, as well as extending the range of repression covered beyond killings (e.g. criminalization, threats, injuries) and levels of impunity for perpetrators. Quantitative and spatially disaggregated studies are needed (see Le Billon & Lujala 2020) to infer root causes of killings, uncover causal linkages, and better identify high-risk places in which protection measures should be deployed to safeguard environmental and land defenders. Predictive models could be refined and tested as more information, especially geo-referenced ones, is added in estimations. Finally, additional research is needed to assess the impacts of repression on environmental and land struggles in terms of social mobilization and project outcomes, as well as the effectiveness of policy reforms on investment criteria and the protection of affected communities.

Notes

1 See, for example, the series of reports in the *Guardian*, www.theguardian.com/environment/series/the-defenders.

2 Even this characterization can be problematic, however, as the impacts of 'traditional' activities depend on their scale, local contexts, and modes of operation. A large number of 'artisanal' miners operating in riverine areas and using mercury, for example, can harm the environment and human health (see Kitula 2006).

3 The EJAtlas documents 'social conflict related to claims against perceived negative social or environmental impacts with the following criteria: 1) Economic activity or legislation with actual or potential negative environmental and social outcomes; 2) Claim and mobilization by environmental justice organization (s) that such harm occurred or is likely to occur as a result of that activity; 3) Reporting of that particular conflict in one or more media stories'; see https://ejatlas.org/about.

4 Some conflicts are only associated with 'Indigenous peoples' (i.e. not with any specific sector), and several sectors can be involved in one killing.

References

Aytaç, S. E., Schiumerini, L., & Stokes, S. (2018). Why do people join backlash protests? Lessons from Turkey. *Journal of Conflict Resolution*, *62*(6), 1205–1228.

Bebbington, A., & Bury, J. (2013). *Subterranean struggles: New dynamics of mining, oil, and gas in Latin America*. Austin, TX: University of Texas Press.

Bob, C., & Nepstad, S. E. (2007). Kill a leader, murder a movement? Leadership and assassination in social movements. *American Behavioral Scientist*, *50*(10), 1370–1394.

Borras Jr, S. M., & Franco, J. C. (2013). Global land grabbing and political reactions 'from below'. *Third World Quarterly*, *34*(9), 1723–1747.

Butt, N., Lambrick, F., Menton, M., & Renwick, A. (2019). The supply chain of violence. *Nature Sustainability*, *2*(8), 742–747.

Campbell, J. M. (2015). *Conjuring property: Speculation and environmental futures in the Brazilian Amazon*. Seattle, WA: University of Washington Press.

Clark, R. D. (2009). Environmental disputes and human rights violations: A role for criminologists. *Contemporary Justice Review*, *12*(2), 129–146.

Conde, M., & Le Billon, P. (2017). Why do some communities resist mining projects while others do not? *The Extractive Industries and Society*, *4*(3), 681–697.

Cruz, J. M. (2011). Criminal violence and democratization in Central America: The survival of the violent state. *Latin American Politics and Society*, *53*(4), 1–33.

Davenport, C. (2007). State repression and political order. *Annual Review of Political Science*, *10*, 1–23.

DeMeritt, J. H., & Young, J. K. (2013). A political economy of human rights: Oil, natural gas, and state incentives to repress. *Conflict Management and Peace Science*, *30*(2), 99–120.

Duffy, R. (2016). War, by conservation. *Geoforum*, *69*, 238–248.
Dunlap, A. (2019). 'Agro sí, mina NO!' The Tía Maria copper mine, state terrorism and social war by every means in the Tambo Valley, Peru. *Political Geography*, *71*, 10–25.
Escobar, A. (2006). Difference and conflict in the struggle over natural resources: A political ecology framework. *Development*, *49*(3), 6–13.
Ghazoul, J., & Kleinschroth, F. (2018). A global perspective is needed to protect environmental defenders. *Nature Ecology & Evolution*, *2*(9), 1340–1342.
Girardi, E. P. (2008). *Atlas da Questão Agrária Brasileira* [Atlas of the Brazilian Agrarian Question]. Presidente Prudente: Unesp/NERA. www.atlasbrasilagrario.com.br/ [last accessed 12 December 2018].
Global Witness. (2014). *Deadly environment: The dramatic rise in killings of environmental and land*. London.
Global Witness. (2017). *Defenders of the Earth: Global killings of land and environmental defenders in 2016*. London.
Global Witness. (2018). *At what cost? Irresponsible business and the murder of land and environmental defenders in 2017*. London.
Grajales, J. (2011). The rifle and the title: Paramilitary violence, land grab and land control in Colombia. *Journal of Peasant Studies*, *38*(4), 771–792.
Grant, H., & Le Billon, P. (2019). Growing political: Violence, community forestry, and environmental defender subjectivity. *Society & Natural Resources*, *32*(7), 768–789.
Hammond, J. L. (2009). Land occupations, violence, and the politics of agrarian reform in Brazil. *Latin American Perspectives*, *36*(4), 156–177.
Hecht, S. (2010). The new rurality: Globalization, peasants and the paradoxes of landscapes. *Land Use Policy*, *27*(2), 161–169.
Hill, D. W., & Jones, Z. M. (2014). An empirical evaluation of explanations for state repression. *American Political Science Review*, *108*(3), 661–687.
Imai, S., Gardner, L., & Weinberger, S. (2016). *The Canada brand: Violence and Canadian mining companies in Latin America*. York, ON: Osgoode Hall Law School of York University.
Jeffords, C., & Thompson, A. (2016). An empirical analysis of fatal crimes against environmental and land activists. *Economics Bulletin*, *36*(2), 827–842.
Kirsch, S. (2014). *Mining capitalism: The relationship between corporations and their critics*. Berkeley: University of California Press.
Kitula, A. G. N. (2006). The environmental and socio-economic impacts of mining on local livelihoods in Tanzania: A case study of Geita District. *Journal of Cleaner Production*, *14*(3–4), 405–414.
Krausmann, F., Gingrich, S., Eisenmenger, N., Erb, K. H., Haberl, H., & Fischer-Kowalski, M. (2009). Growth in global materials use, GDP and population during the 20th century. *Ecological Economics*, *68*(10), 2696–2705.
Le Billon, P., & Lujala, P. (2020). Environmental and land defenders: Global patterns and determinants of repression. *Global Environmental Change*, *65*, 102163. https://doi.org/10.1016/j.gloenvcha.2020.102163
Leitner, H., Sheppard, E., & Sziarto, K. M. (2008). The spatialities of contentious politics. *Transactions of the Institute of British Geographers*, *33*(2), 157–172.
Lynch, M. J., Stretesky, P. B., & Long, M. A. (2018). Green criminology and native peoples: The treadmill of production and the killing of indigenous environmental activists. *Theoretical Criminology*, *22*(3), 318–341.
McNeish, J. A. (2018). Resource extraction and conflict in Latin America. *Colombia Internacional* (93), 3–16.

Mendes, C. (2018). Fight for the forest. In K. Conca & G. D. Dabelko (Eds.), *Green planet Blues: Critical perspectives on global environmental politics* (pp. 76–80). Abingdon, UK: Routledge.

Middeldorp, N., & Le Billon, P. (2019). Deadly environmental governance: Authoritarianism, eco-populism, and the repression of environmental and land defenders. *Annals of the American Association of Geographers, 109*(2), 324–337.

Moore, J. W. (2015). *Capitalism in the web of life: Ecology and the accumulation of capital.* New York: Verso Books.

Muradian, R., Walter, M., & Martinez-Alier, J. (2012). Hegemonic transitions and global shifts in social metabolism: Implications for resource-rich countries: Introduction to the special section. *Global Environmental Change, 22*(3), 559–567.

Peluso, N. L., & Lund, C. (2011). New frontiers of land control: Introduction. *Journal of Peasant Studies, 38*(4), 667–681.

Pierskalla, J. H. (2010). Protest, deterrence, and escalation: The strategic calculus of government repression. *Journal of Conflict Resolution, 54*(1), 117–145.

Rasch, E. D. (2017). Citizens, criminalization and violence in natural resource conflicts in Latin America. *European Review of Latin American and Caribbean Studies, 103*, 131–142.

Scheidel, A., Del Bene, D., Liu, J., Navas, G., Mingorría, S., Demaria, F., Avila, S., Roy, B., Ertor, I., Temper, L., & Martínez-Alier, J. (2020). Environmental conflicts and defenders: A global overview. *Global Environmental Change, 63*. https://doi.org/10.1016/j.gloenvcha.2020.102104

Sylvander, N. (2018). Saneamiento territorial in Nicaragua, and the prospects for resolving Indigenous-Mestizo land conflicts. *Journal of Latin American Geography, 17*(1), 166–194.

Temper, L., del Bene, D., & Martinez-Alier, J. (2015). Mapping the frontiers and front lines of global environmental justice: The EJAtlas. *Journal of Political Ecology, 22*, 255–278.

Totten, S., Parsons, W., & Hitchcock, R. (2002). Confronting genocide and ethnocide of indigenous peoples. In A. Hinton (Ed.), *Annihilating difference: The anthropology of genocide* (pp. 54–91). Berkeley, CA: University of California Press.

UNEP. (2018). *Who are environmental defenders?* United Nations Environment Programme. www.unenvironment.org/pt-br/node/21162

Vadlamannati, K. C., Janz, N., & De Soysa, I. (2019, February 28). *US multinationals and human rights: A theoretical and empirical assessment of extractive vs. non-extractive sectors.* http://dx.doi.org/10.2139/ssrn.3344832

Virdee, S. (2019). Racialized capitalism: An account of its contested origins and consolidation. *The Sociological Review, 67*(1), 3–27.

Watts, M. J. (2005). Righteous oil? Human rights, the oil complex, and corporate social responsibility. *Annual Review of Environmental Resources, 30*, 373–407.

[Part of this chapter first appeared as: Le Billon, P., & Lujala, P. (2020). Environmental and land defenders: Global patterns and determinants of repression. *Global Environmental Change*.]

8 The gendered criminalization of land defenders in Ecuador

From individualization to collective resistance in feminized territories

Melissa Moreano Venegas and Karolien van Teijlingen

Berta Cáceres, Julián Carillo, Isidro Baldenegro López, Amauri Pereira Silva, Raimundo Da Silva, Maria Fernandes, Eraldo Moreira Luz, José Tendetza, Homero Gómez González, Ángel Shingre . . . These are just a few names out of the long list of environmental and land defenders that have been killed across Latin America over the last few years. Unfortunately, this list continues to grow, with new names added to it almost every week.

The violence against land and environmental defenders, activists, and communities that oppose extractive activities is a soaring problem.[1] Worldwide, an average of four defenders died every week in the period 2014–2017 (Butt et al. 2019). These cases generally remain unresolved, as impunity prevails (Maestro 2020). Global Witness (2020) reported that Latin America is the region where most murders are reported: of the 212 murders that occurred globally in 2019, 147 were in Latin America. This figure reflects widespread opposition to the large number of extractive projects in the region (Le Billon & Lujala 2020; Temper et al. 2015), but also strong forms of activism and solidarity in Latin America that allow for better monitoring (Global Witness 2019).

The blatant murders of Latin American environmental defenders are, however, only a fraction of the violence they experience. Generally, these murders are embedded in multiple practices of "passive" violence against activists and leaders, most of which do not make it to the headlines (see Menton et al., this volume). This violence may consist of daily hostilities by public forces or private security guards, bullying, impediment of free mobilization, and destruction of crops, which may escalate to unruly arrests, militarization of communities, criminalization of dissent, sexual assaults, or threats of violence against family members.

Our research focuses on the violence – widely understood – against environmental defenders in Ecuador, particularly its Amazon region. Here, the long-standing oil exploitation and the more recent expansion of large-scale mining have faced fierce resistance (Sawyer 2004; van Teijlingen & Hogenboom 2016). Since the 1980s, Amazonian grassroots organizations and Indigenous and peasant communities have denounced the destructive effects of these activities on the environment, their territories, and their livelihoods. Although the violence against environmental defenders has been less grim compared to neighboring countries, their resistance has not been without risk: over the last decades, various leaders have been killed for defending their territories and land against the expansion of these industries: Ángel Shingre, José Tendetza, Freddy Taish, and Bosco Wisuma. A countless number of leaders have furthermore been living under the threat of violence.

In this context, we have carried out various interviews with the family members, fellow defenders, and community members of two murdered environmental defenders: Ángel Shingre, a *campesino* who organized farmers against oil extraction companies operating in the northern Amazon and was killed in 2003; and José Tendetza, an Indigenous anti-mining leader who was killed in 2014 in southern Ecuador.[2] We have furthermore analyzed recent cases of hostilities against two Amazonian Shuar communities that have been violently evicted to make way for the camp of a mega-mining project in southern Ecuador as part of the work we conduct as members of the *Colectivo de Geografía Crítica del Ecuador* (Critical Geography Collective of Ecuador).[3]

Based on this research, we reflect on two aspects of the violence against land and environmental defenders, anti-extraction activists, and communities that oppose extractive activities. One aspect is the gendered character of this violence, while the other is the tension between individualization vs collectivization of the struggles in relation to this violence.[4] To do so, and as explained in the following, we bring a critical feminist geography perspective to the study of violence against environmental defenders.

The gendered character of violence

Our critical feminist geography perspective highlights the gendered production of territory through everyday power relationships (Ibarra & Escamilla 2016) and across different geographical scales from the body/local to the global (Katz 2001). In the context of the advance of extractive capitalism across Latin America, violence is a vehicle for the capitalist production of space. Spaces of capital are produced "only to be destroyed due to the very dynamics of infinite accumulation of capital, technological change and fierce forms of class struggle" (Harvey 2000: 177). For us, spaces produced by capital are also, without question, gendered, colonial, and racist spaces (Silveira et al. 2017). This is of particular importance in a context of the growing and differentiated violence against men and women (Hyndman & De Alwis 2004) that is being used in Ecuador to territorialize the projects of capital.

Another, related concept we use to understand the practices of environmental defenders and the violence they experience is "body–land–territory". This concept emerged from Indigenous communitarian feminists in Latin American who struggle against the expansion of the extractivist industries (Cabnal 2010, 2019; Paredes 2010). It explicitly positions the body as the first level of territorial autonomy. The concept furthermore affirms the strong interconnectedness between bodily integrity and territorial integrity in struggles against extractivism (Leinius 2020). As Sweet and Ortiz Escalante (2017: 595) argue:

> [we can see] bodies and communities as a continuum . . . revealing feelings and sensations people have as part of space. . . . It is a framework for struggle against sexual violence and mining. It is a political category of indigenous communitarian feminism, a way to suggest and feel the body as territory alive and historical.
>
> (Sweet & Ortiz Escalante 2017: 595)

The concept thus draws analytical attention to the fact that when territories are affected by expropriation or contamination, this directly affects the lives and bodies of those inhabiting these territories. Simultaneously, this connection implies that defending territory is a very bodily practice, too. This then renders the concept of body–land–territory useful in analyzing the violence against environmental defenders, since perpetrators seek to exert power and control over territories by violating the bodily integrity of defenders.

The concept of body–land–territory, moreover, configures differently for women and men (Colectivo Miradas Críticas del Territorio desde el Feminismo 2017). The bodies of women and other feminized bodies especially have historically been arenas of patriarchal domination by state forces, colonialism, and capitalist development models (Colectivo Miradas Críticas del Territorio desde el Feminismo 2017; Falquet 2019). At the same time, as this chapter will discuss in the next sections, the domination of feminized bodies has inspired unique struggles for autonomy of both bodies and territories across the Latin American continent. It is therefore pivotal to recognize the gendered dimension of the concept of body–land–territory when discussing the violence against environmental defenders.

The violence experienced by various Shuar communities in the Ecuadorian Amazon is particularly illustrative to this point. The territory of the Shuar Indigenous nationality occupies a large part of the provinces of Zamora Chinchipe and Morona Santiago, in the southern Amazon of Ecuador, around the Cordillera del Cóndor on the border with Peru. Along with the Shuar population, there is also a peasant settler population who left their lands in other parts of Ecuador decades ago in search of better luck in the Amazon. For several years, the population of some Shuar centers and peasant communities has seen their daily lives affected by the opening of the first mega-mining projects in the country.[5] Violent evictions, deceptive purchase of land, destruction of houses, contamination of rivers, opening of brothels, trials, assassinations of leaders, and militarization are some of the actions that state and company actors have carried out in order to establish the mega-mining projects (van Teijlingen et al. 2017).

In 2015, more than 20 families of the Tundayme parish in the Zamora Chinchipe Province were violently evicted by the police to make way for the infrastructure of the Mirador mega-mining project. In August 2016, the military, police, and personnel of the mining company destroyed various Shuar communities and violently evicted more than 250 people (including children and elderly people) to implement infrastructure of the Panantza-San Carlos project in Morona Santiago, but also to set a precedent to any intention of opposing the mining project in the future. For two months, the provincial territory was militarized for the protection of the mining company (CIAP-Colectivo de Investigación y Acción Psicosocial 2017; Fernández et al. 2019). In what follows, we share testimonies of people evicted from one of the communities in Morona Santiago to illustrate how the physical territory and the Indigenous bodies–land–territories were controlled, producing a violent space according to gender roles.

The first element is that before the violent eviction almost all men of the community were formally accused of "sabotage and terrorism", so they were afraid of being captured by the military and decided to hide in the surrounding forests, coming to the community from time to time. This left the women feeling they would be by themselves when the military arrived:

> The men left us abandoned, our husbands. They were leaving, far away. Where else was the military [going to] look for them, [if] my husband left me abandoned? He told me: stay here with the children and I'm going to hide. Hence I told him: why are you leaving me? I can't be here. He was leaving and I took my things, I took my identity card, my children, I left, I already wanted to leave myself. When I was leaving other people also started to leave. Most of the accused are men.
>
> (interview, 25 May 2019)

Those were times of confusion and fear in which the final decision to leave was indeed made only when the military arrived. As a man told us: "they came firing bursts, advanced to [nearest town], they came shooting. We all listened. There were many rumors that the military is coming to shoot us" (interview, 25 May 2019).

Women also narrated how they faced state terror during the month prior to the eviction, when a helicopter flew over the community, threatening women and children who did not know what to do:

> We were afraid, we only spent time in the house with the children, the children were afraid. But we did not think that the military was going to come. . . . We were already scared. We didn't know what the helicopter was doing. . . . I had my daughter here, when the helicopter flew we ran to hide downstairs with all the children and when it passed we went out. After a little while the helicopter returned, turning around. We didn't even think about eating, we didn't eat anything [for days]. All the women were gathered on that court, we couldn't do anything . . . what to do? . . . We did face [the helicopter]. . . . We didn't stop picking up sticks, we planted our own selves there [in the field] so that [the helicopter] could not land on that field. Because if they landed, I don't know what they would do, they said they were going to rape women, that's what [people] told us.

> That's why we went out with sticks, we stopped there, the helicopter flew low, left again, returned, flew low again, left again and so on. We had no way to be well, or to make lunch. Somebody said: [the helicopter] is already here! And we were running away, and the children were screaming.
> (interview, 25 May 2019)

Sexual threats are a common way to discipline feminized bodies within the capitalist patriarchal system (Segato 2016). Using the body–land–territory concept, our argument is that in the same way that certain bodies are feminized in order to be violated, territories too are feminized. The feminization process started with the fears of men being accused of sabotage and terrorism and women and children being terrorized by the helicopter and rumors that the military was coming. Then, women received the direct violence of the military when, during the eviction, they ran away from the military and, from their hiding places, witnessed the destruction of their community.

Four months after the eviction, women, men, and children started to come back to the community and tried to reconstruct their lives. However, for several months, men hid in the community with fear of being imprisoned and prosecuted. One man told us:

> the charges I think . . . I never left [the community], I didn't go to Gualaquiza (nearest city), I sent people to buy things. [I wondered] why [if] I go to jail, and my children? Who will take care of my wife? For that reason I have not left [the community]. If the charges were dropped, then I could walk around the town. I spent seven months without going out to the town, without leaving the community, I did not move from here. The others who were accused still have not come out. My wife was going out [to the town] . . . I was waiting [for] a bag of bread she brought me.
> (interview, 25 May 2019)

This testimony shows how men and women continued to be affected in different ways even after the eviction; while men were hiding, women had to carry on the legal inquiries and shopping for food in the nearby town:

> [my husband] cannot leave, so far we are not so sure that the charges have been dropped or if they continue, that is why he cannot walk so much in Gualaquiza. Just me. All [those who are accused] do not go out, when they go out they take a taxi, at night, they do not go during the day.
> (interview, 25 May 2019)

> They [the men] do not go out. Sometimes they go out, but soon they return.
> (interview, 25 May 2019)

These testimonies show us how the Shuar communities were feminized territories, as men were neutralized by legal mechanisms (that is, the legal charges against them) and women and children were terrorized by the military siege. Although this feminization of territory placed a disproportionate burden on women to defend their bodies, their land, and reproductive activities, they were not mere victims. Even though the power of capital sees women and children as defenseless, they were the ones who traveled to Quito to claim their communities as their own, creating a trans-territorial network of resistance and support (Fernández et al. 2019). We will return to the emancipatory character of the body–land–territory when discussing the case of *Mujeres Amazónicas* (Amazonian Women).

The individualization of struggles

A second dimension we would like to highlight regarding the violence against, and criminalization of, defenders is that it often has the effect of isolating leaders. Court cases against them target them personally, and fear of infiltrators makes their essential circle of *personas de confianza* – trusted people – shrink. This is illustrated by a quote from an interviewed defender, who has endured various hostilities due to his resistance against a mining project in the south of the Ecuadorian Amazon: "I do not tell people where I am or where I am going. Only a few people are informed about my whereabouts, in order to protect myself" (interview, 20 July 2019). The threats and violence target them and their families, and in an attempt to avoid others being affected by it, leaders move out of their houses, their communities, and reduce their social contacts. The end result: defenders who are often part of collective struggles become isolated and their struggles individualized.

This individualization of environmental defenders is not just related to violence, though. This trend, we suggest, is also reinforced by the patriarchal leadership styles that characterize many of the struggles against extractivism in Ecuador, and extractivism more generally (see Billo 2020; Cirefice & Sullivan 2019; Jenkins 2014). Their leadership is generally organized hierarchically, with a president or a small group of *dirigentes* leading the struggles, as very few people are willing to commit their lives to a cause that seems lost and whose work is very often stigmatized, dangerous, and ill-paid. As one interviewed defender told us: "When we opened the [human rights] office we knew it was a frowned upon job" (interview, 13 August 2019).

Here, we want to argue that defenders who enjoy relative and legitimate positions of power within their organizations or communities are pushed towards a leadership praxis that reflects masculine ideals of the knowledgeable, vocal, and heroic man, lonely at the top (although these positions are held by women, too). This makes them susceptible to further individualization as a result of violence and criminalization. The described leadership praxis is very often reinforced by the defender's growing contacts with external actors like NGOs, public authorities, and journalists, and by the management of increasing resources, such as being able to access and understand large amounts of information about the industry they are facing, the ability to organize public demonstrations, or knowing the "enemy's language and tools".

While this pattern is sometimes an outcome of personal traits and life trajectories, it can also be related to the campaigns launched by national and transnational NGOs in response to increased violence in order to make the violence visible and hence protect threatened defenders. These campaigns aim to give environmental defenders a face and to tell their heroic stories. While these campaigns are well intended, we worry that by doing so they often invisibilize the collective character of struggles, give prominence to the leader(s), and unwittingly contribute to their individualization. At the same time, they may be reinforcing the abovementioned masculine leadership praxis.

Before turning to the perils of individualization for collective action, we want to highlight our research findings, which indicate how isolation and individualization put these leaders at increased risk. Our interviewees, who were closely related to some of the killed environmental defenders of the Ecuadorian Amazon, highlighted the individual character of their way of working as something that contributed to their eventual violent death. A leader from an organization in southern Ecuador, for example, explained the process of individualization that preceded the murder of José Tendetza:

> He started to operate alone, go to meetings without telling us or without us accompanying him. There was no-one to look after him, to have his back or warn him if he would speak up when there were people around one could not trust. On one of those lonely walks, they caught him.
>
> (interview, 23 July 2019)

A colleague of another defender who was murdered told us:

> He did not believe much in protecting himself, he was free, he went to a community, stayed there for 8 days, came back and wrote the reports. His job was to be in the communities, camping and fighting side by side with the communities. . . . He was always alone.
>
> (interview, 13 August 2019)

As these testimonies indicate, the problematic individualizing effect of violence puts the defender in even more danger. Instead of further singling out the defender through awareness campaigns and masculine leadership patterns, we suggest it would be helpful to explore the potential of their opposites: the collective character of environmental struggles and feminist leadership.

The potential of the collectivization of the struggles and of women reclaiming their voices

Individualization puts the defender in danger but also obscures the collective character of environmental struggles. Across Latin America, struggles are enacted and led by collectives: communities, organized neighbors, women or Indigenous organizations, and cooperatives. We argue that highlighting the collective character of the struggles rather than the individuals when communicating or theorizing those struggles is not only a form of protecting the defenders but also a matter of political coherence.

Our research findings on the strategies that environmental defenders have adopted after suffering an episode of violence show "a going back" to the collective as a way to protect themselves and to gather the strength to continue fighting. A colleague of Ángel Shingre explains in this way the formation of the Network of Leaders named after Ángel: "When they killed him, we called the leaders of the 46 communities with whom we were working at that time. We decided to make the Network with 46 people, women and men: Kichwas, Shuar, Waorani, and farmers" (interview, 13 August 2019). Similarly, Shuar people violently evicted from the community, as we reviewed earlier, recovered the "Indigenous guard", a form of communitarian patrolling of the territory, to take care of their people, particularly men with legal charges (interview, 25 May 2019).

Another key strategy to protect individuals is making the collective the leader of a struggle, and by doing so, ensuring the anonymization of its members. This includes making the collective, instead of an individual, the author of demands, press releases, and bulletins; having spokespersons rotate; and appearing in public collectively. This requires, of course, the transformation of the leadership praxis that we described earlier in this presentation. It involves challenging the masculine leadership-figure and promoting another form of leadership.

This explicit collectivization of struggle as well as the challenging of masculine leadership-ideals through women gaining prominence and strengthening their voices are all features of the movement of *Mujeres Amazónicas*. This movement was born in 2013 after large tracts of the Ecuadorian Central and southern Amazon were concessioned and auctioned to transnational oil companies, but soon they also included demands related to the mining industry. Grassroots women from seven Indigenous nationalities gathered to share experiences and build their organizational capacity in order to stand up against the expansion of extraction and the related exploitation of nature and women's bodies. As a result of this weaving of a common subjectivity, they organized women's marches from Puyo, a city in the Ecuadorian Amazon, to the capital city, Quito, to deliver a mandate to the president. The three *Marchas de Mujeres Amazónicas* were a notorious exercise of collective struggle in which more than a hundred women gathered in an assembly, wrote a collective text of demands, and decided to travel to the capital city to present these to the president. There are no apparent leaders, the spokesperson rotates, and their performativity (rooted in their material conditions) is nothing close to that of the lonely hero: women travel with their children, hand in hand, within a collective movement of people.

By means of these actions, we argue, the *Mujeres Amazónicas* transform the masculine form of leadership by challenging the oppressive context in which women decide to become an "environmental defender". Women's roles in environmental struggles are questioned constantly, from both the outside and within their communities. The violence these women defenders suffer is particular to those roles, as a recent report of Amnesty International also evidenced (Amnistía Internacional 2019). The *Mujeres Amazónicas* challenge this in two ways. First, they make a clear connection between the violence used to force extractive projects into Indigenous territories and the violence against women exercised by private security guards and parallel groups (i.e. competing social organizations) created by the government.

Second, the case of the *Mujeres Amazónicas* reveals the increasing gender-based violence that women face with the arrival of extractive projects but also recognizes the structural and historical gender violence they suffer inside the communities. In so doing, they recognize what Lorena Cabnal, a communitarian feminist from Guatemala, has described as the "ancestral patriarchy", a term used to challenge the system of subordination based in gender that exists inside Indigenous societies across Latin America (Cabnal 2010), but also to denounce the "re-patriarchalization of territories" (Colectivo Miradas Críticas del Territorio desde el Feminismo 2017; Hernández 2016) that arises with the arrival of extractive projects. The re-patriarchalization of the territories means that new forms of gender violence overlap with those that previously existed within their communities.

The previous argument is clear when reading the 20th and 21st demands of the "Mandate of the Grassroots Amazonian Women Defenders of the Forest Against Extractivism", written and submitted to the Ecuadorian president in 2018:

> 20 We require a statistical study of the cases of physical and sexual violence, including prostitution, to create a public policy tailored for the Amazon women from all nationalities, in cities and in communities
>
> 21 We demand a deep historical research about sexual and gender violence associated with oil and mining activities and militarization, in order to apply the necessary sanctions and to ensure the NON-repetition [of these violences] in Indigenous territories of the Amazon.

From our perspective, by recognizing the interdependence between the violence exercised against the Indigenous territories and the gender violence they suffer to their bodies, the *Mujeres Amazónicas* appeal to the body–land–territory unity we refer to previously. This concept explicitly ties the individual (that is, the body) to the collective (the territory), locating the body as the first territory of sovereignty, an unavoidable scale from which to fight for their Indigenous territorial sovereignty (Cabnal 2019). And, implicitly, it "pushes us to rethink how bodies have been constructed by multiple oppressions, the historical structure of patriarchy, colonialism, racism, and neoliberal capitalism, which have led to exploitation via different agreements and policies" (Sweet & Ortiz Escalante 2017: 595). In this way, the *Mujeres Amazónicas* incarnate the sense of "putting the body" in the center of territorial defense (Zaragocin 2019), demanding that all forms of violence be eradicated together: those inflicted to the individual body and those imposed on the collective body (the territory). As we said earlier, it could be that this weaving of a common subjectivity reinforces the importance of the collectivization of the struggle.

Conclusions

This chapter focused on the continued violence against land and environmental defenders, anti-extraction activists, and communities that oppose extractive activities in Latin America – and in the Ecuadorian Amazon in particular. We focused on two aspects we deem especially problematic: the gendered character of violence, and the tension between individualization vs collectivization of the struggles in relation to this violence. In order to analyze this, we took a feminist geography perspective and used the concept of body–land–territory.

With the aim of positioning our first argument, we reviewed the case of a Shuar community that was violently evicted in 2016. The argument here is that the state seeks to control Indigenous territories and bodies in different ways according to their gender. Men with legal charges hid, while women who usually perform care work – looking after and feeding the children, the sick, and the elderly – experienced direct violence.

The second argument we put up for discussion is the following. We found that both the criminalization of environmental defenders and the well-intended civil society responses to safeguard threatened defenders tends to isolate and individualize them. Based on our research in the Ecuadorian Amazon, we suggested that this individualization places them at increased risk of suffering violence, whereas the collectivization of struggles provides anonymity and protection. We linked this individualization of struggles to the patriarchal leadership structures that characterize many anti-extraction movements and argued that regaining the collective character of struggle requires the transformation of this leadership. Various women's movements, like the *Mujeres Amazónicas*, may serve as an inspiration for this endeavor.

Notes

1 Although this is an on-going discussion, we think that these three categories encompass the diversity of resistance actions to extractive activities occurring in Ecuador at the moment.
2 This research is part of the project "'Sustainable' development and atmospheres of violence: experiences of environmental defenders", led by the University of Sussex and funded by the British Academy.
3 https://geografiacriticaecuador.org, only in Spanish.
4 We developed these ideas from a previous work: see Moreano et al. (2019).
5 These projects are the Mirador copper mine and the San Carlos-Panantza copper mine, located in the Province of Zamora Chinchipe and the Province of Morona Santiago, respectively. Both projects, of which the first entered the exploitation phase in July 2019, are owned by CRCC-Tonguan Investment Co., a consortium of two Chinese giants in metallurgy and railway construction: Tongling Nonferrous Metals Group Holding (Tongling) and China Railway Construction Company (CRCC).

References

Amnistía Internacional. (2019, April 30). Ecuador: "No nos van a detener". Justicia y protección para las mujeres amazónicas defensoras de la tierra, el territorio y el ambiente. *Amnistía Internacional*. www.amnesty.org/es/documents/amr28/0039/2019/es/

Billo, E. (2020). Patriarchy and progressive politics: Gendered resistance to mining through everyday social relations of state formation in Intag, Ecuador. *Human Geography*, *13*(1), 16–26.

Butt, N., Lambrick, F., Menton, M., & Renwick, A. (2019). The supply chain of violence. *Nature Sustainability*, *2*, 742–747. https://doi.org/10.1038/s41893-019-0349-4

Cabnal, L. (2010). Acercamiento a la construcción del pensamiento epistémico de las mujeres indígenas feministas comunitarias de Abya Yala. *Feminismos diversos: el feminismo comunitario*. Madrid: ACSUR-Las Segovias.

Cabnal, L. (2019, September 25). *Mi cuerpo, mi primer territorio de defensa* [Conference session]. San Cristóbal de las Casas, México.

CIAP-Colectivo de Investigación y Acción Psicosocial. (2017). *La herida abierta del Cóndor*. Quito: Acción Ecológica.

Cirefice, V. C., & Sullivan, L. (2019). Women on the frontlines of resistance to extractivism. *Policy & Practice: A Development Education Review* (29), 78–99.

Colectivo Miradas Críticas del Territorio desde el Feminismo. (2017). *Mapeando el cuerpo-territorio: Guía metodológica para mujeres que defienden sus territorios*. https://miradascriticasdelterritoriodesdeelfeminismo.files.wordpress.com/2017/11/mapeando-el-cuerpo-territorio.pdf

Falquet, J. (2019). Violence against women and (de-)colonization of the "body-territory". In I. Cîrstocea, D. Lacombe, & E. Marteu (Eds.), *The globalization of gender: Knowledge, mobilizations, frameworks of action* (pp. 81–101). Abingdon, UK: Routledge.

Fernández, A., Moreano, M., Gutiérrez, B., Cando, S., Romero Salgado, N., Bayón, M., Murillo, D., & Molina. A. (2019). Violencia estatal, colonialismo interno y despojo: la implantación del proyecto minero Panantza-San Carlos. In M. Bayón & N. Torres (Eds.), *Geografía Crítica para detener el despojo de los territorios: teorías, experiencias y casos de trabajo en Ecuador*. Quito: Abya-Yala.

Global Witness. (2019). *¿Enemigos del estado? De cómo los gobiernos y las empresas silencian a las personas defensoras de la tierra y del medio ambiente*. London.

Global Witness. (2020). *Defending tomorrow: The climate crisis and threats against land and environmental defenders*. London.

Harvey, D. (2000). *Spaces of hope*. Edinburgh: University of Edinburgh Press.

Hernández, D. T. C. (2016). Una mirada muy otra a los territorios-cuerpos femeninos. *Solar*, *12*(1), 45–46.

Hyndman, J., & De Alwis, M. (2004). Bodies, shrines, and roads: Violence, (im) mobility and displacement in Sri Lanka. *Gender, Place & Culture*, *11*(4), 535–557.

Ibarra, V., & Escamilla, I. (2016). *Geografías feministas de diversas latitudes: Orígenes, desarrollos y temáticas contemporáneas*. Mexico: Universidad Nacional Autónoma de México.

Jenkins, K. (2014). Women, mining and development: An emerging research agenda. *The Extractive Industries and Society*, *1*(2), 329–339.

Katz, C. (2001). On the grounds of globalization: A topography for feminist political engagement. *Signs: Journal of Women in Culture and Society*, *26*(4), 1213–1234.

Le Billon, P., & Lujala, P. (2020). Environmental and land defenders: Global patterns and determinants of repression. *Global Environmental Change*, *65*, 102163.

Leinius, J. (2020). From defending body and territory to defending body as territory: Women's politics of translation in eco-territorial conflicts. In A. Daniel, R. Mageza-Barthel, M. Richter-Montpetit, & T. Scheiterbauer (Eds.), *Gewalt, Krieg und Flucht: Feministische Perspektiven auf Sicherheit* (pp. 71–90). Leverkusen: Verlag Barbara Bunrich.

Maestro, G. (2020, February 3). América Latina, una tumba para los defensores de la naturaleza. *Lazarón*. www.larazon.es/internacional/20200202/o544wgvvanfkrhesozqcq3jmwa.html

Moreano, M., van Teijlingen, K., & Zaragocin, S. (2019). El sujeto colectivo en la defensa territorial. *Lasa Forum*, *50*(4), 17–20.

Paredes, J. (2010). *Hilando fino: Desde el feminismo comunitario*. La Paz: Mujeres Creando Comunidad.

Sawyer, S. (2004). *Crude chronicles: Indigenous politics, multinational oil, and neoliberalism in Ecuador*. Durham, NC: Duke University Press.

Segato, R. (2016). *La guerra contra las mujeres*. Madrid: Traficantes de sueños.

Silveira, M., Moreano, M., Romero, N., Murillo, D., Ruales, G., & Torres, N. (2017). Geografías de sacrificio y geografías de esperanza: tensiones territoriales en el Ecuador plurinacional. *The Journal of Latin American Geography*, *16*(1), 69–92. https://doi.org/10.1353/lag.2017.0016

Sweet, E. L., & Ortiz Escalante, S. (2017). Engaging territorio-cuerpo-tierra through body and community mapping: A methodology for making communities safer. *Gender, Place & Culture*, *24*(4), 594–606.

Temper, L., Del Bene, D., & Martinez-Alier, J. (2015). Mapping the frontiers and front lines of global environmental justice: The EJAtlas. *Journal of Political Ecology*, *22*(1), 255–278.

van Teijlingen, K., & Hogenboom, B. (2016). Debating alternative development at the mining frontier: Buen vivir and the Conflict around El Mirador Mine in Ecuador. *Journal of Developing Societies*, *32*(4), 382–420. https://doi.org/10.1177/0169796X16667190

van Teijlingen, K., Leifsen, E., Fernández-Salvador, C., & Sánchez-Vázquez, L. (2017). *La Amazonía minada: Minería a gran escala y conflictos en el sur del Ecuador*. Quito: Editorial USFQ & Abya-Yala.

Zaragocin, S. (2019). Feminist geography in Ecuador. *Gender, Place & Culture*, *26*(7–9), 1032–1038.

9 Insurgent ideas from Indigenous peoples in Brazil

Counter-colonial epistemologies and the defense of life

Felipe Milanez

It is an iconic expression, very popular and common in advertising, that "ideas can change the world". It usually follows the famous Lampedusa quotes, meaning that most of the changes promised are to keep "things to stay as they are". But out of the frightening times Indigenous peoples in Brazil face nowadays, ideas may not be changing the world, but they are at least postponing its end. After Bolsonaro took office in 2019, and during the pandemic of the new coronavirus, two books from Indigenous author Ailton Krenak have been able to "hold the sky" out of the turbulent storm. Both titles are provocative. The first, *Ideas to Postpone the End of the World*, was not just a bestseller in Brazil, but also was shortlisted for the most important literary award in the Portuguese language (Jabuti) and quickly translated into English, French, Italian, and Spanish. The author, who played with the general idea of "change the world" and suggests instead postponing our common end on this planet, is one of most distinguished political leaders of Indigenous movements in the history of Brazil. This book was followed by another, based on different interviews and comments on livestream sessions he engaged in during the pandemic: *Life Is Not Useful* (translated by the author).

The popular recognition of Ailton Krenak's books reveals a new intellectual movement emerging in Brazil: one of Indigenous authors engaged in denouncing the coloniality of power in Brazil and new strategies for conquest and dispossession of territories. Recognized elder leaders, such as Ailton Krenak and Davi Kopenawa, a Yanomami shaman, as well as young academics –representing the generation of Indigenous scholars who entered universities after affirmative actions implemented during the Lula government years (2003–2010) – are fighting a new battle: a battle of ideas inside colonialist Brazilian society. The main Indigenous organization, APIB (Articulation of Indigenous Peoples of Brazil), is led by a bright intellectual woman, Sonia Guajajara, and is associated with many Indigenous leaders who are also great scholars (Eloy Terena, Dinamam Tuxá, and Samara Pataxó, among others – see APIB 2020). Sonia Guajajara has a double BA in literature and nursing, with a master's in education, and is well known for powerful talks at distinguished occasions and international events, frequent interviews, and leading investigative reports; she has an extraordinary political ability to extend alliances. She is leading and reorganizing the Indigenous social movement with great attention paid to social media and new strategies for communication. She is also promoting the emergence of Indigenous women into high posts in the movement, such as her allies Nara Baré, for the Coordination of Indigenous Organizations of the Brazilian Amazon (COIAB), Valeria Payé, and Angela Kaxuyana, who together also coordinate APIB. Guajajara ran for vice-president in the 2018 elections against Jair Bolsonaro, marking the first time an Indigenous candidate ran for the highest posts of the Brazilian republic – her presence engaged new Indigenous candidates all around the country to become involved with institutional politics.

Although the flourishing of these ideas is "raising skies" – published in books, academic theses, or circulating in social media and interviews – these ideas nevertheless arrive into a very adverse situation: there is an extreme right-wing military presidency in power, marked by anti-Indigenous discourse and policies, while the left has been characterized by lethargy after the coup of 2016 and the attacks of *Lava Jato* ('Car Wash') against the Workers Party (PT) and a then-hegemonic left. Moving from ideas to concrete and direct actions, Indigenous intellectuals and Indigenous movements have been able to resist environmental attacks: "Yet the presence of these risks, and long term conflicts related to land grabbing and resource theft linked to said development, has in some cases strengthened community ties and increased capacity for active resistance" (Menton et al. 2021: 1). Among a growing repertoire of actions, "a new cohort of Indigenous youth who had access to higher education, were able to draw on social media and Indigenous led court cases to help counteract the 'genocide by omission' that has been worsened by the pandemic" (ibid.: 1).

500 years of counter-conquest ideas

These ideas are not new. Since the arrival of the first Europeans to the New World, voyagers have reported that the Tupinambá from the Brazilian coast were great speakers; in some of the description of their debates, their sense of war, life, and relationship with nature mark their distinction from European narcissism. As Ailton Krenak (2020: 14) reflects today, "We have to abandon our anthropocentrism. There's a lot more to Earth than us, and biodiversity doesn't seem to be missing us at all. Quite the contrary". Anthropocentrism and the separation of society from nature are the main violences brought by colonizers. For Ailton Krenak, anthropocentrism is a form of violence against collectives that include places supporting life and thereby creates a separation – a form of exile – of kinship linking a community with a river, a forest, a mountain, or a glacier that becomes "violently separated from the human collective, [and thereby] puts an end to a way of life" (Krenak & Milanez 2019). It builds the place of the "other" and the "non-place", the sacrifice zone. It is the mentality of slavery, beyond violence against humans.

The provocations from Krenak, who was a key figure during the debates for the Federal Constitution of 1988 – defending the territorial rights of

Indigenous peoples but also the rights of nature, as well as leading the Union of Indigenous Nations and Forest Peoples Alliances – also dialogue with Davi Kopenawa and his long-term friend's recent book. Both Ailton and Davi claim that the words they use come from long ago, learned through oral history and their political experiences. Davi Kopenawa defines the word ecology as a word that has existed since "the beginning of time"; according to him, "long before these words existed among them and they started to speak about them so much, they were already in us, though we did not name them in the same way" (Kopenawa & Albert 2013: 393). But, as they did not have books that recorded these ideas, the thoughts have been described as "new". As Davi says:

> In the forest, we human beings are the 'ecology' . . . But it is equally the *xapiri* [shamanic spirit], the game, the trees, the rivers, the fish, the sky, the rain, the wind, and the sun! It is everything that came into being in the forest, far from the white people: everything that isn't surrounded by fences yet. (ibid.: 33)

This is why the word ecology, to the Yanomami, is an ancient word.

Amerindian anti-anthropocentrism has long argued against European greed, as French missionary Jean de Lery discovered in dialogue with a Tupinambá elder on the Brazilian coast in the 17th century. Lery (1990: 101–102) wrote the "digression" of his argument in the following, which consists of the oldest and most beautifully written criticism of capitalism from an Amerindian perspective:

> Our Tupinambá are astonished to see the French and others from distant countries go to so much trouble to get their araboutan, or brazilwood. On one occasion, one of their old men questioned me about it: "What does it mean that you Mairs and Peros (that is, French and Portuguese) come from so far for wood to warm yourselves? Is there none in your own country?" I answered him yes, and in great quantity, but not of the same kinds as theirs; nor any brazilwood, which we did not burn as he thought, but rather carried away to make dye, just as themselves did to redden their cotton cord, feathers, and other articles. He immediately came back at me: "Very well, but do you need so much of it?" "Yes", I said (trying to make him see the good of it), "for there is a merchant in our country who has more frieze and red cloth, and even" (and here I was choosing things that were familiar to him) "more knives, scissors, mirrors, and other merchandise than you have ever seen over here; one such merchant alone will buy all the wood that several ships bring back from your country." "Ha, ha!", said my savage, "you are telling me wonders." Then, having thought over what I had said to him, he questioned me further, and said, "But this man of whom you speak, who is so rich, does he never die?" "Certainly he does," I said, "just as others do." At that (since they are great discoursers, and pursue a subject out to the end) he asked me, "And when he is dead, to whom

belong all the goods that he leaves behind?" "To his children, if he has any, and if there are none, to his brothers, sisters, or nearest kinsmen." "Truly," said my elder (who, as you will judge, was no dullard), "I see now that you *Mairs* (that is, Frenchman) are great fools; must you labor so hard to cross the sea, on which (as you told us) you endured so many hardships, just to amass riches for your children or for those who will survive you? Will not the earth that nourishes you suffice to nourish them? We have kinsmen and children, whom, as you see, we love and cherish; but because we are certain that after our death the earth which has nourished us will nourish them, we rest easy and do not trouble ourselves further about it."

Jean de Lery then reflects that this elder, a "poor savage American", from a nation the Europeans considered "so barbarous", was mocking the French for risking their lives only to get rich. He argues that "the Tupinambá mortally hate the avaricious". Out of this great description from Levy, we see a digression in favor of the Tupinambá.

The old Tupinambá's reflection on European greed as opposed to the life of the people in the forest, to life in a broad sense, was made right at the beginning of the expansion of the capitalist system in post-feudal Europe, and is very similar to the analysis made by Yanomami shaman Davi Kopenawa, our contemporary. Kopenawa has been leading an intense struggle against the invasion of their territory by wildcat goldminers supported by President Jair Bolsonaro:

> We are different from the white people and our thought is other. Among them, when a father dies, his children are happy to tell each other: "We are going to share his merchandise and his money and keep them for ourselves!" The white people do not destroy their deceased's goods, for their mind is full of oblivion. As for me, I would not say to my son: "When I die, you will keep the axes, pots, and machetes I happen to own!" I simply tell him: "When I am no longer, you will burn my possessions and you will live in your turn in this forest that I am leaving for you. You will hunt and clear gardens to feed your children and grandchildren on this land. Only the forest will never die!" It is true. We think it is bad to own a dead man's goods. It fills our thought with sorrow. Our real goods are the things of the forest: its waters, fish, game, trees, and fruit. Not merchandise! This is why as soon as someone dies we make all the objects he kept disappear."
>
> (Kopenawa & Albert 2013: 330)

Both elders and shamans, the "old" Tupinambá and the "new" Yanomami, think beyond capitalism, and above all are concerned with the preservation of the forest, the management and care of territories for future generations, knowing that there is what is necessary for them to live. This is precisely how Krenak (2020: 6) defines how "the times we're living in are expert at creating absences: sapping the meaning of life from society and the meaning of experience from life".

Wilma Martins de Mendonça, a Professor of literature, belongs to the Tabajara people, who fought against the violent wars of conquest in the 16th century. In her work, she investigates the discourse of colonialism in the Brazilian literature from both European missionaries and Brazilian writers. Her doctoral thesis investigates the construction of coloniality among missionaries and the first European voyagers, among whom is the Jesuit José de Anchieta, who has been a saint since 2014; she also explores his close relationship with and admiration of Mem de Sá, the third governor of the colony of Brazil, who was responsible for the most violent wars of conquest, comparable to the Spanish conquerors Pizarro and Hernan Cortez.

Mendonça's thesis, "Memórias de Nós" (*Memories of Us*, 2017), discusses the war of her ancestors and the colonial forces of oblivion – forces that seek to consolidate the political project of a singular and linear historical narrative. In 1563, in the midst of the victory over the French in the Guanabara and the cynical peace agreement that the Jesuits brokered with the Tupinambá, José de Anchieta wrote a Portuguese epic, *De gestis Mendi de Saa*, to honor the actions and wars of conquest led and perpetrated by Mem de Sá. As Mendonça observes, Anchieta celebrated the Tupiniquim chief Tibiriçá, ally of the Jesuits, by breaking the skulls of other Indigenous people defending their villages, while also celebrating "the unequal and genocidal wars that this inflicts on our ancestors" (2017: 198). The ferocious Mem de Sá dedicated himself to the crimes of colonialism, and was a murderer made heroic by Anchieta. It was, as Mendonça writes, "the Christian Hell personifying himself in the avenging flames of Portuguese greed" (2017: 200). She identifies a literary duel that followed the genocide and can be seen throughout Brazilian literature, between "voices in tune with colonialism and voices indignant with action and with colonialist vocabulary homilies" (2013: 177). This is a war that has continued in intensity into the contemporary world; one in which, only recently, have Indigenous voices such as Wilma Mendonça's, begun to be heard.

Defending the forest, the mountains, and the rivers provides the opportunity for broad reflection on the meaning of life. Losing the meaning of life promotes "intolerance toward anyone still capable of taking pleasure from simply being alive, from dancing, from singing". Dancing and singing to call for the rain are not exquisite or savage or barbarian, as European racism has classified them, but exist on Earth and with Earth. As Krenak (2020: 33) writes:

> What is being done to our rivers, forests, and landscapes? We get so disturbed by the regional chaos we live in, and so furious over the lack of political policy, that we can't see what really matters to people, collectives, and communities in their ecologies.

Experiences learned from elders in the Indigenous villages are now entering universities through a new generation, as perceived by Luiz Eloy Amado (Eloy Terena), a Terena lawyer and activist with the Indigenous movement. According to him, the trajectories of Indigenous academics are able to break through the coloniality of being and of power, and to challenge the colonial structures of universities (Amado 2020). Indigenous intellectuals like Eloy Terena occupy a social position that requires the ability to bridge not only between Indigenous and non-Indigenous worlds, but also between scientific and traditional knowledges. Life experiences, or "*vivências*", are the foundations of Indigenous epistemological architecture and the political sense of Indigenous academics. And, as Eloy sees, the center of lived experience can be related to the definition of ecology and life from the shaman: with nature, and away from the fences.

Alessandra Korap Munduruku – a law student, leader of the Munduruku women's movement, and recipient of the 2020 Robert F. Kennedy Human Rights Award – published an article in Brazil's national newspaper, *Folha de São Paulo*, denouncing illegal goldmining in the Munduruku territory and the illegal support the mining receives from the Bolsonaro government. She challenged fake news spread by the federal government and denounced Bolsonaro's commitment to the illegal wildcat-mining invaders. Korap denounces Bolsonaro's well-known negationism – denying climate change, the dangerous effects of Covid-19, fires and deforestation in the Amazon: "This negationism reveals the great truth: what does not exists in Brazil is government" (Korap 2020).

APIB released an international report, in partnership with Amazon Watch, denouncing corporations that benefit from the destruction of Indigenous territories: *Complicity in Destruction III: How Global Corporations Contribute to Violations of the Rights of Indigenous Peoples in the Brazilian Amazon* (Amazon Watch & APIB 2020). One of the main targets of the report, the production of cattle ranching and the responsibility of slaughterhouses in spreading coronavirus among the Kaingang and the Kaiowa Guarani Indigenous peoples in South Brazil, was based on a case brought to court by the Indigenous lawyer Fernanda Kaingang. She won an injunction for racism against JBS, Brazil's biggest cattle exporter, after they decided to fire Indigenous peoples because they came from a "vulnerable group".

These resistance movements, both against Covid-19 and invaders, are supported by the sacred songs of the shamans, as Tonico Benites showed in his doctoral thesis at the National Museum in 2014, and are the key elements in "reoccupation of sacred lands". These movements are likewise supported by the strength of women in leading the defense of the collective and their territories, as Elisa Pankararu analyzed in her master's dissertation (Ramos 2019). In the epigraph of her work, Elisa included an excerpt from the Pankararu philosophy:

> In the mythical narrative, in the spiritual and daily orientation of Pankararu People, there is a vision of a female figure seen as the mother of the creator and creation, mother nature, who understands and protects the spaces where there is life. All human and non-human living beings, including stones, waters and sacred female and male spirits. The understanding and knowledge left by our ancestors: traditional knowledge.
>
> (Pankararu Philosophy)

Losing a war is not the same as letting yourself be dominated, as resistances can re-emerge in different dimensions. They emerge in new forms in response to new dimensions of neoliberal territorial wars for conquest and colonialism – Indigenous resistance is continual in its fight to ensure that those attempts at conquest are constantly challenged by creative counter-conquest movements.

References

APIB (Articulação dos Povos Indígenas do Brasil). (2020). *Nossa Luta É Pela Vida*. https://emergenciaindigena.apiboficial.org/relatorio/

Amado, L. H. E. (2020). Para além da Universidade: experiências e intelectualidades indígenas no Brasil. *IdeAs* [Online]. http://journals.openedition.org/ideas/9442. https://doi.org/10.4000/ideas.9442

Amazon Watch, & APIB. (2020). *Complicity in destruction III: How global corporations enable violations of Indigenous peoples' rights in the Brazilian Amazon*. Edited by Luiz Eloy Terena Amado, Sonia Guajajara, & Kretã Kaingang. https://amazonwatch.org/news/2020/1027-complicity-in-destruction-iii

Kopenawa, D., & Albert, B. (2013). *The falling sky: Words of a Yanomami Shaman*. Translated by N. Elliott & A. Dundy. Cambridge, MA: Harvard University Press.

Korap, A. (2020, December 10). O garimpo ilegal existe, o que não existe é governo. *Folha de S. Paulo*. www1.folha.uol.com.br/opiniao/2020/12/o-garimpo-ilegal-existe-o-que-nao-existe-e-governo.shtml?origin=folha

Krenak, A. (2020). *Ideas to postpone the end of the world*. Translated by A. Doyle. Toronto, ON: House of Anansi Press.

Krenak, A., & Milanez, F. (2019). Ecologia Política. *Dicionário Alice*. https://alice.ces.uc.pt/dictionary/?id=23838&pag=23918&id_lingua=1&entry=24271 [last accessed 15 December 2020]. ISBN: 978-989-8847-08-9.

Mendonça, W. M. (2017). *Memória de Nós*. São Paulo: Porto de Idéias.

Mendonça, W. M. (2013). Discursos de discórdia: a temática colonialista nas letras brasileiras. *Cadernos de Estudos Culturais, 1*, 167–182.

Menton, M., Milanez, F., Souza, J. M. A., & Cruz, F. S. M. (2021). The COVID-19 pandemic intensified resource conflicts and indigenous resistance in Brazil. *World Development, 138*, 105222.

Ramos, E. U. (2019). *Mulheres lideranças indígenas em Pernambuco, espaço de poder onde acontece a equidade de gênero*. Master dissertation in Anthropology. Federal University of Pernambuco, Recife.

Tonico Benites. (2014). *Rojeroky hina ha roike jevy tekohape (Rezando e lutando): o movimento histórico dos Aty Guasu dos Ava Kaiowa e dos Ava Guarani pela recuperação de seus tekoha*. PhD thesis. The National Museum/Federal University of Rio de Janeiro.

Part 2

'Dirty' projects

10 The permutations of poverty

Rob Nixon

What does it mean to be poor? And how do people reliant on ecosystem health for their wellbeing resist development agendas imposed from the outside, agendas that may promise to lift the poor out of poverty but instead threaten a community's own definitions of privation and prosperity? To wrestle with these questions is to recognize the role neoliberal development often plays in advancing a rescue narrative that instead plunges the affected community into new levels of destitution.

In the village of Xolobeni, on the South African eastern seaboard, we have witnessed over the past 13 years a sustained struggle against the threat of development-imposed impoverishment. The Australian mining giant MLC campaigned, with the full backing of South Africa's national government, to build a titanium mine in Xolobeni and a highway for transporting the minerals. Together, the mine and highway would have jeopardized the community's future, operating (to adapt a term from sociologist Samer Alatout) as "infrastructures of elimination" (Alatout 2020).

Xolobeni, a village of 10,000, had a deep and successful history of opposing land theft under the old apartheid regime. The community reprised that success when, in 2018, a high court judge blocked the mine in a major victory for customary land rights. The villagers' triumph, however, was exacted at great cost. The proposed mine divided the community, as the company bought out several opponents with free pickup trucks and positions on the corporate board. However, the leader of the resistance, Nonhle Mbuthuma, stood firm against the corporate threats and blandishments. Mbuthuma endured persecution by the company and its surrogates. Hitmen assassinated two of her most prominent allies.

Mbuthuma became a decisive voice on the subject of wealth, poverty, and externally inflicted development.

> We have shelter, we have land, we have livestock. That's all we need. That's why we're saying we are rich. When somebody comes down here telling us "you are poor" because we are staying in a mud house. No! This is our culture and we like it and we enjoy it . . . [This place] will become a desert. They will poison everything. We are living with the plants and nature and we know that without the plants we cannot live. The mine will poison our

> land. And we are nothing without our land. It is our identity. Our way of life will die completely.
>
> (Burke 2016; see also Pearce 2017)

An unnamed elder echoes her views:

> Maybe you think we old men know nothing because we don't know money. Maybe we don't know money, but we do know we need the land. After you sell the land for the tens of thousands that they are promising, the money, unlike the land, will end.
>
> (Grunewald 2015)

Leanne Betasamosake Simpson, a First Nations Leader (Mississauga Nishnaabeg) from Canada, reflects on wealth and poverty in terms that align with Mbuthuma's arguments. Simpson, a founder of Idle No More, observes how:

> Indigenous communities face tremendous imposed economic poverty. Billions of dollars of natural resources have been extracted from their territories, without their permission and without compensation. . . . We have not had the right to say no to development, because ultimately those communities are not seen as people, they are seen as resources. . . .
>
> We need to understand why these communities are economically poor in the first place – and they are poor so that Canadians can enjoy the standard of living they do. I say "economically poor" because while these communities have less material wealth, they are rich in other ways – they have their homelands, their languages, their cultures, and relationships with each other that make their communities strong and resilient.
>
> (cited in Klein 2013)

For Simpson, like Mbuthuma, "economic poverty" is only one among several permutations of poverty. To be poor is not an absolute condition determined by a universal monetary measure. There are forms of wealth that elude official metrics.

Similar refusals of a one-size-fits-all definition of poverty echo across the global South. Carlos Gómez, an El Salvadorian Lutheran minister, insists on the importance of the long-term health of the land in any assessment of indigence and wealth. Gómez testifies to the travesty of economic uplift that doubles as land theft: "Destroying nature so that a few can fill their pockets with money is not justice. . . . The only thing the poor have is the land, and if that is taken, they have nothing" (Barnett 2010).

In November 2010, some strangers arrived in Liberia's Nimba County – men the likes of whom the locals had never seen. The newcomers began clear-cutting the forest that adjoined a village and fencing off the soon-to-be treeless land.

Villagers gathered, angry and aghast. The newcomers, they discovered, were representatives of the Malaysian palm oil colossus, Sime Darby. Government officials in Monrovia, Liberia's capital, had green-lighted Sime Darby's plan to convert forest and farmland into a vast palm oil plantation (BBC 2012).[1] But the residents of this putatively ownerless land were now rising in revolt.

A few years before the Malaysians clear cut the forest, some European researchers had spent time studying this very village (Round et al. 2016). Officially, the people there were poor: statistics showed the community struggled with 35% unemployment. Yet, to the researchers' surprise, they encountered well-nourished poverty. Among the village children, they found zero suffering from malnutrition. (By contrast, in New York, a city that boasts 91 billionaires, 20% of children suffer from malnourishment.)[2] By one measure, the village children were poor. But by another, they were relatively well off, enjoying food security at a critical stage of their physiological development. The mixture of forest commonage and small-scale farming had afforded the children a nutritious, varied diet. The felling of the forest set child development and neoliberal national development on a collision course.

Financial poverty is not the only index of hardship, especially for people dependent on the health of their immediate environment. Before the forest was sacrificed – and before the national government used eminent domain to evict farmers for a pittance – the Nimba community had practiced a blend of small-scale agriculture and agroforestry. They had supplemented what they grew with fruit, nuts, greens, tubers, and bush meat they found and cultivated in the forest. Whatever else the villagers lacked, they had maintained their basic food and water security. And they had health profiles that followed from that.

But now – unconsulted, unforewarned – these Liberian villagers found themselves severed from the conditions that had sustained their modest livelihoods.[3] They had lost their "subsistence security," in James Scott's terms (1977: 5). Like most ecosystem-reliant people, the villagers' dependence on forest fare was part of a precarious survival strategy, a delicate balancing act. Expelled from farms and forests, their balancing act collapsed.[4] Activist Lee Sworh testified to the ensuing impoverishment of his community, as a fertile small farm–forest mix was replaced by a sterile, inaccessible plantation. Sworh struggled to quantify the loss: "the land is our treasury, the land is our bank, the land is our life" (Round et al. 2016).

For generations, the blend of forests and farms had cushioned many villagers from a free fall into penury. But the arrival of the palm oil corporation marked the end of local sovereignty and sustainable possibility. The forest disappeared, replaced by a brutally simplified ecosystem and a labor regime to match. Instead of their customary commons, the villagers stared out at a vast assemblage of regimented trees, a biodiversity desert, a privatized anti-forest.

There is a deeper history to all this. In Liberia and elsewhere, Indigenous land defenders have faced a recurrent dilemma. In communities where possessive individualism had little traction historically, Indigenous communities have had to walk a line between a dominant value system that defines land as owned or ownerless and a customary system resistant to the idea of individual or corporate ownership. Refusing to partake of possessive individualism, Indigenous communities may leave themselves more vulnerable to theft of land that can be officially designated ownerless. Conversely, embracing possessive individualism may leave Indigenous inhabitants vulnerable to the predations of a property system in which land is reduced to a commodity. By transforming a nonproprietary relationship to the land into a proprietary one, Indigenous landholders may risk the desecration of their lifeways. Yet they often have few alternatives. By participating in a system that views land as an alienable commodity, Indigenous communities may leave themselves vulnerable to the predatory logic of the owner class. That class, wielding eminent domain and backed by armed corporate enforcers, can seize property for a pittance. Indeed, as Robert Nichols argues, settler colonial societies have often inducted the Indigenous into a system of land-as-commodity in order to set in motion the machinery of dispossession (2020: 30–33). In this way, all too often possession becomes a mechanism for expropriation rather than a safeguard against it.

Such tensions over what constitutes ownership are linked to another prevalent distinction – between the time frames used to assess poverty and wealth. Indigenous activists tend to stress the intergenerational health of the land as a vital measure of a community's prosperity. By contrast, neoliberals are apt to view wealth through a narrower temporal window. Indigenous land defenders resist a present of resource capture and asset stripping that disregards past practices or long-term prospects. They resist, too, a culture of development that responds to questions about honoring the past and sheltering future lives by revving up the chainsaw.

The number of people who depend on forest access for their livelihood is consequential. In India alone, one-fifth of the population – roughly 275 million people – rely on forests for fuel, water, food, livestock feed, and timber. Most of those forest-reliant people earn less than two dollars a day; many get as much as half their daily caloric intake from foraging (Steinauer-Scudder n.d.). The advance of private plantations and the shrinking of the commons destabilizes a delicate set of survival strategies.

Such concerns are manifest across the global South. For Raoni Metukire, of the Amazon's Kayapó people, wealth and poverty can only be assessed intergenerationally. The onset of plantation culture, he insists, marks regression, not progress. Plantations and mega-ranches weaken the health and spirit of the land:

> Why do you do this? You say it is for development – but what kind of development takes away the richness of the forest and replaces it with just one kind of plant or one kind of animal? Where the spirits once gave us everything we needed for a happy life – all of our food, our houses, our medicines – now there is only soya or cattle. Who is this development for?
> (Metukire 2019)

In the Kayapó language, money is call *piu caprim*, "sad leaves."

Not all Indigenous groups view money in this way. But we can recognize among them a recurrent skepticism toward any assumption that imposed development offers a modern alternative to a "backward" condition of forest dependency. Rukka Sombolinggi, secretary-general of the world's largest Indigenous organization, Aman, which represents 17 million Indigenous Indonesians, puts the matter this way:

> If we want to see the beautiful centuries ahead . . . we need to shift our paradigm of what constitutes wealth or prosperity because too many people see happiness only in terms of material goods and achievements and it is having a devastating impact.
>
> (Taylor 2017)

Indigenous communities across the world are protesting transnational companies that undermine the interests of the many by co-opting the few, buying them first-class, air-conditioned seating on the train called "development." Environmental and land defenders are refusing development without consultation, without prospects, and without equitable distribution. In an era of high speed, extreme extraction, they typically refuse the disastrous dictates of short-term development that inflicts long-term harm on the land and land-dependent people. Such activists refuse the literal clear-cutting of the forest and the metaphoric clear-cutting of environmental time. They refuse terms of redemption that insist on their betterment without their input or against their will. They refuse resource capture and asset stripping. And they refuse the mix of intimidation and blandishments from corporations hostile to transparency.

We see these dynamics at work in a landmark case against Indonesian and Malaysian palm oil conglomerates, when 17 Liberian villages banded together, gathering further allies from Cameroon and Gabon (Round et al. 2016). Together they rose up against the development's enforcers who, as one Liberian lawyer put it, "operate almost as mob gangsters, replete with threats, intimidation, illegal arrests" (ibid.). Across the region, many villagers resented the imposition of a monocultural moonscape; they resented a new order in which the company dictated the terms of labor and survival. By most measures, such communities were not well off, but the combination of forest and small holdings served as an imperfect but substantive protective barrier between a manageable existence and chattel poverty. Silas Siakor, former director of Liberia's Sustainable Development Institute, offered this assessment: "Seen up close palm oil is more often the problem, not the solution. Palm oil seems to be compounding, not alleviating poverty" (Global Witness 2014). Saydee Monboe, an ex-farmer driven off his land, is more forceful: "This is not development, it's modern slavery" (Chaon 2013).

Notes

1 The Liberian government signed away 220,000 hectares of land for conversion into palm oil plantations. Small-scale farmers in the region who depended for their livelihoods on a mix of traditional crops and agroforestry reported being hectored and threatened into relinquishing their land, sometimes at gun point. These intimidating conditions, under

which coercive "development" made its presence felt, resulted in low prices for appropriated land. See also Chaon (2013).

2 In 2019, 1.2 million New Yorkers were – pre-COVID – suffering from chronic food insecurity. See City Harvest (n.d.). www.cityharvest.org/food-insecurity/.

3 In recent decades, much of the expropriation of land and felling of forests in Liberia has constituted a kind of disaster capitalism. Profiteers capitalized on the chaos engendered by two great crises: the Liberian Civil War of 1989–1996 and the Ebola crisis of 2013–2016. Particularly under cover of Ebola, palm oil corporations took advantage of the crisis to press forward plantation culture in unmonitored, unanswerable ways. For two excellent accounts of the rise of plantation culture in Liberia, see *The Land Beneath Our Feet* (Mitman & Siegel 2016), as well as Mitman's *Empire of Rubber: Firestone's Scramble for Land and Power in Liberia* (Mitman 2021).

4 The incursions of the oil palm conglomerates in Liberia after 2010 fits the model of "disaster capitalism," as the companies capitalized on the instability and chaos fostered first by the Liberian Civil War and later by the Ebola crisis. The corporate appropriation of land quickened during both these crises, as the multinationals took advantage of the chaos to intensify their land grabs and the imposition of plantation culture.

References

Alatout, S. (2020, November 16). Technoscience, the continuity of the Zionist settler-colonial project, and infrastructures of elimination. *Jadaliyya*. www.jadaliyya.com/Details/41994

Barnett, T. (2010, June 18). Salvadorian environmental activists put their lives on the line. *Esperanza Project*. www.esperanzaproject.com/2010/latin-america/el-salvador/salvadoran-environmental-activists-put-their-lives-on-the-line/

BBC Today. (2012, December 11). Liberia: Land concessions. *BBC*. http://news.bbc.co.uk/today/hi/today/newsid_9776000/9776648.stm

Burke, J. (2016, June 12). The coastal village, the mining giant and the battle for South Africa's soul. *The Guardian*. www.theguardian.com/world/2016/jun/12/south-africa-titanium-mining-giant-xolobeni

Chaon, A. (2013, January 22). Liberian farmers take on Indonesian palm oil giant. *Phys.org*. https://phys.org/news/2013-01-liberian-farmers-indonesian-palm-oil.html

City Harvest. (n.d.) Facts about hunger. *City Harvest: Rescuing Food for NYC*. www.cityharvest.org/food-insecurity/

Global Witness. (2014, September 10). *Palm oil, poverty and "imperialism": A reality check from Liberia*. www.globalwitness.org/en/blog/palm-oil-poverty-and-imperialism-reality-check-liberia/

Grunewald, R. (Dir.). (2015). *The Shore break* [Film]. Frank Films & Marie-Vérité Films.

Klein, N. (2013, March 6). Dancing the world into being: A conversation with Idle No More's Leanne Simpson. *Yes Magazine*. www.yesmagazine.org/social-justice/2013/03/06/dancing-the-world-into-being-a-conversation-with-idle-no-more-leanne-simpson/

Metukire, R. (2019, September 2) We, the peoples of the Amazon, are full of fear: Soon you will be too. *The Guardian*. www.theguardian.com/commentisfree/2019/sep/02/amazon-destruction-earth-brazilian-kayapo-people

Mitman, G. (2021). *Empire of rubber: Firestone's scramble for land and power in Liberia*. New York: The New Press.

Mitman, G., & Siegel, S. (Dir.). (2016). *The land beneath our feet* [Film]. Alchemy Films.

Nichols, R. (2020). *Theft is property*. Durham, NC: Duke University Press.

Pearce, F. (2017, March 13). A death in Pondoland: How a proposed strip mine brought conflict to South Africa's Wild Coast. *Yale Environment 360*. https://e360.yale.edu/features/titanium-mine-conflict-south-africa-pondoland-rhadebe-caruso

Round, K., Borromeo, L., MacDougall, C., Riley, T., Paddison, L., Weigel, L., & von Harrach, A. (2016, October 20). *Palm oil in Liberia: Hope and anger in one of Africa's poorest countries* [Film]. Banyak Films. www.theguardian.com/sustainable-business/video/2016/oct/20/palm-oil-liberia-hope-anger-one-of-africa-poorest-countries-video?CMP=embed_video

Scott, J. (1977). *The moral economy of the peasant: Rebellion and subsistence in Southeast Asia*. New Haven, CT: Yale University Press.

Steinauer-Scudder, C. (n.d.). One hundred and eleven trees. *Emergence Magazine*. https://emergencemagazine.org/story/111-trees/

Taylor, M. (2017, October 6). Protect indigenous people to help fight climate change, says UN rapporteur. *The Guardian*. www.theguardian.com/world/2017/oct/06/protect-lives-indigenous-people-can-limit-climate-change-says-un

11 Violence and resistance in Indigenous Ceará, northeastern Brazil

Jurema Machado de A. Souza, Mary Menton, Antônia Silva Santos, Daniela Alves de Araújo, and Raquel da Silva Alves

The vast majority of Indigenous lands in northeastern Brazil are characterized by small and insufficient sizes, long and inconclusive demarcation, and regularization processes. They are the target of extensive speculation for agribusiness and commercial exploitation, leading to scarcity of natural resources (Alarcon 2019; Andrade 2020; Carvalho & Reesink 2018; Dantas et al. 1992; Souza 2019). Add to that mega-projects from the Brazilian state itself (Batista 2008; Cruz 2017; Silva & Fialho 2020). Currently, 14 Indigenous peoples are located in Ceará, distributed across 19 municipalities with a population of about 30,000 people (ADELCO 2018: 6). The delays in regularizing and guaranteeing territorial rights to the Indigenous peoples of the state follows the national pattern, with processes that have lasted more than 30 years. Only the Tremembé Indigenous land in the municipalities of Itarema and Acaraú is fully regularized; another four had their declaratory ordinances signed (Pitaguary, Jenipapo-Kanindé, Tremembé de Queimadas, and Tremembé de Itapipoca), while the others remain without measures (ADELCO 2018: 27–28).

It is in this context of the fight for their territories and in defense of their rights, guaranteed by both Brazilian and international laws, that the Indigenous peoples of northeastern Brazil face various forms of violence and violations of their rights. According to the annual report of CIMI (Indigenous Missionary Council), since 2010, 76 Indigenous people have been murdered in the northeast (CIMI 2020). As in other regions of the country, the Indigenous peoples of the northeast have been suffering physical as well as structural (due to the lack of demarcation of their territories), cultural (prohibition of speaking in their native language, performing rituals), and psychological (death threats, criminalization, smear campaigns through social media) violence. Thus, violence comes not only as direct violence against people's bodies but also against their territorial and cultural rights (Fialho et al. 2011). They experience multidimensional violence (Navas et al. 2018) atmospheres of violence (Menton et al., this volume). Navas et al. (2018) outline violence in its multiple dimensions: direct, structural, cultural, slow, and ecological. Thus, in addition to the environmental aspects involved, the analysis must come from an understanding of violence that directly affects social groups and their relationships. In the cases that we will present here, this violence is closely related to the struggle for land and to more structural aspects of inequalities in the northeast region of Brazil. Furthermore, dealing with violence in contexts that involve Indigenous peoples in northeastern Brazil is fundamental for reflecting on the basis of their relations of struggle and resistance for the territory and the recognition of their rights over it (Alarcon 2019; Souza 2019).

In this chapter, we present results from a project, "Mapping Indigenous Rights Abuses in Northeast Brazil," carried out by a network of researchers and Indigenous students from University of Sussex, Federal University of Bahia (UFBA), Federal University of Recôncavo of Bahia (UFRB-BR), and State University of Bahia (UNEB). The first cases point to intense and distinct forms of exploitation that violate Indigenous rights guaranteed in the Brazilian Federal Constitution, harm the autonomy of Indigenous peoples and their ways of life, and prevent their free self-determination. On the other hand, we have seen intense processes of resistance and mobilization, aiming to reverse situations of violation and exploitation of their lands. We focus on three cases in the state of Ceará, Brazil: i) the Lagoa Encantada (Enchanted Lake) Indigenous land, of the Jenipapo-Kanindé people; ii) the Pitaguary Indigenous land, of the Pitaguary people; and iii) the Tremembé de Almofala Indigenous land, of the Tremembé people.[1] All three peoples have experienced different forms of violence, however, all are related to territorial disputes and the exploitation of environmental resources. Three of the five authors of the article are Indigenous researchers who experienced these processes, and for that reason, the choice of the cases presented is directly related to their places of belonging and their daily experiences. We present the cases individually and then reflect on the effects and impacts of these processes on the peoples in question, as well as on conflicts, struggles, political ecology, and resistance.

Theft of water from Lagoa Encantada

Known as "the people of Encantada," the Jenipapo-Kanindé live in the Lagoa Encantada Indigenous land,[2] Aquiraz Municipality, in the state of Ceará. The territory covers 1,731 hectares, with a population of 409 people. The struggle to guarantee rights over their traditional territory began in the 1980s, as a result of conflicts with the company Pecém Agroindustrial Ltd., of the Ypióca business group, a producer of cane spirit. The company started stealing water from the most important lagoon located in the land of the Jenipapo-Kanindé people, Lagoa Encantada (Enchanted Lagoon), for irrigation of their sugarcane monoculture. The theft, which has lasted for almost three decades, still causes many impacts on the community's way of life, such as reduction in water availability, pollution that resulted in fish mortality, and spillage of vinasse, a by-product derived from the manufacture of cachaça. In other words, they suffered from slow violence resulting from this pollution and from a system of structural violence that favors the interests of companies over the territorial and human rights of Indigenous peoples (see Busscher et al. 2020; Galtung 1969; Nixon 2011).

The conflict caused not only the mobilization of the Indigenous people for the regularization of the territory but also reorganized them politically to confront the violations of their rights. The daily life experienced by the community, which before the conflict was centered on agriculture, fishing, and ritual use of the lagoon, started to be determined by efforts to fight the company: blockades, carried out with the support of other Indigenous peoples of Ceará, to prevent water from being removed; and use of sand and tree branches at the end of the lagoon to prevent water from passing through the company's plumbing. In addition, the National Indigenous Foundation – FUNAI; the Brazilian Institute for the Environment and Renewable Natural Resources – IBAMA; and the Ceará Secretariat for the Environment – SEMACE sealed the pump to prevent its operation. Due to the repercussions of the complaint and Indigenous mobilization, the company lost the stamp that allowed for export of its alcoholic beverages. It attempted to criminalize leaders and Indigenous people involved in the denunciation processes through legal cases. In addition, the company has repeatedly questioned the demarcation of Indigenous land in court. Even in 2011, when the declaratory ordinance for the Indigenous territory was signed, the company filed appeals, through injunctions, suspending the demarcation. Only in 2017, after judgment by the Federal Supreme Court (STF), was the Indigenous territory definitively recognized as such. However, the last stages of the regularization process are missing, that is, homologation, followed by the removal of squatters who still remain in the area.

In 2017, through an action brought by IBAMA, Ypióca was forced to remove the pump and pipes that took water from Lagoa Encantada. However, according to Chief Jurema, one of the three chiefs of the Jenipapo-Kanindé people, the company installed the equipment in another lagoon, Lagoa Preta, which despite being located in an area outside the demarcated Indigenous lands is ecologically linked to Lagoa Encantada. Recently, in April 2020, the company reinstalled the pump within the Indigenous area and restarted water withdrawal from Lagoa Encantada, in flagrant disrespect and violation of environmental laws and Indigenous rights over the use of their territory. As soon as the leaders became aware of the fact, they called FUNAI to file a complaint.

The Indigenous leaders affirm that the lagoon ended up at a very low water level due to the lack of rain and excessive daily water withdrawal. The lagoon, in addition to being essential for the survival of Indigenous families, is a sacred place that protects the spirits, and, above all, a space that protects Jenipapo-Kanindé ancestors, like the *Mãe d'Água* (Mother Water), one of the protectors of Indigenous people. During the low water level period, the Indigenous people did not stop visiting the lagoon, and performed *toré*[3] rituals on its banks, especially when it was being polluted. Often these rituals were held to ask for protection for their fight and to spark the return of the spirit *Mãe d'Água*, who, according to the Jenipapo-Kanindé cosmology, might have left the lagoon because of the pollution. However, the leaders are emphatic in saying that the spirit never left, and attribute their victories and progress in claiming their land to her constant presence. "Everything happened with [the spirits'] permission," they say.

Today, after many long years, the Jenipapo managed to see Lagoa Encantada full again, gradually gaining life, so they have gone back to fishing and bathing in the lagoon. Currently, they are struggling to rebuild the lagoon, removing aquatic plants like water hyacinths because they have led to smelly water and skin rashes from bathing in it. The importance of monitoring the situation and reflecting on this conflict are fundamental in the current political moment experienced in Brazil, because even with the demarcated land, it is still intruded upon, and only after this stage may the Jenipapo-Kanindé people reclaim all of their territory.

The industrialization of coconut and the invasion of Almofala territory

The company Ducoco Agrícola SA is the invader of the territory of the Tremembé people, responsible for the most important conflict faced by this people, who inhabit three municipalities in the states of Ceará: Itarema, Itapipoca, and Acaraú. With about 8,000 Indigenous people, 4,000 specifically in the Indigenous land (TI) Tremembé de Almofala, municipality of Itarema, they are divided into 15 villages. The Tremembé de Almofala TI is the territory where most of the Ducoco company is located and where most of the company's production occurs, which therefore prevents Indigenous production, as it directly affects the people's sustainable development. It is also in this territory that the largest monoculture of coconut, which supplies the company, is grown on land grabs acquired in the 1970s, with subsidy support from the now-extinct Superintendence for the Development of the Northeast – SUDENE (Valle 2004).

Unlike what happened with the Jenipapo-Kanindé, the conflict has not yet been resolved in a favorable way for the Indigenous people with demarcation of the TI – on the contrary. According to Chief João Venâncio, the process is paralyzed due to legal proceedings filed by Ducoco. The chief highlights the violence of the conflict beyond the territorial dispute. The company has always provoked a climate of fear and dread among the Indigenous people through actions of physical violence and murders. Yet, the most serious conflict was the poisoning of the land and rivers through chemical products dumped by the company, a slow violence. Children, says João Venâncio, were the biggest victims of the pollution. "There was a house where three children died at once in a single day. Nowadays, it's calmer, but at the beginning it was a huge threat." It is in this sense that poisoning of the water represents both slow violence and an abrupt instance of violence through the death of the children. This scenario of deaths and aggressions has spanned many years.

Despite the relations between Indigenous people and the company being less hostile nowadays compared to years ago, in the sense of reduced violence, João Venâncio emphasizes:

> the violations have not stopped happening, they happen to this day. No matter how much we stand up to them, the results depend a lot on the

> justice system. Justice is too slow to resolve things – even today it has not stopped. It is less intense, but it continues.

In other words, structural violence, due to flaws in the judicial systems, allows atmospheres of violence to continue within the territory.

The company continues to threaten and impede the free movement of Indigenous people on their own land. Nevertheless, the conflicts also led to more active internal organization, such as the Tremembé de Almofala Indigenous Council (CITA). Even so, the lack of definition regarding the regularization of the territory keeps the lands, and consequently the Tremembé people, in a situation of extreme vulnerability. The exploitation and land grabbing continues in spite of the process of regularizing the territory, and new forms of exploitation have been led by other invaders, such as tourism and fish and shrimp farming in nurseries.

Territorial dispute and mineral exploration – the Pitaguary

The Pitaguary occupy a traditional area arising from the fragmentation of old villages near the capital of the state of Ceará, Fortaleza. This proximity caused severe impacts on the people's way of life, mainly due to expansionist projects related to the metropolitan region of the capital. This factor, for a certain period, restricted occupation of the traditional territory. Mobilization for reclaiming the land began in the 1990s, when the Pitaguary challenged the Brazilian state for the demarcation of their lands.

Currently, the Pitaguary reside in a small area between the municipalities of Maracanaú and Pacatuba, with a population of 3,623 Indigenous people, according to 2014 data from the Special Secretariat for Indigenous Health (Siasi/Sesai). The TI is awaiting approval but was declared in 2006, and its limits were recognized by the Ministry of Justice.

In 2018, after years of lawsuits filed by land grabbers who owned a farm called Pouso Alegre, FUNAI revised the limits of the TI, excluding the area of that farm from the area of the TI. However, the greatest territorial and environmental conflict experienced by the Pitaguary people is the exploitation of a quarry in the middle of the Indigenous land. The Britaboa quarry, currently called Canaã Quarry, had been inactive for over 15 years, but after declaration of the TI, the company tried to reactivate it in 2011. In September 2011, fearing the return of mineral exploration and with the support of other Indigenous peoples in Ceará, the Pitaguary held a demonstration against the quarry and its socio-environmental impacts. Two months later, on 15 November 2011, the Pitaguary people reclaimed the area.

In the same year, the company was fined due to the environmental damage caused to the area. Even so, the reactivation process remained in progress. In September 2017, Judge Leonardo Henrique de Cavalcante Carvalho, of the TRF-5, ordered repossession to be carried out in favor of the Canaã Quarry, with the reinstatement attempt being made in November of the same year.

On 2 August 2018, the Federal Prosecutor's Office (MPF) obtained the cancellation of the license that allowed mineral exploration in Pitaguary lands. Since then, the Pitaguary have reclaimed, and occupied, this part of their traditional territory. There is a place referred to by the Pitaguary as having a strong presence of their spirits, where the main rituals of the people are performed. On one of the rocks, the Indigenous people say, you can see a formation similar to the face of an Indigenous person, something that the Pitaguary understand as sacred. Also in this area, in addition to houses, the Pitaguary Indigenous Museum was built, which has a small collection of traditional objects from the people and a workshop for handicrafts. The Museum is seen as a tool to fight for the defense of their rights and the preservation of their memories.

From 2011 to 2020, the Pitaguary people carried out numerous actions to guarantee maintenance of their territorial rights and the social welfare of their villages. However, the violence of the conflict directly impacted their leaders, who suffered court cases and attempts at criminalization as well as direct physical violence. Shaman Barbosa and the leader Ana Clecia were criminalized – judicially accused of trespassing. The leader Maurício Alves Feitosa was beaten and burned with gasoline while sleeping, on 27 August 2017. Chief Madalena was shot in the head on 13 September 2018, but managed to survive. The attack against her is under police investigation. In addition, the Pitaguary report of threats and unknown cars hanging around the homes of leaders. That is, they suffer direct violence and psychological violence manifested in a climate of fear – an atmosphere of violence.

Final considerations

Violations caused by companies and/or land grabbers in lands traditionally occupied by Indigenous peoples are often the main obstacle in the process of demarcating their territories. Undoubtedly, this occurs in large part due to the neglect of the Brazilian state and its strong colonial heritage based on the interests of local elites or the developmental interests of the state itself. As we have seen in the three cases reported here, territorial and environmental exploitation govern violations of the rights of the Jenipapo-Kanindé, Tremembé de Almofala, and Pitaguary peoples. We have no doubt that the absence of demarcation and environmental protection of Indigenous lands by the Brazilian government is largely responsible for the violence experienced daily by these Indigenous peoples.

The immense capacity for political and ethnic reorganization of the peoples caught our attention during the research process, even though they experienced situations of conflict. The creation of Indigenous organizations or the articulation around actions of resistance showed us processes of anti-colonial struggle, as well as forms of socio-political organization guided by mechanisms of resistance, by processes of building memory, and by the search for autonomy (Mendonça 2013).

Our understanding, based on the resistance actions observed in the reported cases and the coping actions engendered by the Indigenous people involved in the described conflict situations, also brings us closer to a political ecology of resistance (Dunlap 2020). From the perspectives of the concepts of struggle and resistance, both the *retomadas* (reclaiming of traditional lands) undertaken by the Jenipapo-Kanindé and Pitaguary and the social and ethnic reorganization strategies of the Tremembé people point to insurgent political practices and emancipatory struggles as everyday acts of resistance (Scott 1986). By also claiming the production of knowledge engaged from research by Indigenous scholars and from politically active researchers, we seek to reinforce the production of situated knowledge (Selister-Gomes et al. 2019). Although exposed to multiple forms of violence, to the atmospheres of violence constituted by companies and land invaders, with the State as an accomplice, their resistance continues to represent the struggle not only for the territory but for the survival of its people. Even in authoritarian political contexts and riddled with anti-Indigenous policies, these peoples have engendered effective forms of struggle, which aim to recover traditionally occupied lands that were usurped by the expansion of the agricultural frontier, industrialization, and mining.

Notes

1 It is important to highlight that in the above-mentioned project, we consider "northeast of Brazil" the large area of coverage of the Articulation of Indigenous Peoples and Organizations of Northeast Brazil (APOINME), which includes, in addition to the states of the northeast region itself, the states Minas Gerais and Espírito Santo. The Indigenous northeast that we consider here is the one articulated by the Indigenous movement, ritualistic connections, the violence suffered by the peoples, the attacks and criminalizations against their leaders, and, especially due to the environmental situation and the regularization of the Indigenous lands, and their area of coverage for regional action.

2 Demarcated through Ordinance No. 184 of the Ministry of Justice, in the Official Gazette, of 24 February 2011.

3 A *toré* is a ritual common among the majority of Indigenous peoples in northeastern Brazil which involves dance, the embodiment of the spirits, and a fermented drink made from the inner bark of the roots of the jurema tree (*Mimosa hostilis*). As a dance, without the ritual aspects, the *toré* can also be performed as a game.

References

ADELCO. (2018). *Violações de Direitos Indígenas no Ceará: terra, educação, previdência, mulheres*. Fortaleza.

Alarcon, D. F. O. (2019). *O retorno da terra: As retomadas na aldeia Tupinambá da Serra do Padeiro, sul da Bahia*. São Paulo: Editora Elefante.

Andrade, L. E. de A. (2020). *Pelejas indígenas: conflitos territoriais e dinâmicas históricas na Serra do Catimbau*. Doctoral thesis. Universidade Federal de Pernambuco, CFCH. Programa de Pós-Graduação em Antropologia.

Batista, M. R. (2008). Os Truká e o impacto da obra de transposição do rio São Francisco. *Reunião Brasileira de Antropologia, 26*. Associação Brasileira de Antropologia.

Busscher, N., Parra, C., & Vanclay, F. (2020). Environmental justice implications of land grabbing for industrial agriculture and forestry in Argentina. *Journal of Environmental Planning and Management, 63*(3), 500–522.

Carvalho, M. do R. de., & Reesink, E. (2018). Uma etnologia no Nordeste brasileiro: balanço parcial sobre territorialidades e identificações. *Revista Brasileira de Informação Bibliográfica em Ciências Sociais (BIB)* (87), 71–104. www.anpocs.com/index.php/bib-pt/bib-87/11594-uma-etnologia-nonordeste-brasileiro-balanco-parcial-sobre-territorialidades-e-identificacoes/file

CIMI. (2020). Violência contra os povos Indígenas no Brasil: Dados de 2019. *Conselho Indigenista Missionário*. https://cimi.org.br/wp-content/uploads/2020/10/relatorio-violencia-contra-os-povos-indigenas-brasil-2019-cimi.pdf

Cruz, F. S. M. (2017). *"Quando a terra sair" Os Índios Tuxá de Rodelas e a barragem de Itaparica: Memórias do desterro, memórias da resistência*. Masters dissertation. Apresentada ao Programa de Pós-Graduação em Antropologia Social da UNB.

Dantas, B. G., Laranjeira, J. A., & Carvalho, M. R. G. de. (1992). Povos indígenas do Nordeste brasileiro: um esboço. In M. Carneiro da Cunha (Org.), *História dos índios do Brasil* (pp. 431–456). São Paulo: FAPESP/Companhia das Letras.

Dunlap, A. (2020). The direction of ecological insurrections: Political ecology comes to daggers with Fukuoka. *Journal of Political Ecology*, *27*(1), 988–1014.

Fialho, V., Neves, R. C., & Figueiroa, M. C. (Orgs.). (2011). *Plantaram Xicão: os Xukuru do Ororubá e a criminalização do direito ao território*. Manaus: UEA Edições.

Galtung, J. (1969). Violence, peace and peace research. *Journal of Peace Research*, *6*, 167–191.

Mendonça, C. F. L. (2013). *Insurgência política e desobediência epistêmica: Movimento descolonial de indígenas e quilombolas na Serra do Arapuá*. Doctoral thesis. Universidade Federal de Pernambuco, CFCH. Programa de Pós-Graduação em Antropologia.

Navas, G., Mingorria, S., & Aguilar-González, B. (2018). Violence in environmental conflicts: The need for a multidimensional approach. *Sustainability Science*, *13*(3), 649–660.

Nixon, R. (2011). *Slow violence and the environmentalism of the poor*. Cambridge, MA: Harvard University Press.

Scott, J. (1986). Everyday forms of peasant resistance. *The Journal of Peasant Studies*, *13*(2), 5–35.

Selister-Gomes, M., Quatrin-Casarin, E., & Duarte, G. (2019). O conhecimento situado e a pesquisa-ação como metodologias feministas e decoloniais: Um estudo bibliométrico. *CS* [Online] (29), 47–72. http://dx.doi.org/10.18046/recs.i29.3186

Silva, W., & Fialho, V. (2020). Povos e Comunidades Tradicionais em confronto com megaprojetos energéticos no Sertão de Pernambuco. *Revista Internacional de Folkcomunicação*, *18*(40), 143–164.

Souza, J. M. A. (2019). *Os Pataxó Hãhãhãi e as Narrativas de Luta por Terra e Parentes, no sul da Bahia*. Doctoral thesis. PPGAS, University of Brasilia.

Valle, C. G. do. (2004). Experiência e Semântica entre os Tremembé do Ceará. In J. P. Oliveira (Ed.), *A Viagem da volta. Etnicidade, política e reelaboração cultural no Nordeste Indígena*. Rio de Janeiro: Contra Capa Livraria.

12 'Land defenders' and the political ecology of coal power in Bangladesh

Paul R. Gilbert and Mohammad Tanzimuddin Khan

The emergence of environmental or land defenders as prominent figures within global environmental politics and multilateral human rights discourse has a complex genealogy. When environmental or land defenders are understood as 'individuals and collectives who protect the environment and protest unjust and unsustainable resource use because of social and environmental reasons' (Scheidel et al. 2020: 1), there are clear analytical parallels with earlier work in political ecology, including on environmentalisms of the poor (Martinez-Alier 2003). The prominent role played by Global Witness and the UN Special Rapporteur on Human Rights in highlighting the violence and threats to life experienced by environmental and land defenders has also situated concern for defenders within a broader human rights discourse.[1] Recent work in what Scheidel et al. (2020) term 'statistical political ecology' has sought to identify cross-national patterns in the political and economic formations associated with violence towards environmental defenders (see also Butt et al. 2019; Le Billon & Lujala 2020; Middeldorp & Le Billon 2019). This work has largely drawn on Global Witness data on the killing of human rights defenders (2002–2017), as well as on the Environmental Justice Atlas, which records socio-environmental conflicts or 'mobilizations by local communities against particular economic activities whereby environmental impacts are a key element of their grievances' in dialogue with affected communities (Temper et al. 2015: 261–262).

This global comparative work has helped to identify a number of political–economic factors that appear to pose a patterned risk to environmental and land defenders, with all studies showing a particular risk of subjection to violence and killings for Indigenous defenders. Most cases of socio-environmental conflict on record relate to mining, and the greatest number of deaths of defenders is associated with mobilization against mining projects (Scheidel et al. 2020). Higher levels of foreign direct investment (FDI) and mineral rents are clearly associated with a higher number of environmental and land defender killings (Le Billon & Lujala 2020). Weak rule of law is identified as an important condition leading to violence against defenders (Butt et al. 2019: 743), and 'semi-authoritarian' regimes are associated with more targeted killings of defenders than authoritarian regimes which are associated with more open repression (Middeldorp & Le Billon 2019: 333). This comparative work has also generated insights into the conditions under which mobilization against mining projects (and other bases of socio-environmental conflict) are more likely to be successful – noting the importance of diversified protest incorporating non-violent resistance and legal action (Scheidel et al. 2020).

In Bangladesh, a number of high-profile socio-environmental conflicts that have resulted in the killing of land defenders appear to fit fairly neatly into these global trends, as discussed in more detail later. However, in this chapter we wish to reflect further from the Bangladesh perspective on some of the measures and assumptions underlying statistical political ecology studies of defenders. In particular, we are concerned with the measures and concepts of governance and rule of law used in the global indices that are used to facilitate comparison and identify patterns in the targeting of defenders. As May (2014) has argued, the rule of law functions as a foundational 'social imaginary' in contemporary global governance, as a set of ideals from which judgements can be made, behind which lies a moral or metaphysical order.[2] The WJP Rule of Law Index, for example, which provides the basis for analysis of risks faced by defenders in Butt et al. (2019), surveys household and expert experience of constraints on government powers, absence of corruption, fundamental rights and regulatory enforcement (among other factors). It is difficult to be opposed to understandings of rule of law that incorporate 'effective enforcement of laws that ensure equal protection . . . the right to life and security . . . [and] due process', as per the WJP measures of fundamental rights. Who could be opposed to fair and equal treatment and due process? It is difficult to articulate opposition to the rule of law, and hard to imagine an expression of Anglo-American political discourse that is as globally prestigious as 'the rule of law' (Mattei 2010).

Yet, as Mattei (2010) argues, the malleability of rule of law is precisely what allows multiple constituencies to see within it what he or she believes, even if at base it is little more than a model in which decision-making is carried out by professional jurists legitimated by legal–technical knowledge. The rule of law has thus as much to do with ensuring due process in the upholding of fundamental human rights, as it has to do with regulatory enforcement. For the WJP Rule of Law Index, this includes assurance that there is 'no expropriation of private property without adequate compensation', but for the private sector wing of the World Bank Group, it can mean requiring enhanced contract protection for foreign investment as part of loan conditionality (Mattei 2010). In Bangladesh, IFC-funded training to enhance the rule of law has involved providing 'capacity building' for lawyers to ensure confidence for foreign investors in extractive (and other) sectors that their assets and revenue streams will not be held up by decisions taken by the domestic judiciary – even though this often undermines legitimate attempts made by the domestic judiciary to hold extractive industry corporations to account for environmental harms and corruption (Gilbert 2020a, 2020b). As D'Souza (2018) has argued, the globalization of concern with *human* rights goes hand-in-hand with efforts to support the rule of law to enforce *property* and *contractual* rights. Initiatives to enhance the rule of law may therefore risk intensifying the 'global environmental structures that require continuous resource extraction' (Scheidel et al. 2020) within which defenders and socio-environmental conflicts are enmeshed.

Measures of governance and democracy are also deployed in statistical political ecology analyses of risks posed to land and environmental defenders – for example, the Combined Index of Democracy utilized by Middeldorp and Le Billon (2019). The Combined Index of Democracy (CID) constructs measures of 'regime quality' by drawing on a number of pre-existing measures of freedom and governance – a relatively common practice in index composition which often introduces opacity since not all underlying methodologies are in the public domain (Thomas 2010). One of the indices upon which the CID draws is the data series produced by Freedom House. The Freedom House rankings have been widely discussed in literature on private authority in global governance and governing by numbers. In particular, critics have noted that the Freedom House rankings' growing influence correlates with the rise in 'neoliberal' authority and the redefinition of freedom in neoliberal terms: away from substantive socio-economic rights and towards procedural freedoms concerned with the protection of private and business freedoms, as much as with the role of public institutions (Giannone 2010). Freedom House rankings also appear to be 'tweaked' to reflect US policy concerns, with allies moved up and rival states moved down (Bush 2017). Nonetheless, these rankings have proved influential and are incorporated into others such as the World Bank's World Governance Indicators – themselves used in other statistical political ecology analyses of defenders (Le Billon & Lujala 2020).

Does it matter that the indices available for statistical political ecology analyses of the risks faced by defenders incorporate broadly 'neoliberal' measures of democracy, governance, freedom and the rule of law that valorize enhanced property protection and contract protection? It certainly does not simply invalidate global comparisons of land and environmental defenders. As shown in the next section, there are many resonances between the findings of these large-scale comparisons and the experiences of defenders and mobilizations against mining in Bangladesh. However, rooting analyses of land and environmental defenders in this rule of law 'social imaginary' can risk reproducing liberal developmental norms that work to further intensify violent forms of extractive development in Bangladesh and elsewhere. As such, the body of this chapter examines the mobilization and killing of land defenders around coal mines and coal power plants in Bangladesh, highlighting how these patterns of violence fit within the global trends identified by Scheidel et al. (2020) and Le Billon and Lujala (2020). It situates these patterns of violent development in terms of a move towards 'developmental centrism' (Khan 2020) that sees the Energy Ministry and Bangladesh Power Development Board take an increasingly prominent role in shaping the nation's developmental landscapes.

The final section of the chapter draws together this critical discussion of land in Bangladesh with the notion of 'developmental centrism' and identifies some of the main transnational actors who work with and through the Bangladeshi state to cultivate atmospheres of violence around coal power projects. Inspired by efforts to build a 'counter-index' to Transparency International's Corruption Perceptions Index which have sought to move away from grounding perceptions of corruption in countries of the Global South and instead identified the role that key nodes in global movements of capital play in facilitating corruption, we argue for the development of a 'violence footprint' measure that can facilitate global comparative studies of land and environmental defenders in relation to the involvement of donors and foreign investors, rather than domestically rooted measures of governance, corruption and rule of law. Firstly, however, we will briefly review some dominant approaches to the study of land, dispossession and resistance in Bangladesh.

Land, law and dispossession in Bangladesh

Land is of particular significance in rural Bangladesh, with land transmission occurring through a lineage or *gusthi* central to identity and livelihood, and landless *gusthi* often reliant on kinship patronage or selling labour as seasonal sharecroppers (Gardner 1995: 66–72). A great deal of attention given to land ownership and dispossession in Bangladesh has been focused on the *char* lands of Noakhali – the shifting alluvial lands that generate particular problems of ownership. Adnan's (2013) landmark work on shrimp farming in the *char* lands highlights the extent to which land grabbing by political–business elites is shaped by World Bank/IMF-led prescriptions – in this case, a push to export-led shrimp farming – and creates class-based alliances that at times cross Bangladesh's otherwise rigid lines of party affiliation. In a different geographical context, Ahasan and Gardner (2016) have highlighted the extent to which dispossession of land is *functional* to development in Sylhet's gas fields, and not merely a side effect. Unocal (the former owners of a gas field now operated by Chevron) were supported by police to clear land for development, and local political elites were pressured not to oppose the 'development' by national leaders. The state worked in concert with transnational corporations – and indeed with local NGOs contracted to bring 'development' to the dispossessed.

Other writers, however, reproduce troubling tropes that sit with the rule of law 'social imaginary' and appear to attribute failings to endemic characteristics of Bangladeshi territory and social life, rather than to dynamic political–economic conjunctures. Feldman and Geisler (2012: 975–978), for instance, refer to 'lawlessness' and the character of the *char* lands as a 'stateless place' in a country with a 'disturbing record of violence . . . and corruption'. This is not to deny that violence and corruption take place in Bangladesh – but it is curious to at once refer to lawlessness and 'statelessness' while also invoking the *excessive* thickness of elite business–government connections. To invoke an absence of the law and the state seems to imply the need for externally imposed 'rule of law' development. In his recent work on the political ecology of climate adaptation in the *char* lands, which opens up important lines of analyses around how elites enclose land in the context of adaptation policies, Sovacool

(2018) also reproduces understandings of weak governance as embedded in and confined to Bangladesh's national territory, arguing that 'the most pernicious sets of consequences do not arise from the forces of global capitalism or neo-liberalism . . . it is local actors – community leaders, criminals, state officials, businesspersons, political elites – who perpetuate classism, racism, elitism, and chronic poverty' (Sovacool 2018: 184). There can be no doubt that local business and political elites contribute to the perpetuation of poverty and inequality, but it seems analytically questionable to separate the environments in which they operate (cf. Adnan 2013) and the alliances they make (see the following) from global, transnational and multilateral actors.[3]

Not only are global, transnational and multilateral actors instrumental to the processes through which land defenders become exposed to atmospheres of violence, but the framing of parts of Bangladesh as 'stateless' or 'lawless' – or the ranking of Bangladesh's governance and rule of law as inadequate – neglects the complexity of legal arrangements in rural Bangladesh. Village shalish courts, as Berger (2020) shows, are recognized by the state for the trial of minor offences *and* have become central to the advancement of 'good governance' agendas by international donor agencies who have identified village courts as potential sites from which to advance human rights. And yet, as noted earlier, donors are equally involved in 'rule of law' and governance reform initiatives designed to enhance protection of *property* rights and contract enforcement in territories deemed to be insufficiently 'civilized' (Anghie 2006; see Gilbert 2020a) to adequately arbitrate commercial disputes with transnational corporations. As Le Billon and Lujala (2020: 7) observe, 'contemporary killings of environmental and land defenders are part of a long history of colonialization and resource exploitation. . . . Propelled by accumulative economic regimes . . . and often underpinned by racial and socio-economic hierarchies'. We must perhaps be wary of the rule of law imaginary and ensure that the tools we utilize to measure, compare and make visible global patterns of violence towards environmental defenders do not partake of or give license to those same accumulative economic regimes.

Land defenders, 'developmental centrism' and coal power

The widely used Global Witness dataset of land and environmental defenders killed globally 'seeking to protect land, community and environment' (Middeldorp & Le Billon 2019: 325) records at least 1,570 killings between 2002 and 2017. A particular spike was recorded for Bangladesh in 2016, when four residents of Gondamara village near the planned Banshkhali power plant – Mortuza Ali, Anowarul Islam Angur, Zager Ahmend and Zaker Hossain – were killed by police shooting. These killings, in April 2016, followed the government giving the go-ahead to a power plant joint venture between S. Alam Group (an increasingly prominent Bangladeshi power company) and Chinese and US partners in February of that year. Also in 2016, three Santal men – Shyamal Hembrom, Ramesh Tudu and Mangal Mardi – were shot dead by police in

Gaibandha while occupying land that they claim was forcibly acquired without compensation during the Pakistan period. These killings fit into some of the prominent patterns detected by Scheidel et al. (2020) and Le Billon and Lujala (2020): mining and power-related protests are often at the centre of violence towards defenders, and Indigenous people appear to be particularly targeted.[4] Whether Bangladesh – which registers as authoritarian on the index used in Middeldorp and Le Billon's (2019) analysis – is here engaged in targeted killing (associated with semi-authoritarian regimes) or open repression (associated with authoritarian regimes) is perhaps difficult to parse.

The targeting of Indigenous people has also been widely documented in Modhupur, where Koch and Mandi, protesting against an eco-park that would be built on their ancestral land, were fired upon in 2004. Piren Snal, a young Koch man, was killed instantly, and his comrade Utpal Nokrek was shot, partially paralysed and had a 'forest case' (alleged infringement of forestry laws) filed against him while in hospital. The criminalization and violence towards Indigenous forest users has roots in the prohibition upon Independence in 1971 of Koch and Mandi people from using forest lands they had occupied for over 300 years. Indigenous people were restricted from forest uplands (*chala*) and confined to *khas* or government land. Land to which permission had been granted ('Record of Rights') was treated informally as titled land, with ownership passed down – but when in 1999 plans were released for a series of eco-parks on land occupied by 25,000 Mandi and Koch, the land was treated as 'unsettled' *khas* land; no compensation would be given (see Ahmed & Low 2020; Luthfa 2017: 239). Compensation was also not forthcoming when, with Asian Development Bank funding, land was transferred from Mandi and Bengali families to 'social forestry' rubber plantations in the 1980s (Pfoffenberger 2000: 99). Both Indigenous forest users and Bengali forest users have been subjected to heavy criminalization. Hasan Ali, for example, has been appearing before Tangail Forest Court since 1998 in relation to the same forest case – one of more than 65 he is currently facing. Another 35 have been settled, incurring significant costs. Each court appearance is a further financial burden, costing around 20,000 taka per month and permanently indebting Ali. Ali, and others in Modhupur, Bangladesh, whose livelihoods have been criminalized insist that the Forest Department's 'social forestry' scheme is the real driver behind depletion of the Modhupur *sal* (*Shorea robusta*) forests (Gain 2018).[5]

This pattern of criminalization, killing and targeting of Indigenous people fits the global patterns documented by recent work in statistical political ecology, underscoring the importance of identifying cross-national trends and drivers of environmental violence. The Forest Department was also at the centre of corruption narratives when during the 2007 emergency and military caretaker government, the Chief Conservator of Forests became emblematic of corruption when he was discovered with mattresses and jars stuffed full of money in one of his (many) houses (Chowdhury 2020: 326). But the period of emergency highlights precisely why we should perhaps be cautious using measures of governance 'quality' and rule of law derived from Anglo-American neoliberal social imaginaries. During the emergency, there was considerable urban middle-class support in Bangladesh for the military, based on a circular logic whereby 'withholding of democratic rights *for the sake of democracy*' could be justified, and a 'repressive, corrupt, and undemocratic governmental apparatus is blamed for the underdeveloped political rationality of its citizens' (Chowdhury 2014: 34). The exercise of sovereignty as a form of domination is thereby justified as a way to protect the 'masses' from themselves – from acting as crowds rather than rights-bearing citizens.

Mass action was, however, at the centre of resistance to a (still) planned coal mine in Phulbari, north-west Bangladesh, a few months before the emergency. On August 26, 2006, some 40–50,000 protestors converged on the offices of Asia Energy, a wholly owned subsidiary of London-listed GCM Resources plc., to organize a *gherao* or sit-in. Police, paramilitaries and the Bangladesh Rifles fired upon the protestors, resulting in the deaths of three young men aged 11 to 18: Al Amin, Mohammad Salekin and Tarikul Islam. A diversified campaign enrolling transnational support resulted in the seeming success of the resistance to a project that could have displaced 55,000–250,000, dewatered considerable agricultural land and exported coal while Bangladesh remained chronically underserved by electricity supply (Luthfa 2017).[6] Some scholars have announced the 'death' of the project, in part due to the successful mobilization, and in part due to repeated failures in corporate attempts to engage with the community (Faruque 2018). In March 2018, however, the license-holders for the Phulbari project (stalled now for over a decade) announced an agreement with Energy China to develop a 2000MW mine-mouth coal-fired power plant, suggesting that the mine may yet go ahead, even though transnational campaigns against it continue to mobilize.

The attempt to re-invigorate the mine by partnering with Chinese state-owned enterprises under the banner of the Belt-and-Road initiative is also an attempt to resuscitate the mine's speculative value with reference to government plans to increase domestic coal extraction and coal power capacity. In their 2016 Power System Master Plan, the Government of Bangladesh have declared their intention to become a 'high-income country' by 2041. This would entail radical social, economic and infrastructural change, and result in Bangladesh no longer being eligible for donor assistance from OECD countries and the World Bank. Central to this plan, termed 'Vision 2041', is a radical overhaul of the country's energy and power sector. Reliance on natural gas is set to be reduced, while reliance on domestic coal will be increased (from 0.7mt to 11mt per annum), along with reliance on imported coal (0 to 60mt per annum). It is in this context that the revival of a project which has already cost three lives, as well as the loss of further lives in opposition to up-and-coming power plants, is taking place. The potential for these coal power developments to further provoke atmospheres of violence and put land and environmental defenders at risk is palpable and cannot be captured by broadly static measures of governance quality, democracy or rule of law. Instead, we need to turn to the historically specific global and national political–economic shifts which have laid the ground for the re-emergence of coal power and the potential for violence against defenders.

The Bangladeshi state has made manifest its sector-specific developmental priorities and elevated a dominant bureaucratic agency in line with Vision 2041 and the Power Sector Master Plan. Their energy ministry-centric approach to development, and the enormous power and influence afforded the Bangladesh Power Development Board (BDPB), reflects the institutionalization of 'developmental centrism' (Khan 2020: 295), whereby party support for decisions taken by this specific ministry cascade down through party-affiliated bureaucratic cadres and local activists. The presence of BDPB bureaucrats on the board of an Indo-Bangladeshi joint-venture power plant at Rampal in the Sundarbans highlights the increasing scope of their influence. The Rampal power plant has been controversial from the outset, since land acquisition took place in advance of environmental impact assessment (EIA), and the EIA has been particularly lax about the risk of pollution from the plant and coal transport (Khan 2020). Approximately 4.72m tons of coal will pass through the Passur river annually, and 9,150 cubic metres of water will be withdrawn and half of that pumped back in (Mookerjea & Misra 2017). In addition, while 1834 acres was handed to BPDB, only half of this was *khas* or government land (Mahmud et al. 2020),[7] and 3,500 claims for compensation were refused. In a now familiar pattern of intimidation of land defenders, numerous false court cases have been filed against leaders of this campaign – although, in line with Scheidel et al.'s (2020) identification of successful mobilization strategies, the landowners have entered into alliances with litigation-focused middle-class environmental NGOs, as well as international NGOs concerned with the UNESCO World Heritage status of the Sundarbans. Still, the landless tenants who were first driven off their land remain invisible to the subsequent campaign – and particularly invisible to the 'consultancy firms like Price Waterhouse Coopers and the McKinsey Group who develop policy for hapless governments' (Mookerjea & Misra 2017), with Price Waterhouse Coopers having authored the feasibility study for one of the planned coastal industrial zones due to be powered by Rampal. Notwithstanding the protest movement, landowners affiliated with the ruling party were pressed into support for Rampal and have been able to benefit from *khas* allocations under elite-biased laws that deny compensation or resettlement for small landowners and the landless (Mahmud et al. 2020).

As Mahmud et al. (2020: 11) note, the bureaucratic administration is not a 'neutral counterforce to local structures' in the Rampal affair, but neither can this be dismissed as, or assimilated to global indices that characterize Bangladesh in terms of weak rule of law or ineffective institutions. International donor agencies are implicated in pushing Bangladesh towards coal power, as are the weakly governed financial centres which make extractive operations like Phulbari possible (see Conclusion). The 2016 Power System Master Plan was sponsored by JICA, the Japanese bilateral aid agency, and has been referred to by JICA officials as 'our PSMP 2016'.[8] The PSMP pushed, as noted earlier, for an increase in domestic and imported coal use, based on seemingly poor performance in natural gas exploration. Defending the focus on coal, a JICA energy specialist in Dhaka argued that the PSMP focus was on finding the right energy mix to deal with an expected 60,000MW demand in 2040. Post-Fukushima,

> as Japan cannot support coal too much, but you have to understand the Bangladesh situation . . . gas is depleting so if Bangladesh has no coal, they may further go for nuclear. The situation of Bangladesh is very much do or die.[9]

Japan has also secured 143 billion Yen financing for the Matarbari Phase 2 development, led by Sumitomo, which includes a 1200MW coal-fired power plant and a deep-sea port. This loan financing will not only indebt Bangladesh and create pressure on public finances, but forms part of a chain of 17,944MW of coal power plants planned for a 25 square kilometre stretch of coastline. This planned string of coal plants is a reflection of the BPDB's 'developmental centrism' and a coal-based growth strategy centred upon coastal economic zones, and will create the world's largest coal cluster in a delicate ecological zone. Thousands of families have been, or risk being, further displaced. Employment opportunities and livelihoods have not been provided for displaced fishermen and farmers, evictions have been widespread, and where compensation has been paid it has been highly variable and arbitrary, especially where landowners have not been able to produce required documents (Bangladesh Poribesh Andolon & Waterkeepers Bangladesh 2019: 27). Opposition to the project continues along the 'diversified' lines outlined by Scheidel et al. (2020) for successful projects, with NGOs and Supreme Court barristers working to give voice to displaced families in Dhaka. These initiatives have highlighted the loss in income from salt cultivation and shrimp farming, the loss of thousands of jobs in the salt supply chain, including on cargo vessels, lack of notice for displacement and non-payment of compensation.

While there are clear failings in terms of the letter and application of laws governing land acquisition, settlement and compensation, it is difficult to ascribe these only to governance in Bangladesh. External donor agencies are implicated too, as are World Bank group consultants who have advised in favour of the privatization and deregulation of the power sector since the 1990s[10] – though anti-corruption campaigns never seem to focus on the 'inflated value of (expatriate) "expert knowledge"' (Chowdhury 2020: 326). One clear example of the need to expand our lens on governance, corruption and rule of law beyond the nation state relates to quick rental power plants. In 2012, the Government provided support for 49 'quick-rental' power plants that were established to cover a daily power shortfall of 1,500MW. Subsidies for high-sulphur fuel oil imported to run the quick-rental plants ran to an annual cost of 231.25 billion taka or US$2.93 billion (Ahamad & Tanin 2013: 280), with significant consequences for Bangladesh's foreign exchange reserves. Certain groups affiliated with the ruling Awami League – including Summit Group, headed up by a former Awami League Minister of Commerce, and S. Alam, involved in the Banshkhali plant discussed previously – benefitted unduly from the quick rental scheme (Mirza 2020). Land and infrastructure were provided by the state, and substantial subsidies paid, amounting to US$774 million in 2008–14. While this might seem to constitute a clear example of corruption in the terms of the World Governance Index (i.e., capture of the state by elites and private interests), Summit has also received financing from the UK's CDC Group, a taxpayer-funded, 'development'-focused private equity-style fund. Should this enter into measures of UK corruption, too? Equally, while Bangladesh fares poorly on the Combined Index of Democracy and WJP Rule of Law Index (authoritarian and 115th respectively), Japan scores highly (functioning democracy and 15th). Yet Japanese expertise, capital and indebtedness engineering is indispensable to ongoing displacement and coal power expansion in coastal ecological zones. In the final section, we will discuss implications of this for developing new measures through which to compare the risks faced by environmental and land defenders.

Conclusion: violence footprints[11]

This chapter began by bringing recent work on 'statistical political ecology' to bear on the experiences of land and environmental defenders in Bangladesh. We have shown that many of the patterns found globally seem to resonate with Bangladesh – including an association with mining and the capacity for diversified movements to have some measure of success (Scheidel et al. 2020); levels of foreign direct investment and landholding inequality (Le Billon & Lujala 2020); and a particular risk to Indigenous people (Butt et al. 2019). However, we have also encouraged reflection upon the use of global governance, corruption and rule of law rankings to conduct comparisons of the experiences of defenders across contexts. In particular, we noted the degree to which archetypal 'neoliberal' rankings produced by Freedom House, which privilege protection of private over public property and economic over civil rights, have come to underpin rankings widely used by academic analysts.[12] We also highlighted the tendency for the promotion of human rights to be part of the same discourse and apparatus as the promotion of property and contractual rights (D'Souza 2018). While social movements and international donors may be interested in promoting frameworks for protection of human rights, donors also participate in promoting rule of law reforms designed to increase the ease with which FDI in extractive industries – two significant risk factors for land and environmental defenders – can flow in and out of countries like Bangladesh. In some cases, these reforms actively seek to undermine the domestic judicial system and empower transnational corporations over state sovereignty (Gilbert 2020b).

We have also outlined patterns in the political economy surrounding the targeting of land defenders in Bangladesh, in particular the 'developmental centrism' exercised by the increasingly powerful energy ministry. Attempts to develop and intensify coal power through domestic coal extraction and coastal coal-fired power plants tied to new economic zones have been intimately connected to the violence exercised on land defenders in Phulbari and Banshkhali. Threats of further displacement continue along the coastal belt, incorporating the Matarbari development. Clearly, organs of the Bangladeshi state are implicated in this violence: police firing is a brutal demonstration of sovereign power over defenders' lives. Our concern here is more with the approach to 'writing and envisioning global space' (Ó Tuathail 2005), which sees troubling measures of rule of law and governance written (sometimes literally, in the case of certain indices) on to the map of Bangladesh. This has the effect of locating atmospheres of violence in domestic arrangements which – implicitly, within the rule of law social imaginary – emerges from a perceived 'lack' and failure to conform to an Anglo-American ideal (Mattei 2010).

What is needed, perhaps, is an approach to mapping the extraterritorial actors involved in provoking atmospheres of violence – for instance, London-listed companies like GCM Resources Plc., or bilateral donors like JICA. Such an approach could draw for example on the work of the Tax Justice Network, which has long been critical of the degree to which Transparency International's Corruption Perceptions Index locates (perceived) corruption in countries of the Global South, even while the *facilitation* of corruption through financial secrecy jurisdiction primarily takes place in OECD countries – notably London and the UK's network of offshore territories and crown dependencies. The Tax Justice Network proposed an alternative measure that takes account of the *volume* of capital flow as well as the secrecy legislation adopted in facilitating territories (Cobham et al. 2015). We propose a shift in emphasis from using neoliberally inflected indicators that write governance shortcomings into national territories, towards a tool for mapping the 'violence footprint' of specific transnational corporations, domestic corporations, para-statals and bilateral or multilateral development agencies. This approach would differ from existing 'carbon footprint' and 'ecological footprint' measures, which are concerned with life-cycle carbon accounting or aggregated indicators or resource and land use. It would complement the work of the Environmental Justice Atlas (Temper et al. 2015, 2020), but would focus on specific firms or agencies, and the oversight (or lack of oversight) to which they are subjected.

For instance, the near non-existent oversight of extractive industry companies listed on London's Alternative Investment Market (including GCM Resources Plc.) is undoubtedly an aspect of governance that has intersected with domestic political wrangling in order to cultivate an atmosphere of violence around Phulbari. Similarly, a violence footprint mapping project might also draw attention to governance failures in Japan – such as support for the Matarbari coal-fired project being at odds with the Cabinet's Paris Agreement strategy and Strategic Energy Plan. A lens on violence footprints would not only help defenders and their supporters target involved parties more effectively, it would also avoid the pitfalls of fixing the risks that defenders face within national or domestic space and encourage the development of social imaginaries that do not risk endorsing facilitation of the expansion of violent extractive industry investment by excessively valorizing the rule of law.

Notes

1 While the Special Rapporteur on Human Rights has cultivated the language of environmental defenders within the UN system since 2007, there have also been earlier and parallel initiatives that use partially overlapping language. For instance, the Sierra Club/ Amnesty International collaboration 'Defending the Defenders' (1999–2002), which was framed as providing US support for victims of US corporations overseas, partly in response to the killing of Ken Saro-Wiwa. In this campaign, however, language centred on 'defending the rights of environmentalists', rather than framing land and environmental defenders as a particular subset of *human rights* defenders, as per the UN discourse.

2 May (2014) draws on Charles Taylor's concept of social imaginary, but as several scholars have argued, such diffuse notions of imaginary risk reproducing unmoored notions of holistic 'culture' that are not grounded in relations or practice. Although the idea of social imaginary is a useful heuristic, to avoid its analytical pitfalls, the rankings themselves can be viewed as *technologies of imagination* (Gilbert 2020a), which do not determine but do afford particular imaginative effects. That is, the circulation of rankings of rule of law or global governance *provoke* geographical imaginations in which certain territories are associated with violence because of a putative lack of (nominally Anglo-American) liberal institutions and legal frameworks encompassing both human rights and property rights.

3 Sovacool (2018) even goes so far as to argue in severe instances, *lathiyals* or stick-wielding enforcers 'can kill protestors or activists' (Kotikalapudi 2016). In fact, the paper referred to by Kotikalapudi discusses the *police* shooting of protestors at the Banshkhali power plant.

4 The language of indigeneity is complex in Bangladesh – see Uddin (2019).

5 Chalesh Ritchil also fell victim to extrajudicial killing in 2007. Chalesh, a Mandi leader from Modhupur, was arrested on March 18, 2007, by joint military and police forces and died in custody on the same day. His family and other activists who were arrested along with him allege that he was tortured at an Army camp, punishment for the role he played organizing against plans for the Modhupur eco-park that would have displaced up to 25,000 Mandi and Koch. The day he was killed was due to be the first meeting of a committee established with the Government Forest and Environment Advisor to discuss the potential impact of the proposed eco-park. Modhupur police ruled in September 2007 that his death was due to a heart attack. After journalist Tasneem Khalil reported that Chalesh Ritchil's body was mutilated in 15 different places, he was detained, interrogated and tortured by the Directorate General of Forces Intelligence in May 2007. See Khalil (2015).

6 The tension between nationalist–developmentalist support for coal power and resistance to ecologically and socially destructive coal extraction projects often emerges in struggles over Bangladesh's energy future.

7 Mookerjea and Misra actually report a much lower figure of only 86 acres.

8 Interview with JICA official, 16 July 2019.

9 Ibid.

10 Interview with former Power Cell chairman, 19 July 2019.

11 Thanks to Fran Lambrick and Mary Menton for prompting this terminology.

12 See also Gilbert (2020c) for a discussion of how these and similar rankings underpin 'criticality assessments' with regard to critical raw materials or 'green minerals'.

Bibliography

Adnan, S. (2013). Land grabs and primitive accumulation in deltaic Bangladesh: Interactions between neoliberal globalization, state interventions, power relations and peasant resistance. *The Journal of Peasant Studies*, *40*(1), 87–128.

Ahamad, M., & Tanin, F. (2013). Next power generation mix for Bangladesh: Outlook and policy priorities. *Energy Policy*, *60*, 272–283.

Ahasan, A., & Gardner, K. (2016). Dispossession by 'development': Corporations, elites and NGOs in Bangladesh. *South Asia Multidisciplinary Academic Journal* (13). https://doi.org/10.4000/samaj.4136

Ahmed, F., & Low, N. P. (2020). Environmental justice dialogues and the struggle for human dignity in the deciduous forest of Bangladesh. *Journal of Political Ecology*, *27*(1), 300–316.

Anghie, A. (2006). The evolution of international law: Colonial and postcolonial realities. *Third World Quarterly*, *27*(5), 739–753.

Bangladesh Poribesh Andolon, & Waterkeepers Bangladesh. (2019). *The tourist capital of Bangladesh endangered by plans to build the largest coal power hub in the world*. Dhaka.

Berger, T. (2020). The logic of non-enforcement: Entanglements between state and non-state law in Bangladesh. *Contributions to Indian Sociology*, *54*(2), 152–172.

Bush, S. S. (2017). The politics of rating freedom: Ideological affinity, private authority, and the freedom in the world ratings. *Perspectives on Politics*, *15*(3), 711–731.

Butt, N., Lambrick, F., Menton, M., & Renwick, A. (2019). The supply chain of violence. *Nature Sustainability*, *2*, 742–747.

Chowdhury, N. S. (2014). Picture-thinking: Sovereignty and citizenship in Bangladesh. *Anthropological Quarterly*, *87*(4), 1257–1278.

Chowdhury, N. S. (2020). The Taka, transparency, and an alternative politics of seeing from Phulbari, Bangladesh. In M. T. Khan & M. S. Rahman (Eds.), *Neoliberal development in Bangladesh: People on the margins* (pp. 321–350). Dakha: University Press Limited.

Cobham, A., Janský, P., & Meinzer, M. (2015). The financial secrecy index: Shedding new light on the geography of secrecy. *Economic Geography*, *91*(3), 281–303.

D'Souza, R. (2018). *What's wrong with rights?: Social movements, law and liberal imaginations*. London: Pluto Press.

Faruque, O. (2018). The politics of extractive industry corporate practices: Anatomy of a company-community conflict in Bangladesh. *The Extractive Industries and Society*, *5*(1), 177–189.

Feldman, S., & Geisler, C. (2012). Land expropriation and displacement in Bangladesh. *Journal of Peasant Studies*, *39*(3–4), 971–993.

Gain, P. (2018, April 27). The man with 100 forest cases. *The Daily Star*. www.thedailystar.net/star-weekend/spotlight/the-man-100-forest-cases-and-why-he-claims-he-innocent-1568197

Gardner, K. (1995). *Global migrants local lives: Travel and transformation in rural Bangladesh*. Oxford: Oxford University Press.

Giannone, D. (2010). Political and ideological aspects in the measurement of democracy: The Freedom House case. *Democratization*, *17*(1), 68–97.

Gilbert, P. R. (2020a). Speculating on sovereignty: 'Money mining' and corporate foreign policy at the extractive industry frontier. *Economy & Society*, *49*(1), 16–44.

Gilbert, P. R. (2020b). Expropriating the future: Turning ore deposits and legitimate expectations into assets. In K. Birch & F. Muniesa (Eds.), *Assetization: Turning things into assets in technoscientific capitalism* (pp. 173–201). Cambridge, MA: MIT Press.

Gilbert, P. R. (2020c). Making critical materials valuable: Decarbonization, investment & 'political risk'. In A. Bleicher & A. Pehlken (Eds.), *The material basis of energy transitions: Interdisciplinary perspectives on renewable energy and critical materials* (pp. 91–108). Amsterdam: Elsevier.

Khalil, T. (2015). *Jallad: Death squads and state terror in South Asia*. London: Pluto Press.
Khan, M. T. (2020). Developmental centrism, Rampal power plant, EIA, and the Sundarbans. In M. T. Khan & M. S. Rahman (Eds.), *Neoliberal development in Bangladesh: People on the margins* (pp. 293–320). Dakha: University Press Limited.
Kotikalapudi, C. (2016). Corruption, crony capitalism and conflict: Rethinking the political ecology of coal in Bangladesh and beyond. *Energy Research & Social Science*, *17*, 160–164.
Le Billon, P., & Lujala, P. (2020). Environmental and land defenders: Global patterns and determinants of repression. *Global Environmental Change*, *65*, 102163.
Luthfa, S. (2017). Transnational ties and reciprocal tenacity: Resisting mining in Bangladesh with transnational coalition. *Sociology*, *51*(1), 127–145.
Mahmud, M. S., Roth, D., & Warner, J. (2020). Rethinking 'development': Land dispossession for the Rampal power plant. *Land Use Policy*, *94*, 104492.
Martinez-Alier, J. (2003). *The environmentalism of the poor: A study of ecological conflicts and valuation*. Cheltenham: Edward Elgar.
Mattei, U. (2010). Emergency-based predatory capitalism: The rule of law, alternative dispute resolution and development. In D. Fassin & M. Pandolfi (Eds.), *Contemporary states of emergence: The politics of military & humanitarian interventions*. New York: Zone Books.
May, C. (2014). *The rule of law: The common sense of global politics*. Cheltenham: Edward Elgar.
Middeldorp, N., & Le Billon, P. (2019). Deadly environmental governance: Authoritarianism, eco-populism, and the repression of environmental and land defenders. *Annals of the American Association of Geographers*, *109*(2), 324–337.
Mirza, M. (2020). State-business nexus in Bangladesh: Quick rental power plants in perspective. In M. T. Khan & M. S. Rahman (Eds.), *Neoliberal development in Bangladesh: People on the margins* (pp. 113–136). Dakha: University Press Limited.
Mookerjea, S., & Misra, M. (2017). Coal power and the Sundarbans: Subaltern resistance and convergent crises. In D. Kapoor (Ed.), *Against colonization and rural dispossession: Local resistance in South/East Asia-Pacific and Africa* (pp. 164–186). London: Zed Books.
Ó Tuathail, G. (2005). *Critical geopolitics: The politics of writing global space*. Abingdon, UK: Routledge.
Pfoffenberger, M. (2000). *Communities and forest management in South Asia*. Gland: IUCN.
Scheidel, A., Del Bene, D., Liu, J., Navas, G., Mingorría, S., Demaria, F., Avila, S., Roy, B., Ertor, I., Tmeper, L., & Martínez-Alier, J. (2020). Environmental conflicts and defenders: A global overview. *Global Environmental Change*, *63*, 102104.
Sovacool, B. (2018). Bamboo beating bandits: Conflict, inequality, and vulnerability in the political ecology of climate change adaptation in Bangladesh. *World Development*, *102*, 183–194.
Temper, L., Avila, S., Del Bene, D., Gobby, J., Kosoy, N., Le Billon, P., Martinez-Alier, J., Perkins, P., Brototi, R., & Walter, M. (2020). Movements shaping climate futures: A systematic mapping of protests against fossil fuel and low-carbon energy projects. *Environmental Research Letters*, *15*(12), 123004.
Temper, L., Del Bene, D., & Martinez-Alier, J. (2015). Mapping the frontiers and front lines of global environmental justice: The EJAtlas. *Journal of Political Ecology*, *22*(1), 255–278.
Thomas, M. A. (2010). What do the worldwide governance indicators measure? *European Journal of Development Research*, *22*, 31–54.
Uddin, N. (2019). The local translation of global indigeneity: A case of the Chittagong Hill Tracts. *Journal of Southeast Asian Studies*, *50*(1), 68–85.

13 Manifestations of violence

Case study of Moolampilly eviction for a development project in Kerala

Chitra Karunakaran Prasanna

The modernization project in India initiated by the British during the colonial period continued on a greater scale after the country gained independence in 1947. The ensuing industrialization and infrastructure development has brought the nation into the trajectory of economic growth without considering the social and ecological dynamics that have resulted in large-scale displacement of people and disruption of natural eco-systems. Between 1950 and 2005, nearly 60 million people were displaced within India in the name of development (Cernea & Mathur 2008). The land acquired for development projects during these years totalled 20 million hectares (ha.), which include 7 million ha. of forest area and 6 million ha. of other common property resources (Fernandes 2004). Due to the absence of strong policy, legal frameworks and political will for social and environmental protection, the processes of land acquisition in India continue to be enmeshed with people's protests and conflicts over forced eviction, displacement, livelihood loss and ecological disruption.

In the wake of massive industrialization and infrastructure development processes initiated by the adoption of liberalization, privatization and globalization policies in India in the 1990s, conflicts surrounding development projects are on the rise (Guha 2014). Unfortunately, most of these conflicts are addressed not by the state through development dialogues or consensus building, but rather through direct and indirect means of physical and ideological suppression. While physical violence is a tool used by the state for subjugation of the individuals or communities fighting for environmental and land rights, the communities also experience subtle forms of violence through multiple manifestations of state power and coercion. This chapter attempts to explore both visible and invisible forms of violence experienced by the people fighting for land, environment and livelihood rights in the wake of large-scale land acquisition and displacement of Indigenous communities happening in India. The arguments will be substantiated primarily with a case study of the controversial Moolampilly eviction, which occurred during the implementation of the Highway Connectivity for Vallarpadam International Container Transhipment Terminal Project (ICTT) in Cochin, Kerala, India. The case study was undertaken using a qualitative approach involving in-depth interviews and focus group discussions with the affected community, officials and activists, as well as content analysis of project-related documents.

Project and area profile

In 2004, a license agreement was signed between the Cochin Port Trust (CPT) and Dubai Port World (DPW), a multinational company, for initiating an International Container Transhipment Terminal (ICTT) project in Cochin, Kerala. The project components included the construction of a rail connectivity and a four-lane highway connectivity to ensure high-speed cargo transportation linkages with the Port. The focus of this chapter is on the displacement and environmental insecurities caused by the road project. The National Highway Authority of India (NHAI), which comes under the Central Ministry of Road, Transport & Highways, monitored the construction of the road connectivity project. The 17-kilometre NH project corridor passed through seven villages in the Ernakulam District of Kerala, three of which fall into the backwater belt of the Vembanad wetland, which is a designated *Ramsar* site and covers an area of more than 2,000 square kilometres. The acquired area is a network of islands intersected by large waterbodies, low-lying wetland regions, paddy fields, ponds and canals. The region is also known for *pokkali* cultivation, a traditional practice of cultivating rice and shrimp alternatively, which can survive in saline and tidal water conditions and which ensure food and income security for the local community. In the region, 521 landholders were affected, of which 183 families lost their houses. Though only about 43 ha. of land were acquired by the state, unscientific conversions and illegal parallel conversions by real estate groups resulted in larger-scale land use changes in the project area (Chitra 2013). During and after the acquisition and conversion of land for the road project, the region including the *pokkali* lands underwent major social and ecological transformation. The process of land acquisition faced severe hurdles due to people's struggles against eviction without rehabilitation. In this context, rehabilitation would include provision of alternative land and compensation for loss of houses. The period 2005 to 2008 saw people's protests across the region, which escalated after the infamous Moolampilly forced eviction that took place in February 2008. Following the strong protests that emerged and public outcry against eviction without rehabilitation, the government was forced to announce rehabilitation measures for all the evictees in the project. Though issues related to rehabilitation continued, Moolampilly eviction and the protests that ensued determined the course of the implementation of the development project.

The case of Moolampilly eviction in Kerala

Moolampilly is an island within the Kadamakkudy village area that was earmarked for acquisition of land for the highway connectivity project. Though people's struggles against acquisition of land had started in 2005, the protests were not strong enough to stop the process. Severe pressure and verbal commitment of rehabilitation from the land acquisition authorities led 22 households, earmarked for eviction in Moolampilly by the land acquisition office, to submit their permission notes for surrendering their property in 2007. But with the intervention of the Coordination Committee, which spearheaded the struggle against eviction without proper rehabilitation, ten families withdrew their permission for acquisition in 2007. It was at this juncture that forceful acquisition of land was undertaken by the authorities. On 4 February 2008, the land acquisition officials came with the police force and demolished the houses of 12 families whose permit letters they had previously obtained. They attempted to demolish the house of one family that had withdrawn their permission as well but backed out due to community protest. On 6 February, the officials returned with sufficient police force and started demolishing the remaining ten houses using earth movers despite people's protests. The local community tried to prevent the demolition, but more police arrived and it became impossible for them to defend themselves. The house of the Moolampilly struggle committee convener was demolished first. Many evictees, especially women, protested by remaining inside their houses, yet the demolition continued and they were forcefully evicted. Many of those who protested outside their houses, including the local parish priest, were beaten up by the police. Using police force, all the houses were demolished, and the families were forced to surrender their land to the authorities. The following narrative by the Moolampilly struggle committee leader, who is also a *pokkali* farmer, throws light on the exhibition of the authorities' power in order to facilitate the acquisition of the land.

> After demolishing my house, they went to the house of a family headed by a woman. They demolished her house also. She had fainted. They forced her to sign the paper in that condition holding her hand. They asked me to sign also. But we said that we won't agree without a decision on rehabilitation. After two houses, they started demolishing the rest. What they did in other places was devilish. The children who went to school, they left seeing the house intact. When they came back they saw the house demolished. Even the food kept for them was thrown away. They threw away the books. . . . We didn't get an opportunity to shift our things to a suitable place. We demanded them to give us time but that also they didn't allow. They were attacking us as if . . . we were opponents of India and we were not the citizens of India. Such was their audacity. We are not against development. But what we were saying was that our development cannot stand affected by their atrocities. We were dealing democratically. But they did this cruelty. But luckily no one was provoked. Otherwise murders would have happened here. The Government is functioning as a cheating fake company.
>
> (interview with farmer evictee and struggle committee leader, Moolampilly, 27 October 2010)

This narrative is explicit about the visible forms of violence inflicted on the community through the use of police force and bureaucratic power to smoothen the process of eviction. The various tactics employed by the authorities resulted in an atmosphere of fear and insecurity building among the evictees. The narrative also shows how the violence inflicted by the authorities affected the morale of the community, which was democratically protesting against forceful acquisition of land. The forceful eviction made evictees feel that they were not treated as rightful citizens of India. The following narrative shows how the belief of the people in a democratically elected government was violated after the incident.

> No one thought that they will come and demolish the houses like this. Since this is a democratic government and they work for the welfare of the people, the people had a belief that the houses will be demolished only after their security is ensured. But that belief got collapsed.
>
> (interview with local parish priest and struggle committee member, Moolampilly, 27 October 2010)

Though the political leadership and senior bureaucrats denied any involvement in the incident, the protest leadership did not accept this. The following narrative shows how they perceived the role of the political leaders and bureaucrats in the Moolampilly violence. At a higher level, bureaucracy and political leadership knew about that operation.

> To say that they didn't know is a lie. They thought by taking this step, the people will be defeated, they will be deserted, and they will be demoralized. They thought people will get dispersed. They never thought that people will stand back and question them. The Moolampilly eviction became a turning point in the course of struggle.
>
> (interview with coordination committee convenor, 14 September 2010)

The narrative argues that the violence was intended to demoralize the community and defeat their opposition to land acquisition. The authorities expected that forceful acquisition tactics would create fear and insecurity among those few of the project affected and create a split in the people's struggle. The Moolampilly demolition, due to its explicit violence and public humiliation of the evictees, became a decisive factor in the people's struggle against forced eviction without rehabilitation.

> The demolition was a mistake done by the Government. If the demolition was not done this struggle would not have occurred and there would not have been any issue.
>
> (interview with struggle support committee member, 9 February 2010)

Ten families were left without any alternate housing facilities after the demolition and were put up in the local parish hall for 42 days. The families with children and elderly persons stayed there and continued their protest against the authorities, demanding a reasonable rehabilitation package. Every day for 42 days, the families sat in a tent put up in a public place during the daytime, cooking food and thus exhibiting a show of protest against the atrocities of the state. The protest continued until the Moolampilly rehabilitation package was announced. The demolition of houses and eviction of families in Moolampilly for the development project became a landmark event in the history of Kerala. Though it was not the first event of its kind to happen in Kerala or India, the live media broadcast of the demolition of the houses and of people's protests had a lasting impact on the minds of the people of the state. Because of the impact of this event, the people's protests surrounding the struggles came to be referred to as pre-Moolampilly and post-Moolampilly struggles. The strong protests that ensued culminated in the formulation of a 'Moolampilly package', the name by which rehabilitation package for the evictees was known; this consisted of rehabilitation plots for those who lost their homesteads along with compensation in cash for the land acquired.

Manifestations of invisible/subtle forms of violence

The forced eviction in Moolampilly was not an isolated incident but rather a strategic move by the state to create fear and to force communities in other project locations to surrender their land without resistance. The violence experienced by the local community in the project area in general during the development interventions did not confine itself to physical forms. Manifestations of violence can also have a 'subtle' nature that is not as easily visible as the open violence meted out to the people during a forced eviction. Even after the incident, families with children and elderly persons, including female-headed families, were forced to live in socially insecure conditions without any efforts for rehabilitation from the state. The right to live with dignity and the right to property, as enshrined in the constitution of India, were denied to them in the name of development. This violation of citizens' fundamental rights also needs to be seen as an example of a subtle form of violence. The apathy of the political leadership and state bureaucracy in matters concerning land acquisition, compensation and rehabilitation resulted in denial of these fundamental rights. The subtle forms of violence experienced by the affected community can be seen in the following day-to-day manifestations of development politics during the implementation of the project.

Objectification as violence

The development project is always presented as a nation-building project, the argument for which is utilitarian in nature and seeks to highlight only the economic benefits of implementation. Any form of social and ecological violence during the process is considered to be a normal phenomenon as per the utilitarian argument. The utilitarian argument does not consider the social and ecological ethos in the development process and treats the people and common property resources (here traditional agricultural land, backwater) as objects to be sacrificed in the name of 'development'. This objectification, which is normal in the current development discourses, is a violence meted out against the rights to resources of the people as well as right of nature to sustain itself. Here, land acquisition for development becomes a technical project to be completed by the state rather than a social/ecological project.

Legitimization and coercive forms of ideological internalization

The mainstream 'development' narrative focussing on the development project is given recognition by the state over the local narratives of social and environmental insecurities generated by the project. The prospective economic development in the state upon implementation of the development project focussed repeatedly on legitimizing the indiscriminate acquisition and conversion of *pokkali* land/backwaters and the displacement of people. The social and ecological disruption caused by the project was undermined and further legitimized by constant reference to the project as 'prestigious' by state representatives, bureaucrats, court and the media. Due to this legitimization drive, the people's protests were often seen and projected in the public sphere as illegal or anti-development. Along with the fundamental rights to live with dignity and right to livelihood, the right to expression was also denied to the local community affected by the project.

Media and political parties act as ideological apparatuses (Althusser 2006) of the state, carrying out obsessive campaigns for development projects that only focus on prospective economic advancement and employment generation. These mainstream development narratives are constructed and propagated by these ideological apparatuses without any critical reflection. These campaigns are also tactics to enable the general community to internalize the language of the mainstream development narrative and to silence any form of opposition against the project. They act as counter-mobilizing forces against any forms of protests that erupt against the majoritarian development narratives. These coercive ideological apparatuses create fear and insecurity within the community that stands affected by the project. A favourable opinion of the ICTT project was created among the general public through continuous media updates and political party narratives at the local level. Any form of dissent was treated with an anti-development/Maoist label to uproot it from the beginning, which also helps to create a split within the community. This labelling is done to create psychological pressure on the dissenters of wronging the nation and acting against its development aspirations. Media act as an ideological apparatus of the state, promoting the development propaganda and using its platform to label the dissenters as holding an anti-development position and thus to coerce the dissenters into giving consent for acquisition. In the Moolampilly struggle and the ensuing protests, people had to take up the slogan 'We are not against development' as a consequence. While after the Moolampilly eviction, media in general reversed this stance and stood with the community, support continued to exist within the confines of the existing development framework. The process of manufacturing consent (Chomsky 1999) through indirect propaganda and coercion by media and political parties also can be seen as exemplifying subtle forms of violence in the development process.

Manipulation of existing legal framework

The ICTT project also saw manipulative interpretations of existing legal frameworks for land acquisition and environment conservation laws in order to facilitate the development project. This included manipulation of the National Highway Act, coastal zone regulations, quarrying norms and wetland conservation regulations to speed up the acquisition of land, address environmental restrictions and smooth construction activities. Usage of the Land Acquisition Act 1894 instead of the National Highway Act to introduce one-time settlement was made to speed up the land acquisition procedures. This brought out interim orders to supersede environmental regulations related to coastal zone, quarrying, wetland regulations, in addition to a biased EIA report, with a declaration that land was to be acquired under the urgency clause, even before environment clearance had been obtained, with a District Level Purchase Committee set up for land acquisition and declaring double compensation to lure people to part with their land are a few instances. These legal manipulations were legitimized by the judiciary through rulings in favour of the mainstream development narrative, calling ICTT a 'prestigious' project and rendering judgements after consultation with state agencies against whom the petitions were filed.

Withdrawal from participatory governance procedures

The environmental protection rules and procedures – Environment Impact Assessment (EIA) and Public Hearing – for the ICTT and auxiliary projects were treated by the state machinery as merely a procedural requirement, without considering their real intent of environment conservation and social protection. Many issues could be observed around the environmental clearance procedures undertaken for project implementation: the same agency functioned as project and EIA consultant, raising questions regarding the credibility of the assessment; the conduct of public consultation and Public Hearing occurred without the people's participation; reductionist assessment techniques failed to focus on real issues of social and ecological disruption; a time gap of four years separated the EIA and Public Hearing; the EIA report was manipulative; and there was a lack of transparency and proper information flow. In addition, there was a complete absence of democratic dialogue in the processes, which were interspersed with threats to the people with binding dates to surrender their land, announcement of fake promises, offers of lucrative advance payments to split the community, etc. The right of the community to participate in decision-making in matters affecting their lives and livelihood was denied in all stages of project implementation.

Denial of rehabilitation

Even after announcement of the Moolampilly package, bureaucratic delay could be observed in the provision of compensation/rehabilitation to the evictees, handing over of rehabilitation plots with legal issues or with construction constraints due to particular ecological features or absent basic facilities. The rehabilitation benefits also were not equally distributed initially, exhibiting differential treatment of the evictees of the project. Rehabilitation benefits were denied to the families under additional acquisition, and families were forced to move legally against the injustice. Coastal zone regulations were relaxed for the project, but the same regulations were used by the authorities to prevent many evictees from constructing houses in the Coastal Regulation Zone (CRZ) area provided as rehabilitation plots. The concept of rehabilitation was treated as 'charity' rather than the right of the people being evicted by the state.

Environmental violence

The land under acquisition included *pokkali* lands and backwaters which, in addition to being sources of livelihood, were also ecologically sensitive regions. As part of the Vembanad wetland designated as a *Ramsar* site, the land under acquisition was ecologically significant, but the environmental violations were not focussed on in the struggles. The protests were mainly against loss of private property and largely remained silent on loss of common property resources. A state of helplessness and fear contributed to the dispossession of the common property resources. Reclamation of backwaters/*pokkali* lands were legitimized by the state through the argument of 'development' and 'rehabilitation' needs. Another example of environmental violence is the dumping of the project's construction waste into waterbodies and *pokkali* lands with permission of the bureaucracy, overriding existing state laws for wetland/paddyland conservation. Parallel land transactions and conversions by real estate groups also resulted in accumulation of wetlands and their destruction.

Conclusion

The Moolampilly incident and the general land acquisition procedures for the project show how the state coerced the common people to surrender their land using the physical apparatuses (e.g. police and bureaucrats) and ideological apparatuses (e.g. media, political parties). The bureaucratic system, which includes both the police and officials, was formulated for ensuring social security and the welfare of the people, but was here used as a tool of the state to threaten, coerce and evict people through physical and psychological pressure. The process of accumulation by dispossession (Harvey 2005) of both private property and common property resources was facilitated by these physical apparatuses since they work in subordination to the political executive. It can also be seen that the process was engineered through strategies, such as propaganda, manipulation, coercion, discipline and legitimization (Chomsky 1999; Gramsci 2006; Abrams 2006; Foucault 2006) and through the use of ideological apparatuses manifested in the day-to-day politics of mainstream development processes. The functioning of governmentality (Foucault 2006) – defined as the collective of all those institutions, procedures, calculations and tactics that help to exercise power on its target populations – was demonstrated in the tactics adopted by the political executive and bureaucrats, using their discretionary powers to move forward with land acquisition, with its manifestations taking the shape of legal manipulations, development propaganda and verbal and physical coercion used for acquisition of land.

The engagement of the state in visible and invisible forms of violence during development processes exhibits the state's apathy towards the democratic rights of the people. The consequence of these physical and ideological manifestations of power is that the state cannot be seen as a 'socio-political system' for ensuring social and environmental justice but rather comes to be defined as a 'cheating fake company' by the aggrieved communities. Rather than an organic entity that responds to the requirements of the population, the state is seen as a technocratic machine without any human or environmental ethos. The concept of citizenship remains questionable when fundamental rights to life, livelihood, expression and participation in development governance are denied to the people affected by the project. In this scenario, local communities experience not only social and ecological displacement, but also political displacement as rightful citizens of a democratic state.

References

Abrams, P. (2006). Notes on the difficulty of studying the state. In A. Sharma & A. Gupta (Eds.), *The anthropology of the state: A reader* (pp. 112–130). Hoboken, NJ: Blackwell Publishing.

Althusser, L. (2006). Ideology and ideological state apparatuses (Notes towards an investigation). In A. Sharma & A. Gupta (Eds.), *The anthropology of the state: A reader* (pp. 86–111). Hoboken, NJ: Blackwell Publishing.

Cernea, M. M., & Mathur, H. M. (Eds.). (2008). *Can compensation prevent impoverishment: Reforming resettlement through investments and benefit sharing*. Oxford: Oxford University Press.

Chitra, K. P. (2013). *Politics of land acquisition and conversion with reference to two development projects in Kerala*. Unpublished doctoral dissertation. Tata Institute of Social Sciences.

Chomsky, N. (1999). *Profit over people: Neoliberalism and global order*. Delhi: Madhyam Books.

Fernandes, W. (2004). Rehabilitation policy for the displaced. *Economic and Political Weekly*, *39*(12), 1193–1194.

Foucault, M. (2006). Governmentality. In A. Sharma & A. Gupta (Eds.), *The anthropology of the State: A reader* (pp. 131–143). Hoboken, NJ: Blackwell Publishing.

Gramsci, A. (2006). State and civil society. In A. Sharma & A. Gupta (Eds.), *The anthropology of the State: A reader* (pp. 71–85). Hoboken, NJ: Blackwell Publishing.
Guha, R. (2014). *Environmentalism: A global history*. London: Penguin UK.
Harvey, D. (2005). *A brief history of neoliberalism*. Oxford: Oxford University Press.
Sharma, A., & Gupta, A. (Eds.). (2009). *The anthropology of the State: A reader*. Hoboken, NJ: Blackwell Publishing.

14 Land defenders and struggles against agro-industrial and mining projects

Louisa Prause and Philippe Le Billon

Introduction

Environmental and land defenders are key actors in attempts to stop or slow down land transformations for agro-industrial and mining purposes. Driven by high commodity prices and financial market uncertainties, a wide range of resource companies, financial firms, and governments have gained control over lands and resources, especially in the Global South, over the past two decades (Deloitte 2018; Nolte et al. 2016). These shifts in land control have often been met with resistance by environmental and land defenders, including local communities seeking to hold on to their lands and livelihoods, to be better compensated for their losses, or to be incorporated into projects (Bebbington 2012; Borras & Franco 2013). As such, they play a crucial role in addressing power inequalities and environmental change related to large-scale land transformations. Instances of resistance against agro-industrial or large-scale mining projects have received much academic attention (Conde & Le Billon 2017; Hufe & Heuermann 2017). However, few studies have compared the actions of land defenders against agro-industrial with those against large-scale mining projects (but see Brent 2015; Prause 2019; Prause & Le Billon 2020), even though both types of conflicts relate to 'land' (Li 2014), land uses, and associated power relations and property rights (see Peluso & Lund 2011).

This gap reflects the limited attention given within the Agrarian Studies literature to resistance against mining, as research on the diversity of 'land grab' drivers and processes is largely focused on variations within the agricultural sector (e.g. Borras et al. 2012; Cotula 2012; but see Andrews 2018). Similarly, the growing field of 'Extractive Studies' has given relatively little attention to agro-industrial activities and associated resistance movements, with research framed by the resource curse (e.g. Di John 2011) or (neo-)extractivism (e.g. Acosta 2013; Bebbington & Bury 2013) being mostly focused on mineral sectors, including their impact on agrarian livelihoods (e.g. Perreault 2013). The narrower field of studies on environmental and land defenders likewise focuses mostly on case studies of conflicts around a particular type of land transformation or broader statistical analysis, thus lacking an understanding of the specificities and commonalities of land and environmental defenders' actions and contexts within different sectors.

As such, we know relatively little about the similarities and differences of resistance movements to agro-industrial and mining projects, despite their potential significance for environmental and land struggles. Resistances not only shape the trajectories of specific land investments, but may also contribute to slowing down processes of environmental degradation and prevent a further deepening of social inequalities in the countryside. Since the agricultural and mining sectors are increasingly spatially and financially interwoven (see Cuba et al. 2014), there is also a need to bridge some of these differences to build alliances between environmental and land defenders active in different sectors.

This chapter systematically compares resistance motives, practices, and outcomes related to agro-industrial and large-scale mining projects, focusing on their differences and similarities, and discussing how these might be explained. First, we use the Global Atlas of Environmental Justice (EJ Atlas) to statistically describe patterns within a broad universe of agro-industrial and mining conflict cases. Second, we review some of the case study literature on resistance against agro-industrial and mining projects to identify and explain likely patterns of similarities and differences in resistance motives, physical and narrative practices, and outcomes. Third, we present findings from a specifically designed comparative field study in Senegal.

Global patterns of resistances against agricultural and mining projects

Instances of resistance to agro-industrial and mining projects have risen considerably in the midst of the recent commodity boom and 'global land grab' (EJ Atlas 2019a; Grain 2016). By May 2019, EJ Atlas had recorded 331 agro-industrial conflicts and 576 mining conflicts around the world.[1] The occurrence of socio-environmental struggles reflect many factors: project-related investment factors rendering land investable, such as the geological or agricultural characteristics of project sites, as well as investment policies and state-based extractivist logics (see Le Billon & Sommerville 2017; Li 2014); project impacts and perceptions, levels of grievances, and demands for alternative development models by communities in opposition to industrial-scale land exploitation and control of territory (Haslam & Tanimoune 2016; Temper et al. 2018); and different forms of social contestation, including contentious modes of politics (Engels & Dietz 2017; Grant & Le Billon 2019). According to information collected through EJ Atlas, conflicts relating to agro-industrial and mining projects show relatively similar frequency of forms and outcomes of resistance (see Figure 14.1; Prause & Le Billon 2020). Among the differences, movements opposing mining projects were more frequently in the form of popular referendums, street protests, blockades, and especially sabotage, than those resisting agro-industrial projects. Conflicts over mining were reported to be more frequently of 'high intensity' (e.g. mass mobilization, violence, arrests), and resulting in the violent targeting and deaths among environmental and land defenders. Finally, movements opposing mining projects would more frequently yield a 'success'

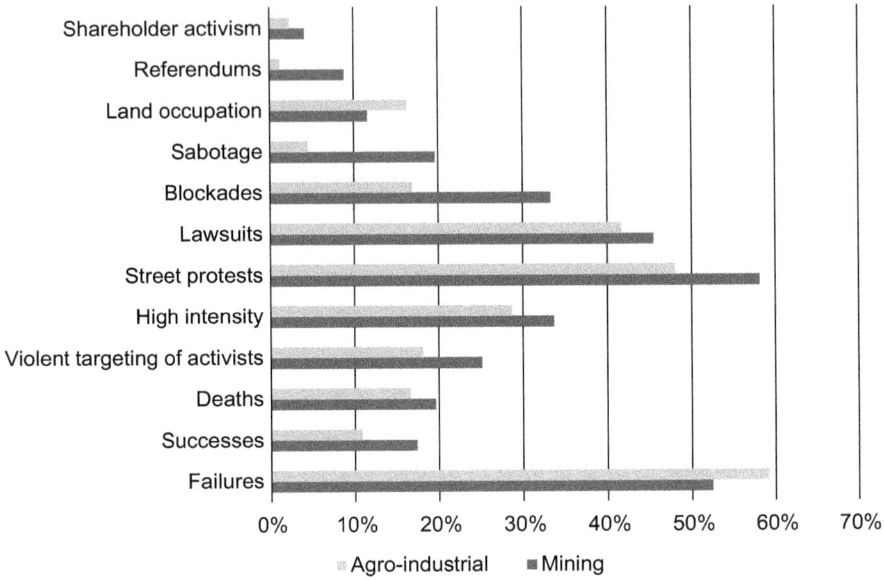

Figure 14.1 Comparison of resistance to agro-industrial and mining projects[2]

(i.e. with key demands addressed, or project suspended or canceled) than those opposing agro-industrial projects. This preliminary data analysis suggests broad similarities across the two sectors, with some differences regarding outcomes and the intensity of conflicts. Yet, we should approach these results with caution given uneven levels of reporting and possible biases.

Analytical framework

To examine factors responsible for differences in resistances to large-scale land investments by land and environmental defenders, our analytical framework draws on the literature on land and mining conflicts from the fields of contentious politics and political ecology. Both fields extensively engage with processes of resistance, which we define here as verbal and physical actions undertaken in opposition to existing power relations (Hollander & Einwohner 2004). Physical actions of resistance can take the form of invisible or covert actions that are often part of everyday forms of resistance (Scott 1985), or visible and overt actions taking the form of lawsuits, protests, or blockades (Taylor & Van Dyke 2004). Verbal actions encompass the formulation of specific claims and the development of narratives: strategic stories constructed by resistance movements to articulate claims or grievances, promote the interests of the resisting group, and oppose the narratives of their antagonists (Poletta & Chen 2012). Resistance is usually motivated by some form of grievance and/or perceived injustice (Hollander & Einwohner 2004). The literature on conflicts around industrial mines and agro-industrial projects outline several factors that explain in which instances perceived grievances lead to resistance, what form resistance takes, and whether it is successful. Much of the literature hereby emphasizes the importance of contextual political factors, particularly political opportunities.

Motives of resistance

Motives for resistance are generally understood to be rooted in some form of grievance or perceived injustice (Hollander & Einwohner 2004). Studies of resistance to mining projects have generally considered potential motivation factors (see Conde & Le Billon 2017), such as environmental impacts, loss of land and population displacement, the relative deprivation of affected communities, and the behaviors of resource corporations and governments, such as authorities openly favoring corporate interests (CNRG 2015), lack of community participation in project approval (e.g. Middeldorp & Le Billon 2019; Owen & Kemp 2014), or counter-productive corporate social responsibility (CSR) programs (e.g. Dougherty & Olsen 2014; Warnaars 2012). Motives of resistances to large-scale agricultural projects predominantly point to the loss of cultivated or grazing lands as drivers of resistances (e.g. Alonso-Fradejas 2015; Gerber 2011), as well as the loss of religious places or ancestral lands (e.g. Gingembre 2015; Kandel 2015). Significantly fewer studies link resistance with unfulfilled promises or general disappointment with a company's development projects, job creation, and CSR programs (e.g. Famerée 2016; Ndi & Batterbury 2017). Again, few studies also show that resistance can seek to achieve better terms of incorporation into large-scale agricultural projects, especially over contract farming conditions, and within areas with a long history of political and economic marginalization and little alternative livelihood options (e.g. Hall et al. 2015; Larder 2015).

Studies on resistances against agro-industrial movements stress that local communities are more likely to resist land grabs – and thus 'become' land defenders – if they can build upon experiences of previous agrarian struggles (see Gingembre 2015; Martiniello 2015). Studies on resistances against mining projects suggest that previous community experiences of economic dispossession and/or marginalization (Haslam & Tanimoune 2016) or feelings of relative deprivation exacerbated by high commodity prices and perceptions of unfair compensation or access to jobs (see Bond & Kirsch 2015; Christensen 2019) seem to be drivers for resistance, while dependence on mining revenues and entrenched political marginalization can hinder resistance (Horowitz 2010). Contributions to both literatures mention that the motives of resistance can be in part shaped by the framing of issues and range of options presented by informants and (potential) allies – whether media, NGOs, or other communities already affected by similar projects – as well as political opportunities and challenges presented, for example, by a change in government. (e.g. Engels 2018; Middeldorp & Le Billon 2019).

Practices and narratives of resistance

Resistances against land investments play out in part according to the political and cultural contexts in which they take place, including what forms of repression will be used and how counter-narratives will be formed (Dietz & Engels 2020). The actors involved and their positions vis-à-vis the political institutions likewise shape resistance practices, including narratives. Marginalized groups that lack access to institutionalized channels of participation and resources are more likely to use less organized and more disruptive forms of protests (Engels 2018). It is noticeable that the literature on resistance against agro-industrial projects places a stronger emphasis on hidden forms of resistance, such as illegal harvesting of fields (e.g. Martiniello 2015; Moreda 2015). It is unclear whether this is due to everyday forms of resistance being more prominent in conflicts over agro-industrial projects, or to the theoretical approaches informing Agrarian Studies. Similar to the findings of the EJ Atlas, the literature also points to popular referendums being highly important in the mining sector (Conde 2017), while they are not mentioned as resistance practices in the studies analyzed on resistances in the agrarian sector.

Contentious actors in conflicts over mining and agro-industrial projects employ a range of narratives to mobilize support for their resistances and articulate their claims and grievances. Studies on conflict over agro-industrial projects identify narratives constructed around the concepts of food sovereignty (e.g. Alonso-Fradejas 2015; Prause 2019), the defense of land as territory (e.g. Alonso-Fradejas 2015; Brent 2015), the depiction of land grab as a violation of human rights, existing property, and ancestral or Indigenous rights (e.g. Gingembre 2015; Grajales 2015), as well as the framing of an agro-industrial project as a threat to food security (e.g. Gingembre 2015). For the mining sector, authors also identified the defense of territory and the violation of rights as important narratives (see Conde 2017). Furthermore, several authors have stressed the importance of the environmental justice frame for resistances against industrial mines, including with regard to water pollution issues (e.g. Martinez-Alier et al. 2016; Rodríguez-Labajos & Özkaynak 2017).

Outcomes of resistance

The EJ Atlas suggests that resistance movements against mining projects were more often successful that those opposing agro-industrial projects. In both literatures, success is largely attributed to favorable political opportunities, particularly the successful forging of alliances. Successful outcomes of resistance, which we understand as the claims of land and environmental defenders being fully or partly fulfilled, such as the cancellation or the temporary halting of a project, are often achieved through the assistance of national and international NGOs, the scaling-up of protests beyond the local level (see Anyidoho & Crawford 2014; Temper 2018), and divisions among influential national elites (see Engels 2018; Renauld 2016). Furthermore, studies on resistance against agro-industrial projects identify divisions among local land users based on class, ethnic differences, or distinct rights to land as reducing successful resistance (see Hall et al. 2015).

Building on the literature, we can identify some key factors shaping the actions of environmental and land defenders in conflicts around agro-industrial and mining projects. In terms of *motives*, defending access to land is a key motivation for resistance in conflicts over both agro-industrial projects and industrial mines. Both sectors also share motives relating to dissatisfaction with the options for 'incorporation' into projects, and in particular with the availability and allocation of project-related jobs. However, motives related to participation and consent for project approval, perceived injustice with regard to the distribution of revenues and compensation payments, and one-sided support by governments for corporate interests seem to be more prominent in mining-related conflicts than in agro-industrial ones. In terms of *practices*, forms of resistance seem to be largely similar in both sectors, with the notable exceptions of popular referendums, which are predominantly used in the mining sector. The violation of rights and the defense of territory are prominent *narratives* of resistance in both sectors among local communities, civil society actors, and social movements opposing projects. Agricultural sector narratives, however, tend to rely on subsistence livelihood issues, associated with concepts of food security and food sovereignty, while in the mining sector narratives are often built around environmental harm and health risks associated with water pollution, dust, noise and traffic, or the desecration of cultural sites. Finally, *outcomes* seem to be largely determined in both sectors by political opportunities; however, studies on resistances against agro-industrial projects additionally stress divisions based on class or land rights as hindering factors to resistance movements' success. Overall, the literature on resistances against agro-industrial projects and industrial mines suggests that despite many similarities, there can also be important differences in the motivations, practices, narratives, and outcomes of resistance.

Resisting land-based agro-industrial and mining investments in Senegal

To verify whether or not the differences regarding resistances of land defenders suggested by the literature can be identified in our empirical data, we selected two land investments in Senegal, the Sabodala gold mine and the Senhuile agro-industrial plantation, which produced rice, corn, and groundnuts (see Prause & Le Billon 2020), to systematically compare resistances. Our selection was based on the high level of opposition they had been facing, as well as their relatively similar characteristics in terms of land area affected, time frame, and constellation of actors. The conflict around the Senhuile agro-industrial project might be best captured as a struggle against expulsion; however, some groups also engaged in acts of resistance to achieve better incorporation, mainly in the form of jobs. The struggle against the Sabodala gold mine has been dominated by demands for better incorporation, even if local communities also frequently resisted instances of expulsion and displacement.[3]

Motives of resistance

Differences between the two cases included a larger variety of motives in the conflict around the Sabodala gold mine (Table 14.1). While loss of land and water, along with poor job inclusion, were prevalent motives in both sectors, the displacement of the village Sabodala and the desecration of religious sites were specific motives for the Sabodala gold mine conflict. As a villager explained: 'If you want to displace Sabodala, this will only be possible using violence. Our grandparents are buried here. Someone who is from Sabodala, who is born in Sabodala will never leave here' (interview, Sabodala village, 2016). Displacement is also a key motivation for resistance in other mining conflicts in Senegal, such as around the ICS phosphate mine and the Grand Côte zircon mine (EJ Atlas 2019b). This suggests the importance of place-based meanings for the actions of land defenders. Compensation payments as well as the distribution of profits from the mine have also been major issues in the conflict around the Sabodala gold mine, as in conflicts around the phosphate mines in Senegal (Diallo 2017; Boidin & Simen 2016), but had no significance in the conflict around the Senhuile project. Table 14.1 offers a comparison of the motives of resistance in both cases. This echoes our findings from the literature, which likewise suggest a larger variety of motives in the mining sector.

While some differences reflect the values and expectations of different actors that are involved in these conflicts, sector-specific components – including structural ones – also help explain these differences. One is related to the specific materiality of the resources targeted by both land investments. Mining projects, especially those extracting high-value minerals, such as gold, are strongly place-bound. Land, as Li (2014: 589) reminds us, 'is not like a mat.

Table 14.1 Comparison of motives between the Sabodala mine and Senhuile project[4]

	Motives	*Sabodala gold mine*	*Senhuile project*
Motives related to integration and profit sharing	Lack of jobs and training opportunities	Yes	Yes
	Lack of effective corporate social responsibility	Yes	No
	Low royalties and unfair distribution	Yes	No
Motives related to expropriation	Loss of access to crop and livestock lands	Yes	Yes
	Loss of access to artisanal mines	Yes	No
	Low compensation for loss of land	Yes	No
	Lack of participation of local population in land allocation and project implementation	Yes	Yes
	Displacement of villages	Yes	No
	Threats to and loss of important cultural places	Yes	No

You cannot roll it up and take it away', yet its use for agro-industrial projects is slightly more flexible than for mining projects. Mineral deposits are location-specific, and especially in the case of gold, they have a high value. Village relocation and the destruction of culturally significant sites are, from the perspective of mining companies, imperative when they are located *above* underground deposits and deep-shaft mining is not a commercially viable option: a situation that led TGO to decide to relocate the village of Sabodala and the hamlet of Dambankhoto. In contrast, Senhuile planted its crops *around* the villages in the project area, and as a result, resettlement has not been an issue in the conflict. Differences in terms of displacement and resettlement as resistance motives can thus be not only project-specific but also sector-specific, as they reflect particular operational constraints and imperatives shaping the practices of companies and grievances of communities.

Structural aspects related to institutional and legal dimensions also differentiate the mining and agricultural sectors, providing different opportunities for the actions of environmental and land defenders, as these are usually regulated by a different set of laws and regulations. In Senegal, the Mining Act contains clear provisions on compensation payments to be made by companies (code minier, loi n° 2016–32, 08 November 2016, article 93). There are also fixed levies that companies must pay on the mineral or metal they extract (code minier, loi n° 2016–32, 08 November 2016, article 77). In contrast, the Senegalese Land Law remains ambiguous regarding the payment of compensation, and there is no legally binding levy in the form of rent payments or something similar. As a result, local populations losing their access to land face a situation where adequate compensation payments, as well as financial payments to the community, are unlikely. Opposition to agro-industrial projects in Senegal is therefore less concerned with the way certain projects are implemented than is the case in the mining sector. Claims against agro-industrial projects are mostly straightforward in demanding their cancellation since local communities are aware of the lack of options to benefit from these investments and instead focus on the defense of their access to and control over agricultural land. In the focus group discussions with the communities involved in the conflict around the Senhuile project, women especially stressed the point of defending the land access for the local communities in order to protect the land and land-based livelihoods of the future generations (focus group discussion, village opposing the Senhuile project, January 2015). This shows how different grievances and motives for resistance are produced through the differing project materialities and institutionalization of land and resource rights.

Physical acts of resistance

Physical acts of resistance were relatively similar across both sectors, yet there were notable exceptions regarding the more confrontational forms of resistance practices (see Table 14.2 providing an overview of the differences and similarities regarding the physical actions of resistance).

Table 14.2 Comparison of the physical actions of resistance between the Sabodala mine and Senhuile project

	Physical actions of resistance	*Sabodala gold mine*	*Senhuile project*
Confrontational forms	Undeclared demonstrations and violent clashes with security forces	Yes	Yes
	Blocking project development through attacks on workers and machinery	No	Yes
	Blockading of roads access to impede production process	Yes	No
	Targeted attacks on rural council	No	Yes
	Targeted attacks on national government and administration buildings, insults to national government officials	Yes	No
	Collective sabotage of equipment and machinery	No	Yes
Conventional forms	Communication and press campaign	Yes	Yes
	Lobbying political, economic, and religious leaders (meetings, petitions, etc.)	Yes	Yes
	Reporting on land allocation (e.g. critical reports on consequences for local population)	Yes	Yes
	Refusal to communicate with the company	Yes	Yes
	Mobilization and organization of the local population	Yes	Yes
	Boycott of social and environmental impact assessment	Yes	Yes
	Participation in international events	No	Yes

A first difference in registers of resistance is again related to the materialities of the projects and their consequences in terms of impacts and opportunities for resistance. While in both cases, members of local communities tried to impede production, only in the case of the Sabodala gold mine did they repeatedly use road blocks. Members of KODEN instead attacked workers and sabotaged machinery or fences to stop the clearing of the land, resistance strategies also observed in other cases in Senegal such as the 'Ferme Mame Tolla Wade' and the 'Ranch de Ouassadou' (Gagné 2019). This could reflect the different infrastructures involved in the two projects and related options for resistance. Agro-industrial projects often rely on existing transport infrastructure, which provide an important incentive for the decision to invest in a specific location (Nolte et al. 2016). As a result, agro-industrial projects are often well connected and have different access points to the road infrastructure, and this infrastructure is public and used by diverse groups.[5] Mining projects, in contrast, often need to invest in new transport infrastructure as their investment decision is primarily based on the location of the deposits. As a result, mining operations are often only accessible via one or very few access roads or railways, with access being sometimes restricted to company use. This increases the effectiveness of a blockade, while reducing its cost to non-mining interests. Both forms of confrontational resistances were met with strong reactions from the state, and several of the protesters involved have been arrested. Additionally, both companies tried to secure the relevant infrastructure using private security personnel.

A second difference in registers of resistance is related to the institutional dimensions of the land appropriations. Opposition to both projects was directed at different state authorities. A key addressee of the resistance practices in the conflict around the Senhuile project, especially in its beginnings, was the local council, *Conseil Rural*. But also, after the relocation to Ndiael, the local population lamented the complicity of the elected local authorities (Open letter, KODEN 2012). This targeting of the local authorities eventually paid off, and the maire of the community recently joined forces with protestors (telephone interview, speaker of KODEN, 01.11.2020). Resistance practices in the conflict around the Sabodala gold mine were directed at representatives of the central state at the regional and district levels, the *Préfet* and the *Sous-Préfet*, and were more often confrontational, mainly as a result of the greater social disconnection between local communities and central authorities, compared to elected local authorities. One of the villagers from Faloumbou stated during a focus group discussion: 'The state and in the form of its local representative the Sous-Préfet . . . is indeed responsible for the problems experienced by the population of Sabodala vis-à-vis the mine'.

This mirrors the different responsibilities of the central state and the local governance structures in both sectors, which offers distinct political opportunities for certain forms of resistance. The local level in the form of the *Conseil Rural* plays a greater role in land allocation in the agricultural sector. In the mining sector, land use rights are granted exclusively by the central, or 'national', government. This shows the impact of different state structures on resistance in different sectors. Furthermore, NGOs were more directly involved in the Senhuile conflict and provided more institutionalized channels of resistance, thereby directing efforts towards negotiations rather than physical confrontation.

Finally, there were also differences in temporalities. While Senhuile project opponents used public demonstrations and forceful tactics only at the very beginning of the project, such practices have been used throughout the entire period of opposition to the Sabodala gold mine. This partly reflects the success of the Senhuile project opponents in getting the project relocated to another area, one where the state holds a clearer 'claim' of property (i.e. natural reserve). Furthermore, while resistance in both cases was largely motivated by a sense of economic or cultural loss, these losses were more immediate in the Senhuile case as they directly followed the rapid clearing of land by the company, a strategy used by investors to strengthen their claim to the land that has also been observed in other cases of agro-industrial investments (e.g. Moreda 2015). In contrast, losses resulting from mining reflect the slower physical transformation of the landscape for mining purposes due to the different requirements that gold's materiality poses on the appropriation process. In the case of the Sabodala gold mine, local communities thus lost their livelihoods progressively over a longer period of time and experienced chronic impacts – such as blasting – rather than a one-time shock. This, in turn, continuously motivated some degree of opposition to the presence and extension of the mine. Sector-specific differences in some temporal aspects can thus influence resistance practices.

Resistance narratives

In both cases, opponents of the land investments built their narratives around the violation of human and ancestral rights by the company, as well as livelihood loss and the defense of the land for future generations (Prause 2019; see Table 14.3). As one chef du village involved in the resistance against the Senhuile-project put it,

> as a person one can only live on the land. It is on the land that you do your activities and when you die, it is this land where they will put your body. So, they will have to kill us, if they want to take the land.
>
> (focus group discussion, village opposing the Senhuile project, January 2015)

These narratives can also be found in many other instances of resistance against land investments in Senegal (Gagné 2019; EJ Atlas 2019b). Additionally, protest actors in the conflict around the Senhuile project framed the project as a danger for the food security of the population and for the national development goal of achieving food self-sufficiency (*autosuffisance alimentaire*), thereby linking their protests to the vision of food sovereignty, narratives that are likewise found in other conflicts around agro-industrial projects in Senegal (see Gagné 2019).

The opponents of the Sabodala gold mine stress social injustices related to the mine and especially the failure of industrial mining to enhance the local economy and create new and sustainable livelihoods for local communities (Prause 2019). Table 14.3 provides an overview of differences and similarities with regard to the narratives used. These resonate with some of our literature review, although narratives around environmental justice and the defense of territory that are often prominent in case studies of resistance against mining projects do not play an important role in Senegal. This suggests that while variations may be not just project-specific but also sector-specific, they can remain influenced by local and national contexts, including relative sensitivities and discursive mobilization around specific issues.

These differences reflect not only grievances directly perceived by affected communities but also the different discourses mobilized by the companies and

Table 14.3 Comparison of the narratives constructed by the resistance movements in the conflicts around the Senhuile project and the Sabodala gold mine

Narratives	*Sabodala gold mine*	*Senhuile project*
Violation of ancestral use rights	Yes	Yes
Violation of human rights	Yes	Yes
Loss of livelihoods and increase of poverty	Yes	Yes
Danger to food security and food self-sufficiency	No	Yes
Danger to local economic development	Yes	No
More productive use of land through local land users and vision of food sovereignty	No	Yes
Low share of revenue for Senegalese nation and local communities	Yes	No

state authorities to legitimize and enable investments. In order to delegitimize project proponents and recruit supporters, project opponents created counter-narratives linked to the dominant discourses shaped by state and investors, as seen in the case of food security interpreted through self-sufficiency rather than industrial productivity. On the mining side, the Senegalese government advanced an extractivist and neoliberal development discourse, declaring the Sabodala gold mine as a key pillar to economic growth and development, a typical discourse for many mining investments beyond Senegal (Le Billon & Sommerville 2017). Protest actors tried to delegitimize this narrative by referring to the lack of local economic development related to the mine, the risks that it poses to the existing ASGM-based local economy, and the low share of revenues to be received by the state. While both agricultural schemes and mining operations vary in scale and economic significance, the mining sector will often raise higher economic stakes, shaping both legitimization and resistance narratives.

Outcomes

Resistances to these two projects have produced very different outcomes. The Senhuile project was stopped in 2011 in the community of Fanaye, where the company had first leased 20,000 hectares of land. It has since been relocated to the Ndiael, the main investor has pulled out, and the company has so far been unable to cultivate more than a few hundred hectares. Nevertheless, the government has not yet officially returned the 20,000 hectares to the local community. Instead, it seems that the lease for the land has now been taken over by a Rumanian businessman who has been allegedly involved in a corruption scandal surrounding the Senegalese offshore oil and gas concessions (Cissé 2020, August 24). The Sabodala gold mine, on the other hand, has continually expanded its operations, and state authorities have approved the resettlement of the Sabodala village and the exploration and extraction of the *Niakafiri*-deposit, which necessitates work within culturally and religiously important sites.

Two main factors influenced these contrasting outcomes. First, there is a wide gap in economic value between these two projects. Total government revenues in 2016 from the Sabodala gold mine accounted for $49.7 million, representing a quarter of all extractive sector revenues, including concrete (EITI 2019). The mine also created about a thousand jobs. In contrast, the Senhuile project yielded very little revenue to the government and only created a few hundred jobs. According to a manager of Senhuile, the land has been given by the government without demanding any payments in return (interview, 2016, Dakar) and the company benefited from tax exemptions (interview, investment agency, 2015, Dakar). Whereas there is very little dispute among state officials about supporting the Sabodala gold mine, support for the Senhuile project is more contested, as was shown by the temporary memorandum on land transactions imposed by Macky Sall in 2012, which offered an important political opportunity for resistance.

Second, there is a stronger coalition opposing the Senhuile project than the Sabodala mine, in terms of both the number and reach of organizations, which provides a strong political opportunity structure for the resistance movement against Senhuile. Relatively different civil society actors became involved in both conflicts. The Senhuile case benefited from well-established NGOs defending smallholder agriculture (see more broadly, McKeon et al. 2004), but there was no equivalent for small-scale gold miners. In a country where (post) colonial economic attention was mostly directed at the commercial agricultural sector, agrarian organizations have long drawn attention to the danger posed by the expansion of agro-industrial projects for small-scale agriculture. In contrast, there is only a relatively thin history of industrial mining in Senegal, and agrarian organization have until recently not given much attention to large-scale mining projects, including the Sabodala gold mine, partly because the affected local population mostly consisted of artisanal gold miners whose interests have not been defended by agrarian organizations. Support for artisanal miners thus comes in a more limited way through international NGOs focusing on the impacts of extractive industries in the form of brief field visits and reports.

Conclusion

Resistance to agro-industrial and mining projects has received much attention, yet few studies have systematically compared resistance movements across these two sectors. Based on an analysis of conflicts reported in EJ Atlas, our literature review, and the comparison of two specific cases in Senegal, this study points to some differences and possible explanations.

Regarding motives, both the literature and our case study point to a larger variety of motives in the mining sector. In terms of physical practices of resistance, our case study found some variations, especially with regard to confrontational tactics (e.g. road blocks against mining vs. attacks on worker and machinery for farming), but otherwise much similarity across both sectors, which is in line with findings from the EJ Atlas and the literature. In terms of discursive forms of resistance, we found that protest actors used different narratives to oppose mining (i.e. economic development) than against agro-industrial projects, which supports findings from the literature (i.e. food security and food self-sufficiency). Finally, outcomes also differed across our two case studies, with resistance against large-scale farming being more successful than against mining. Yet, no generalization can be made as a result of this preliminary study. A more systematic and in-depth analysis of the literature and EJ Atlas may yield some more general patterns of differentiation across sectors, including in terms of the motives – or 'project impacts' – reported in the EJ Atlas dataset, which were not examined here at this global level.

Given the limited scope of this study, it is difficult to point at systematic explanations for the differences so far observed, but our case study analysis suggests several hypotheses relating to the specific material, institutional, and discursive characteristics of the two sectors.[6] While we summarize each in turn, we emphasize that project- or even sector-specific characteristics often relate to and can influence each other, and that we do not seek to argue for sectoral characteristics as the key or only determinant of resistance, but rather to stress the importance of considering sector-specific factors while trying to understand the actions of environmental and land defenders.

First, the *materialities* of land investments can influence motives and physical modes of resistances. For example, agro-industrial projects are generally less place-bound than mining projects and can thus more easily avoid resettlements or religious site desecration than mining projects, and thereby avoid some key motives for resistance. Second, variations in *institutions* and *financial stakes* across sectors can provide distinct political opportunities and thus influence resistance. Third, differences in the legitimating *discourses* of project proponents can influence the distinctive counter-narratives adopted by resistance movements and demands placed upon project proponents. Differences in outcomes of resistance against the Senhuile agro-industrial project and the Sabodala gold mine were linked in part to factors reflecting the history and position of the sector in the country, rather than broad or site-specific sector characteristics. Successful outcomes were largely due to the availability of powerful allies such as national and international NGOs, wavering state support for the land investment in question, and the way different resistance movements were able to make use of political opportunities. This confirms findings from the literature.

Approaches from contentious politics, such as narratives and political opportunities, allowed us to identify institutional structures and discursive contexts as important factors that differ in the mining and agrarian sector and thus offer distinct incentives or impediments for resistances. However, only in accounting for the distinct forms of land appropriations and socio-environmental relations produced through the materiality of different resources and project sites, including the spatialities of government institutions managing the project and engaging with resistance practices, were we able to more fully account for differences in the physical acts and narrative forms of resistance, as well as some of the initial motives. Thus, our study points to the relational importance of biophysical materialities with social discourses and practices to understand the resistance practices of land defenders.

One important finding from our case studies is that struggles for land appropriation for industrial mining and agribusiness are still largely separate. Alliances between protest actors in both sectors seem to be rare. Different institutional settings governing land allocations in the mining and agricultural sectors mean that movements often face different political opportunities and material conditions of resistance at different times. Yet, our comparison has also shown that loss of land and water, along with poor job inclusion, are prevalent in both sectors, thereby providing shared motives. More broadly, defending rights principles – for example relating to property, health, consultation, and physical integrity – might serve as a master frame for resistances against both agricultural and mining projects. However, in order to build such common frames, social ties between land defenders are key, which are often lacking due to different histories of social struggles in both sectors. Here international NGOs and social movements could play an important role as 'bridge builders or 'brokers' in facilitating communication and coordination among land defenders from both sectors, but also in linking seemingly disparate struggles. In Senegal, agrarian NGOs such as Enda Pronat are slowly starting to get active in struggles against industrial mines, suggesting a growing linkage between struggles in both sectors. Globally, the platform of the Global Convergence of Land and Water Struggles attempts to bring different kinds of rural struggles together. Thus, further research on resistance movements across land-based sectors could help build further synergies between land defenders active in conflicts around different types of land transformations.

Funding

This work was supported by the Freie Universität Berlin within the Excellence Initiative of the German Research Foundation as well as the German Federal Ministry of Education and Research under its funding line 'Global Change'.

Acknowledgements

We are very grateful to all interview partners for their time and invaluable input. Further thanks to IPAR, particularly Dr. Aminata Niang, the West Africa office of the Rosa-Luxemburg Foundation, ENDA PRONAT and Dr. Lamine Diallo for their logistical support, and to the GLOCON team for their feedback and input.

Notes

1 Agro-industrial conflicts include 'intensive food production (monoculture and livestock)', 'agro-fuels and biomass energy plants', and 'plantation conflicts (including pulp)' projects; mining conflicts include 'mineral ores and building materials extraction' projects.

2 Figure 14.1 was compiled by the authors, using the filter functions of the EJ Atlas database.
3 For Senegal, EJ Atlas (2019a) reports four cases of resistance to agro-industrial projects, including Senhuile (characterized as a conflict of 'high intensity' that by the last update in February 2018 was not seen as a success for the movement), and six cases of resistance to large-scale mining, including Sabodala (characterized as a 'medium intensity' conflict that by the last update in February 2018 was not seen as a success for the movement). These intensity levels reflect the 2011–13 period, rather than later years.
4 Tables 14.1–14.3 were compiled by the authors on the basis of the primary data collected during five months of fieldwork in Senegal.
5 However, crops requiring rapid processing after harvesting, such as sugar-cane or palm oil, often have only one access road connecting fields and processing plant.
6 The networks of actors getting involved have also contributed to the differences; however, their configuration seemed rather case-dependent and linked to the different history of agrarian and mining struggles in Senegal.

References

Acosta, A. (2013). Extractivism and neoextractivism: Two sides of the same curse. In M. Lang, L. Fernando, & N. Buxton (Eds.), *Beyond development: Alternative visions from Latin America* (pp. 61–86). Amsterdam: Transnational Institute.

Alonso-Fradejas, A. (2015). Anything but a story foretold: Multiple politics of resistance to the agrarian extractivist project in Guatemala. *The Journal of Peasant Studies, 42*(3–4), 489–515.

Andrews, N. (2018). Land versus livelihoods: Community perspectives on dispossession and marginalization in Ghana's mining sector. *Resources Policy, 58*, 240–249.

Anyidoho, N. A., & Crawford, G. (2014). Leveraging national and global links for local rights advocacy: WACAM's challenge to the power of transnational gold mining in Ghana. *Canadian Journal of Development Studies/ Revue canadienne d'études du développement, 35*(4), 483–502.

Bebbington, A. (2012). *Social conflict, economic development and extractive industry*. Abingdon, UK: Routledge.

Bebbington, A., & Bury, J. (Eds.). (2013). *Subterranean struggles: New dynamics of mining, oil, and gas in Latin America*. Austin, TX: University of Texas Press.

Boidin, B., & Simen, S. F. (2016). Industrie minière et programmes de développement durable au Sénégal. *Développement durable et territoires*, 7(2).

Bond, C. J., & Kirsch, P. (2015). Vulnerable populations affected by mining: Predicting and preventing outbreaks of physical violence. *The Extractive Industries and Society, 2*, 552–561.

Borras, S. M., & Franco, J. C. (2013). Global land grabbing and political reactions 'from below'. *Third World Quarterly, 34*(9), 1723–1747.

Borras, S. M., Kay, C., Gómez, S., & Wilkinson, J. (2012). Land grabbing and global capitalist accumulation: Key features in Latin America. *Canadian Journal of Development Studies/ Revue canadienne d'études du développement, 33*(4), 402–416.

Brent, Z. W. (2015). Territorial restructuring and resistance in Argentina. *The Journal of Peasant Studies, 42*(3–4), 671–694. doi:10.1080/03066150.2015.1013100

Christensen, D. (2019). Concession stands: How mining investments incite protest in Africa. *International Organization, 73*(1), 65–101.

Cissé, F. (2020, August 24). Frank Timiss bénéficie d'une titre foncier de 25.000 ha dans le nord du Sénégal. *PressAfrik*. https://farmlandgrab.org/post/view/29805 [last accessed 11 November 2020].

CNRG. (2015). *Communities, companies and conflict*. Harare: Centre for Natural Resource Governance.

Conde, M. (2017). Resistance to mining: A review. *Ecological Economics, 132*, 80–90.
Conde, M., & Le Billon, P. (2017). Why do some communities resist mining projects while others do not? *The Extractive Industries and Society, 4*, 681–697.
Cotula, L. (2012). The international political economy of the global land rush: A critical appraisal of trends, scale, geography and drivers. *The Journal of Peasant Studies, 39*(3–4), 649–680.
Cuba, N., Bebbington, A., Rogan, J., & Millones, M. (2014). Extractive industries, livelihoods and natural resource competition: Mapping overlapping claims in Peru and Ghana. *Applied Geography, 54*, 250–261.
Deloitte. (2018). *Tracking the trends 2018: The top 10 issues shaping mining in the year ahead.* www2.deloitte.com/content/dam/Deloitte/global/Documents/Energy-and-Resources/gx-TTT-report-2018.PDF
Diallo, L. (2017). L'industrie du phosphate de Taïba au Sénégal: Front minier et tensions locales. *Vertigo, 28.*
Dietz, K., & Engels, B. (2020). Analysing land conflicts in times of global crises. *Geoforum, 111*, 208–217.
Di John, J. (2011). Is there really a resource curse? A critical survey of theory and evidence. *Global Governance, 17*, 167–184.
Dougherty, M. L., & Olsen, T. D. (2014). Taking terrain literally: Grounding local adaptation to corporate social responsibility in the extractive industries. *Journal of Business Ethics, 119*(3), 423–434.
EITI. (2019). *EITI progress report 2019.* Extractive Industry Transparency Initiative. https://eiti.org/files/documents/eiti_progress_report_2019_en.pdf [last accessed 1 December 2020].
EJ Atlas. (2019a). *Environmental Justice Atlas.* https://ejatlas.org
EJ Atlas. (2019b). *Projet Grande Côte for zircon and ilmenite mining.* Senegal. https://ejatlas.org/conflict/diogo-zircon-mining-niayes-senegal [last accessed 4 October 2019].
Engels, B. (2018). Nothing will be as before: Shifting political opportunity structures in protests against gold mining in Burkina Faso. *The Extractive Industries and Society, 5*(2), 354–362.
Engels, B., & Dietz, K. (Eds.). (2017). *Contested extractivism, Society and the State: Struggles over Mining and Land.* London: Palgrave Macmillan.
Famerée, C. (2016). Political contestations around land deals: Insights from Peru. *Canadian Journal of Development Studies/Revue canadienne d'études du développement, 37*(4), 541–559.
Gagné, M. (2019). Resistance against land grabs in Senegal: Factors of success and partial failure of an emergent social movement. In T. Barley (Ed.), *The politics of land* (pp. 173–203). Bingley: Emerald Publishing.
Gerber, J. F. (2011). Conflicts over industrial tree plantations in the South: Who, how and why? *Global Environmental Change, 21*(1), 165–176.
Gingembre, M. (2015). Resistance or participation? Fighting against corporate land access amid political uncertainty in Madagascar. *The Journal of Peasant Studies, 42*(3–4), 561–584.
Grain. (2016). *The global farmland grab in 2016? How big, how bad?* Barcelona: Grain.
Grajales, J. A. (2015). Land grabbing, legal contention and institutional change in Colombia. *The Journal of Peasant Studies, 42*(3–4), 541–560.
Grant, H., & Le Billon, P. (2019). Growing political: Violence, community forestry, and environmental defender subjectivity. *Society & Natural Resources, 32*(7), 768–789.
Hall, R., Edelman, M., Borras, S. M., Scoones, I., White, B., & Wolford, W. (2015). Resistance, acquiescence or incorporation? An introduction to land grabbing and political reactions 'from below'. *The Journal of Peasant Studies, 42*(3–4), 467–488.
Haslam, P. A., & Tanimoune, N. A. (2016). The determinants of social conflict in the Latin American mining sector: New evidence with quantitative data. *World Development, 78*, 401–419.

Hollander, J. A., & Einwohner, R. L. (2004). Conceptualizing resistance. *Sociological Forum, 19*(4), 533–554.

Horowitz, L. S. (2010). 'Twenty years is yesterday': Science, multinational mining, and the political ecology of trust in New Caledonia. *Geoforum, 41*, 617–626.

Hufe, P., & Heuermann, D. F. (2017). The local impacts of large-scale land acquisitions: A review of case study evidence from Sub-Saharan Africa. *Journal of Contemporary African Studies, 35*(2), 168–189.

Kandel, M. (2015). Politics from below? Small-, mid- and large-scale land dispossession in Teso, Uganda, and the relevance of scale. *The Journal of Peasant Studies, 42*(3–4), 635–652.

Larder, N. (2015). Space for pluralism? Examining the Malibya land grab. *The Journal of Peasant Studies, 42*(3–4), 839–858.

Le Billon, P., & Sommerville, M. (2017). Landing capital and assembling 'investable land' in the extractive and agricultural sectors. *Geoforum* (82), 212–224.

Li, T. M. (2014). What is land? Assembling a resource for global investment. *Transactions of the Institute of British Geographers, 39*(4), 589–602.

Martinez-Alier, J., Temper, L., Del Bene, D., & Scheidel, A. (2016). Is there a global environmental justice movement? *The Journal of Peasant Studies, 43*(3), 731–755.

Martiniello, G. (2015). Social struggles in Uganda's Acholiland: Understanding responses and resistance to Amuru sugar works. *The Journal of Peasant Studies, 42*(3–4), 653–669.

McKeon, N., Watts, M., & Wolford, W. (2004). *Peasant associations in theory and practice.* Geneva: United Nations Research Institute for Social Development.

Middeldorp, N., & Le Billon, P. (2019). Deadly environmental governance: Authoritarianism, eco-populism, and the repression of environmental and land defenders. *Annals of the American Association of Geographers, 109*(2), 324–337.

Moreda, T. (2015). Listening to their silence? The political reaction of affected communities to large-scale land acquisitions: Insights from Ethiopia. *The Journal of Peasant Studies, 42*(3–4), 517–539.

Ndi, F. A., & Batterbury, S. (2017). Land grabbing and the axis of political conflicts: Insights from Southwest Cameroon. *Africa Spectrum, 52*(1), 33–63.

Nolte, K., Chamberlain, W., & Gige, M. (2016). *International land deals for agriculture. Fresh insights from the Land Matrix: Analytical Report II.* Bern: Centre for Development and Environment (CDE), University of Bern; Centre de coopération internationale en recherche agronomique pour le développement (CIRAD); German Institute of Global and Area Studies (GIGA); University of Pretoria.

Owen, J. R., & D. Kemp. (2014). 'Free prior and informed consent', social complexity and the mining industry: Establishing a knowledge base. *Resources Policy, 41*, 91–100.

Peluso, N. L., & Lund, C. (2011). New frontiers of land control: Introduction. *The Journal of Peasant Studies, 38*(4), 667–681. doi:10.1080/03066150.2011.607692

Perreault, T. (2013). Dispossession by accumulation? Mining, water and the nature of enclosure on the Bolivian Altiplano. *Antipode, 45*(5), 1050–1069.

Poletta, F., & Chen, P. C. (2012). Narrative and social movements. In J. Alexander, R. N. Jacobs, & P. Smith (Eds.), *The Oxford handbook of cultural sociology*. Oxford: Oxford University Press.

Prause, L. (2019). Success and failure of protest actors' framing strategies in conflicts over land and mining in Senegal. *Canadian Journal of Development Studies/Revue canadienne d'études du développement, 40*(3), 387–403.

Prause, L., & Le Billon, P. (2020). Struggles for land: Comparing resistance movements against agro-industrial and mining investment projects. *The Journal of Peasant Studies*. doi: 10.1080/03066150.2020.1762181

Renauld, M. (2016). The Esquel effect: Political opportunity structure and adaptation mechanisms in anti-mining mobilisation in Argentine Patagonia. *Canadian Journal of Development Studies/Revue canadienne d'études du développement, 37*(4), 524–540.

Rodríguez-Labajos, B., & Özkaynak, B. (2017). Environmental justice through the lens of mining conflicts. *Geoforum* (84), 245–250.

Scott, J. C. (1985). *The weapons of the weak: Everyday forms of peasant resistance*. New Haven, CT: Yale University Press.

Taylor, V., & Van Dyke, N. (2004). 'Get up, stand up': Tactical repertoires of social movements. In D. A. Snow, S. A. Soule, & H. Krisie (Eds.), *The Blackwell companion to social movements*. Wiley Online Library.

Temper, L. (2018). From boomerangs to minefields and catapults: Dynamics of trans-local resistance to land-grabs. *The Journal of Peasant Studies, 46*(1), 188–216.

Temper, L., Demaria, F., Scheidel, A., Del Bene, D., & Martinez-Alier, J. (2018). The global Environmental Justice Atlas (EJAtlas), ecological distribution conflicts as forces for sustainability. *Sustainability Science, 13*(3), 573–584.

Warnaars, X. S. (2012). Why be poor when we can be rich? Constructing responsible mining in El Pangui, Ecuador. *Resources Policy, 37*(2), 223–232.

[Part of this chapter first appeared as: Prause, L., & Le Billon, P. (2020). Struggles for land: Comparing resistance movements against agro-industrial and mining investment projects. *The Journal of Peasant Studies*.]

15 How violence is justified in 'democratic countries'

Justine Taylor

One of the most effective means of defense against the climate and biodiversity crisis we all face is the Indigenous peoples and local communities defending nature and protecting their lands. Indigenous environmental defenders are highly effective at protecting our natural resources; Indigenous peoples manage at least ~38 million km^2 globally (Schleicher et al. 2017), about a quarter of the world's land surface, which covers about 40% of all ecologically intact landscapes (Garnett et al. 2018). Environmental defenders risk their well-being by putting their bodies on the frontlines. Notwithstanding the massive scale and grave risks environmental human rights defenders face in countries where rule of law is weak (Butt et al. 2019; Le Billon & Lujala 2020), targeted attacks against environmental defenders are a global phenomenon, occurring even in democratic countries (Simpson and Le Billon 2021). While often defenders are easily exoticized, perceived to be living in far-away places, increasing data shows people's lives are permanently changed due to abuse from state actors in the UK. Examining the violence, surveillance, and intimidation by United Kingdom (UK) police and private enforcement companies inflicted on defenders helps us to cast an eye upon how those here who are fighting against the environmentally destructive will of the government have a different experience of what living in a democracy means. The focus of this chapter on the UK does not intend to diminish the experience of international defenders who are murdered and harassed for risking their lives to protect land. Global Witness's 2019 report *Enemies of the State?* records that three people are murdered a week while standing up for the environment. The UK has not reported a death of any environmental protectors. Yet, in the UK the level of harm done to peaceful protestors is increasingly at odds with the notion of democracy and so-called British values[1] that sustain our international reputation.

In 2019, Not1More started to work on the issue of environmental defenders (self-identifying as 'protectors') in the United Kingdom. Our involvement with data gathering on environmental protest is likely to continue as testimonies continue to roll in. *High Speed 2* (HS2) is a railway building company which has been given political and economic significance in the UK. Protectors are fighting the building of the highspeed railway's first stage, which is due to run from London to Birmingham through 108 ancient woodlands and 33 legally protected scientific sites (Barkham 2020), and destroy 900 homes and 1000 businesses (Gabbatiss 2018). Staff members, subcontractors and police connected to HS2 physically harm environmental defenders outside of work-sites, and in their homes and tree houses along the route of the railway on a daily basis.

Our research began shortly before a moratorium on fracking was called in the UK. In a moment of cautious optimism, we thought our research might only be a historic examination of controversial policing on sites where local communities resisted fracking projects. Communities in the UK have been fighting fracking for over a decade; impacts upon the life of a local community following fracking include contaminated ground water and earthquakes, while general concerns about fracking are intrinsically linked to the climate crisis (Friends of the Earth 2017). We gathered data that scratched the surface, spanning eight years of peaceful resistance with protectors being systematically subjected to police violence and other forms of repression. Various publications and organizations have reported beatings (Thompson 2014),[2] sexual violence (Gilmore et al. 2019: 33),[3] a disproportionate number of arrests (Gilmore et al. 2016: 36),[4] and inappropriate arrests, unfair bail conditions, and intimidation during the protests (BBC 2013; Devlin 2017).[5] Conduct outside of the actual protests contributes to the chilling effect on environmental activism. Continuing surveillance (The Guardian 2019),[6] sharing activists' details with the disability benefit system (Pring 2018), the persistent qualification of climate activists as terrorist threats (Jones & Scott 2018), and the distribution of fake news (Pidd 2018; Rowell 2018)[7] are part of a system that induces fear of participating in environmental action and creates a chilling effect – threatening to slow down or even stop the environmental movement as the fear of persecution, harm, or benefit loss makes the cost of going to a protest disproportionate to the perceived benefit of attending.

In just less than a year of research, we recorded over four hundred incidents against peaceful protestors. While testimonies were gathered from both historical and live protests, the proportion of incidents that happened concurrent with our data gathering shows there is no slowing of human rights violations. Our historical work included three main anti-fracking sites: Preston New Road, Barton Moss, and West Newton. We gained supplementary testimonies from anti-fracking sites at Missen Springs, Kirby Misperton, and Balcombe as well. The ongoing incidents were recorded at live sites for anti-HS2 protests, including Harvill Road, Denham Country Park, Crackley Woods, and Steeple Claydon. We know that the number of incidents we recorded are only a small proportion of those that have happened. To fulfil the need for a more complete dataset, we are working with grassroots groups to improve recording of incidents for those on the frontlines. At the very beginning of our research period, while we were preparing for an interview at Preston New Road, we asked if violence was a regular thing, and the interviewee said that at the peak, there were around 20 beatings per day. The problem is, with such an astounding number of violations, only a few stand out, and the rest blur into one. It has thus been challenging to accurately gauge the state of human rights abuses relating to protests in the UK: while the protectors experience regular and life-changing violence, it becomes normalized and difficult to count. This normalization of violence is toxic both within and outside of the movement. Most of the public are too far removed to understand the complexity, and unrelatable stories of violence make the stories of protestors' lives even more alien. The evidence we have gathered and examine in this chapter shows that beyond misunderstanding, people outside the protest think protectors are harmed because they deserve to be harmed.

In the UK, the violence and psychological intimidation against environmental human rights defenders surpasses what the general public would normally see as unacceptable. Examples from our own anonymized interviews at the HS2 sites during COVID-19 lockdown include: strip searching children (interview, 23 July 2020),[8] stalking and pressuring family members of protectors (interview, 11 August 2020), broken fingers and broken noses (interview, 14 August 2020), a broken jaw (Courtney-Guy 2020), and racial and sexual discrimination. There is concern that people who have experienced violence in the past months will be permanently unable to work or walk properly. Those deprived of their hearing and other long-term physical damage to their bodies, including head wounds (interview, 17 July 2020; Shadwell 2020) or tendon and nerve damages (interview, 6 April 2020) will have their long-term well-being negatively affected.

Not1More's research findings, analysed by Global Diligence LLP, showed that violence is perpetrated by police, security guards, and bailiffs (Global Diligence LLP 2020). The UN General Assembly guides us to understand that when private bodies are acting in state interest, they can be seen as acting in the interest of the state:

> The conduct of a person or entity which is not an organ of the State under article 4 but which is empowered by the law of that State to exercise elements of the governmental authority shall be considered an act of the State under international law, provided the person or entity is acting in that capacity in the particular instance.
>
> (UN General Assembly 2008)[9]

When bailiffs use violence against defenders, it is often ignored by police evidence collectors, offering private bodies a level of impunity. HS2 staff and security, including bailiffs, are employed through contracts made for HS2, which is a company overseen by the UK Department of Transport, answering to the government's select committee on the topic (House of Commons Select Committee on High Speed Rail 2018). HS2 enforcers, including the police, are answerable to the government, and yet attempts by protestors to report the aforementioned harms to the police, or to HS2's complaints-line, have been ignored. Protestors who contact the police to report when bailiffs have beaten them up have found themselves being arrested instead (interview, 14 August

2020). This was a strategy also typical for the anti-fracking protests (interview, 1 June 2020), despite there being a larger degree of plausible deniability on the topic of collusion between the police and the company.[10]

This form of violence is not condoned by the police's own guidelines. The College of Policing states that: 'the police have a duty to take reasonable steps to protect those who want to exercise their rights peacefully' (n.d.). The 'human rights compliant policing policy' (ACPO 2010; HMIC 2009) illustrates how the police tried to rein in their own behaviour following the death of a newspaper vendor, Ian Tomlinson (Jackson et al. 2018).[11] Tomlinson, attempting to commute during the G20 protests, was struck by an officer mistaking him for a protestor; Tomlinson died shortly after (ibid.). Our laws, aforementioned policing policy papers, and also international human rights law, which the UK has been instrumental in writing,[12] condemn violence against those who are peacefully protesting and aim to have a human rights compliant approach to policing.

There are very narrow stipulations for organs of the state being allowed to cause harm. The violence we have gathered evidence of is far beyond the scope of lawful enforcement. It is hard to identify the basis given to legitimize inflicting permanent and serious harm on protestors. Theory behind who is, and who isn't, allowed to cause harm may shed light on how the law does not protect protestors from the police. Power theory argues that whoever has power gets to decide who legitimately can cause harm to others (Brunkhorst et al. 2017). Gaventa's (2006) analysis of power illustrates how power-holders can choose what is or isn't an issue. The violence experienced by protestors in the UK is barely reported in the news, even when the injury is potentially life threatening.[13] This shows that violence can be framed in the media as a non-issue and can continue no matter how outrageous because it is aligned with the political understanding of humanness (Ferrarese 2015). With Gaventa's workshops on locating power (Powercube n.d.), it is possible to see how a culmination of interactions illustrate who is agenda-setting, what is normative (Kagan 1998), who has hegemony (Crehan 2016), and what makes people deserving of human rights. This is achieved by understanding how the impacts of observable phenomena denote the intent and situation of power.

From our research, we gathered instances where protestors were treated as subhuman – not by the police, who have been established as orientated to enforce the proceeding of company's expansion, but rather by local people whose own lives would be negatively impacted by these infrastructure and energy projects. People who stand to lose homes and businesses appear to wish harm on protestors who are sticking up for all of our rights to water, nature, and a healthy environment as well as their local community's right to self-determination. Our 430 recorded incidents indicated the normalization of violence and dehumanizing of protestors. A lack of public outcry on stand-out cases, such as the aforementioned broken bones and stalking of family members, indicate the normalization of repression of people who protest to protect the environment and members of the public internalizing this new lesser status of protestors. Project Justice gathered evidence of targeted bullying by adults of children whose parents were involved in anti-fracking protests, and, more specifically, of targeted hate attacks online aimed at a trans child whose parent was involved in a local campaign (Protest Justice 2020). A mother explained to Not1More how distressing it was that police would not investigate these incidents. General attacks participation spanned a wide range of incidents of protestors in their everyday life, including a member of the public known to protestors who booked a beauty appointment with the spouse of a protestor, assaulted her, and then left. Police said there was not enough evidence to investigate, even though continuing threats were made over social media before the attack (interview, 1 June 2020). Members of the public purposefully try to run over activists (Project Justice 2020), often clipping people and more regularly carrying them down the road on the hood of a car (interviews, 9 April, 6 May 2020). Protestors would describe how, while standing with their picket, often children would lean out of car windows, spitting and yelling profanities. This particular action was upsetting because the parent driving the car was allowing, and in effect condoning, abusive behaviour.

Despite their diversity of backgrounds and motivations, protestors in the UK are known as 'the great unwashed' rather than as human rights defenders. How this has come to pass can be observed in many ways. Power analysis is possible when we observe how the police ingratiate themselves into normative culture while forcing protestors out. Many of the strategies in what follows fall into the category of 'astroturfing': the battle for 'hearts and minds' version of greenwashing, where companies win over communities who might otherwise be opposed to damaging projects. A key way that both corporations (Brock 2020; interview, 9 September 2020)[14] and the police influence communities is by going into schools (NETPOL 2018) and taking time to influence both children and teachers. Both anti-HS2 campaigners and anti-fracking campaigners said that schools had run programs in support of these respective projects, which had been extremely biased and therefore unhelpful (ibid.). Many areas in which these projects exist are in a funding void. Company funding for essentials in the local area, such as football teams, demonstrate the care companies take to be well situated in the community (Brock 2020).

In the police's own training materials, activists such as anti-fracking activists or animal rights activists are aligned closely with extremist groups, such as 'Irish Related – Republican & Loyalist' [sic] and 'Infidels' [sic] (Gilmore et al. 2019). Perhaps the training materials which align peaceful protestors with actual terrorist groups (with a track record for killing people) has something to do with the manner in which policing is carried out. The data that Not1More gathered highlights that it is normal for officers to adopt high levels of violence as policing techniques of people who have turned up to exercise their democratic rights.[15] An additional impact of heavy-handed policing is that the greater police presence actually makes protestors look like they are doing something wrong. Many protestors have said they are sure that huge police presence makes them look even more criminal; often 60 police will turn up for a two-person lock-on or a three-person eviction. As a protestor told us, 'it looks like we are doing something wrong, even though we aren't. It is so obvious they are doing this on purpose, to make us look like criminals. It's such a waste of public money' (anonymous interview, 9 September 2020). Protestors feel that police generally side with security guards and company staff. Throughout the years that fracking policing was at its most violent, people's Freedom of Information requests uncovered a Memorandum of Understanding that police would be actively supportive of the corporation with regard to the activists. This happened so often that the police had to stop agreeing to them on paper (Brock 2020).

Case study 'dog gate'

The unfathomable incident 'dog gate' began on 3 June 2018, when police violently arrested an activist (breaking his partner's bones) for allegedly poisoning a dog belonging to a security guard (Patterson 2018; Rowell 2018). The protestor did not poison the dog – vet records show that no poisoned dog was ever admitted to them. As they carried out the arrest, police sent out press releases about the incident. People living on site received huge amounts of abuse, accused as dog murderers. Members of the public were horrified and set about to hurt and humiliate the protestors online and in the camp. The accused protestor applied for a Freedom of Information request to understand where the story had come from. They discovered that the healthy dog had been taken to the vet and the vet had been confused as to why the dog was there (interview, 1 June 2020). However, the damage done by the police's own false press release was never repaired. Horrible reports about how vile protestors remain live on the internet (ibid.).

Conclusion

The level of harm that is being done to people who attend the frontlines of environmental protest in the UK is indisputable. The rub comes in trying to understand how there is such a huge disconnect between human rights standards, to which we legally subscribe, as do our police, and the way environmental protestors are treated. There is nothing unique about the demographics of people who protest. The background of these protestors is diverse, from students, parents, and the elderly to mid-career professionals. Our data shows that any of these attendees might expect to be choked or sexually assaulted. While the root of this violence is attached to policy instructions and training materials, it is much harder to understand the complacency of the general public. Examples given here are only a snapshot of the current normative state at this moment. There is much such violence in the UK, belying a hegemony of ideas about who is and isn't worthy of harm or protection by the law. Rehumanizing protestors requires more than simply legal acknowledgement; as we saw, 'human rights compliant policing' did not have the desired impact. While defenders cannot return to work, some feel a public inquiry is now the only means to vindicate the hundreds of people traumatized and smeared as a result of their conscientious objection to habitat destruction.

Notes

1 'British Values' is a part of the 'Prevent' strategy and is taught in schools (Department for Education & Lord Nash 2014).
2 See, for example, the assault of Vanda Gillett (Thompson 2014).
3 See, for example, Gilmore et al. (2019: 33) about the targeted approach to (younger) women at the protests.
4 At the Barton Moss protests, within 20 weeks, 231 individuals were arrested (Gilmore et al. 2016: 36).
5 See, for example, the kettling and subsequent removal of 'tealady' Jackie Brooks (Devlin 2017). More generally, anti-fracking protests are characterized as being over-policed, with around 150 police officers deployed for a small protest (BBC 2013).
6 See the *Guardian* (2019), for example, on surveillance at the Balcombe protests.
7 See, for example, Pidd (2018) on accusations that anti-fracking protestors groomed a 14-year-old boy, or Rowell (2018) on the distribution of a police press release that an anti-fracking protestor poisoned a guard dog.
8 The interviews referenced throughout this chapter were conducted for Not1More research between April and September 2020.
9 *Article 5 Conduct of persons or entities exercising elements of governmental authority*, UN General Assembly, *Responsibility of States for internationally wrongful acts: resolution/adopted by the General Assembly.*
10 For an assessment of the collusion between police and private companies, I defer to Brock (2020):

'Corporate-state security collaborations are institutionalised in Memoranda of Understanding (MoUs), more common in the earlier years of anti-fracking policing. "People . . . kept making FoI request so they stopped doing these [MoUs]", a legal expert who monitors fracking policing explains. The MoU between the Greater Manchester Police (GMP) and drilling company IGas, for instance, showed that the company had insider access to police command meetings at the highest (gold command) level, with "daily briefings or video conferences" to discuss information and intelligence'.

11 An officer struck a commuter (who was actually a newspaper vendor), who died shortly after the blow. The officer claims he believed the vendor was a protestor. https://en.wikipedia.org/wiki/Death_of_Ian_Tomlinson.
12 See United Nations, *Charter of the United Nations (1945*).
13 Not1More research for documentary, 11 August 2020.
14 Research interview conducted by both Not1More and University of Sussex.
15 Not1More research results publication, forthcoming.

Bibliography

ACPO. (2010). *The manual of guidance on keeping the peace*. London: National Policing Improvement Agency.

Barkham, P. (2020). HS2 will destroy or damage hundreds of UK wildlife sites. *The Guardian*. www.theguardian.com/uk-news/2020/jan/15/hs2-will-destroy-or-damage-hundreds-of-uk-wildlife-sites-report

BBC. (2013, December 3). Barton Moss: MP says anti-fracking protest 'over-policed'. www.bbc.com/news/uk-england-manchester-25196705

Brock, A. (2020). 'Frack off': Towards an anarchist political ecology critique of corporate and state responses to anti-fracking resistance in the UK. *Political Geography*, *82*, 102246.

Brunkhorst, H., Kreide, R., & Lafont, C. (2017). *The Habermas handbook*. New York: Columbia University Press.

Butt, N., Lambrick, F., Menton, M., & Renwick, A. (2019). The supply chain of violence. *Nature Sustainability*, *2*, 742–747. https://doi.org/10.1038/s41893-019-0349-4

College of Policing. (n.d.). *Public order: Core principles and legislation*. www.app.college.police.uk/app-content/public-order/core-principles-and-legislation/

Courtney-Guy, S. (2020, October 9). H2 bailiffs investigated for 'breaking protester's jaw' while off duty. *Metro News*. https://metro.co.uk/2020/10/09/hs2-bailiffs-investigated-for-breaking-protesters-jaw-while-off-duty-13395738/

Crehan, K. (2016). *Gramsci's common sense: Inequality and its narratives*. Durham, NC: Duke University Press.

Department for Education & Lord Nash. (2014, November 27). *Guidance on promoting British values in schools published*. www.gov.uk/government/news/guidance-on-promoting-british-values-in-schools-published

Devlin, A. (2017, October 10). Shocking moment tea lady, 79, is forcibly removed by 12 cops from serving refreshments to anti-fracking protesters. *The Sun*. www.thesun.co.uk/news/4653092/shocking-moment-tea-lady-79-is-forcibly-removed-by-12-cops-for-serving-refreshments-to-anti-fracking-protesters/

Ferrarese, E. (2015). Habermas: Testing the political. *Thesis Eleven, 130*(1), 58–73.

Friends of the Earth. (2017). https://friendsoftheearth.uk/climate/fracking-facts

Gabbatiss, J. (2018). Destruction from HS2 'far worse' than previously thought as hundreds of homes set for demolition. *The Independent*. www.independent.co.uk/news/uk/home-news/hs2-railway-route-houses-demolish-woodland-destroy-environment-impact-london-birmingham-manchester-a8582316.html

Garnett, S. T., Burgess, N. D., Fa, J. E., et al. (2018). A spatial overview of the global importance of Indigenous lands for conservation. *Nature Sustainability, 1*, 369–374. https://doi.org/10.1038/s41893-018-0100-6

Gaventa, J. (2006). Finding the spaces for change: A power analysis. *IDS Bulletin, 37*(6), 23–33.

Gilmore, J., Jackson, W., & Monk, H. (2016). *Keep moving! Report on the policing of Barton Moss community protection camp, November 2013–April 2014*. York: Centre for the Study of Crime, Criminalization and Social Exclusion, LJMU, & Centre for URBan Research.

Gilmore, J., Jackson, W., Monk, H., & Short, D. (2019). *Protesters' experiences of policing at anti-fracking protests in England, 2016–2019: A national study*. London: NETPOL.

Global Diligence LLP. (2020). *Systemic violations of environmental protesters' right to peaceful assembly in the United Kingdom*. Report to Not1More. [Available upon request].

Global Witness. (2019). *Enemies of the state? How governments and business silence land and environmental defenders*. Washington, DC.

The Guardian. (2019, June 17). 'Domestic extremism' is no way to describe peaceful protest. www.theguardian.com/world/2019/jun/17/domestic-extremism-is-no-way-to-describe-peaceful-protest

HMIC. (2009). *Adapting to protest: Nurturing the British model of policing*. London: HM Inspectorate of Constabulary.

House of Commons Select Committee on High Speed Rail. (2018). *Promoter's select committee's second special report of session 2017–2019*. Department for Transport. https://assets.publishing.service.gov.uk/government/uploads/system/uploads/attachment_data/file/893698/hs2-phase-2a-promoters-response-select-committee-second-special-report.pdf

Jackson, W., Gilmore, J., & Monk, H. (2018). Policing unacceptable protest in England and Wales: A case study of the policing of anti-fracking protests. *Critical Social Policy, 39*(1), 23–43.

Jones, M., & Scott, R. (2018, September 9). Why are counter terrorism police still spying on the anti-fracking movement? *Spinwatch: Public Interest Investigations*. https://spinwatch.org/index.php/issues/climate/item/6006-why-are-counter-terrorism-police-still-spying-on-the-anti-fracking-movement

Kagan, S. (1998). *Normative ethics*. Boulder, CO: Westview Press.

Le Billon, P., & Lujala, P. (2020). Environmental and land defenders: Global patterns and determinants of repression. *Global Environmental Change, 65*. https://doi.org/10.1016/j.gloenvcha.2020.102163

NETPOL. (2018, May 24). Police maintain absolute secrecy over counter-terrorism surveillance on anti-fracking campaigners. *The Network for Police Monitoring*. https://netpol.org/2018/05/24/information-tribunal-prevent/

Oxford Probono Publico. (2020). *The law on policing peaceful protests*. Oxford. [Available upon request].

Patterson, S. (2018, January 13). Two anti-fracking protesters are arrested after a guard dog was poisoned by aniseed balls at a drilling site in Yorkshire. *Daily Mail*. www.dailymail.co.uk/news/article-5266355/Protesters-arrested-dog-poisoned-aniseed-balls.html

Pidd, H. (2018, July 30). Anti-fracking activists falsely accused of 'grooming' boy, 14. *The Guardian*. www.theguardian.com/world/2018/jul/30/anti-fracking-activists-falsely-accused-grooming-boy-14

Powercube. (n.d.). www.powercube.net/analyse-power/what-is-the-powercube/

Project Justice. (2020). Report to Not1More. [Available upon request].

Pring, J. (2018, December 20). Police force admits passing footage of disabled protesters to DWP. *Disability News Service*. www.disabilitynewsservice.com/police-force-admits-passing-footage-of-disabled-protesters-to-dwp/

Rowell, A. (2018, June 13). The curious case of the guard dog, anti-fracking protesters and North Yorkshire police. *openDemocracy*. www.opendemocracy.net/en/opendemocracyuk/curious-case-of-guard-dog-anti-fracking-protestors-and-north-yorkshire-police/

Schleicher, J., Peres, C. A., Amano, T., Llactayo, W., & Leader-Williams, N. (2017). Conservation performance of different conservation governance regimes in the Peruvian Amazon. *Scientific Reports*, 7(1), 1–10. https://doi.org/10.1038/s41598-017-10736-w

Shadwell, T. (2020, May 4). HS2 protester filmed with blood dripping from head 'after being assaulted by worker'. *Mirror*. www.mirror.co.uk/news/uk-news/hs2-protester-filmed-blood-dripping-21971339

Simpson, M., & Le Billon, P. (2021). Reconciling violence: Policing the politics of recognition. *Geoforum, 119*, 111-121. https://doi.org/10.1016/j.geoforum.2020.12.023

Thompson, D. (2014, March 20). Fracking protester who complained about GMP charged with assaulting an officer. *Manchester Evening News*. manchestereveningnews.co.uk/news/fracking-protester-who-complained-gmp-6856852

UN General Assembly. (2008, January 8). *Responsibility of states for internationally wrongful acts: Resolution/adopted by the general assembly*. Article 5 A/RES/62/61. www.refworld.org/docid/478f60c52.html

United Nations. (1945, October 24). Charter of the United Nations. 1 UNTS XVI. www.refworld.org/docid/3ae6b3930.html

Part 3

'Green' projects

16 Resist or comply?

Experiences of violence around dams in Cambodia

Sarah Milne

The Cardamom Mountains region of Southwest Cambodia has witnessed dramatic and ongoing environmental struggles over the last decade. Conflict has especially emerged around the planning and construction of hydropower dams, which have generated high profile but distinct dynamics of resistance and violence. In this chapter, I explore and compare two prominent examples of these dynamics: first, the assassination of environmental activist Mr. Chut Wutty in 2012, who was trying to expose illegal forest exploitation around the Atay dam; and second, the rise and violent dispersal of the Areng Valley anti-dam campaign, which involved intimidation of campaigners, along with a success story of sorts. Notably, public resistance to the Areng dam appears to have played a role in the Cambodian government's suspension of plans to build the dam in 2017, but it was not a factor in the Atay dam, which became operational in 2014.

The dramatic examples of violent suppression of environmental activism and community resistance may be thought of as "punctuated moments" (Ahmann 2018) or flash points in what has been a long period of struggle in the Cardamom Mountains. This ongoing drama has involved different forms of violence, mainly in relation to the enclosure and appropriation of forest and land resources by Cambodian elites. This violence has played out across scales and between multiple actors, including: government officials at all levels, local villagers, Indigenous people, foreign activists, Cambodian urban activists, foreign non-government organisations (NGOs), local NGOs, youth networks, religious organisations, political parties, Cambodian tycoons, Chinese state-owned enterprises (SOEs), foreign researchers, the Cambodian diaspora, carbon market proponents, eco-tourists, illegal logging gangs . . . the list goes on. This is a complex and multi-sited struggle, involving competing interests and perspectives: in many ways it is a classic environmental conflict of the Global South (see Temper et al. 2015).

At the heart of the struggle is a core domain of contention: forested land. The forests of the Cardamom Mountains are valued by many, but for different reasons. For example, they are the homeland and customary territory for local Indigenous people (Martin 1997), the target of global biodiversity interventions since 2000 (Milne 2009), the source of vital "ecosystem services" for "sustainable" hydropower (Killeen 2012), and now a source of illicit finance for elites who plunder the forest (Milne 2015). Not surprisingly, these differing claims, agendas, and perspectives have fuelled many of the environmental conflicts that I observe in this chapter. These conflicts exhibit violence that is material, symbolic, and structural: what emerges is a complex web of power relations and contests over forested land, akin to what Peluso and Watts describe as a "violent environment" (2001).

I begin by characterising the Cardamom Mountains landscape and its hydropower developments. I then compare the two sites where conflict and violence occurred: (i) the Atay dam, where there was no overt local opposition, but environmental campaigners from outside the area *did* contest illegal logging and other illicit activity that emerged around the dam; and (ii) the proposed Areng dam, where a remarkable campaign mobilised local resistance, transnational advocacy networks, and a newly politicised urban youth to contest the dam. In comparing these sites, I explain why they experienced such divergent outcomes. I also explore the different forms of violence that emerged at each site. Here, I especially point to the consequences of resistance in the Areng Valley, which saw campaigners being subject to deportation, social isolation, intimidation, and arrest from 2013 onwards. While conflicts over the dam have now eased, tensions remain as other environmental threats emerge.

Ultimately, I argue that the combination of uncertainty and fear that prevails among villagers in the Cardamom Mountains is a form of slow violence. This resonates with observations of similar land and resource struggles in contemporary Cambodia (e.g. Beban et al. 2017; Grant & Le Billon 2020; Schoenberger & Beban 2018), which together reflect an increasingly constrained and authoritarian political landscape, in which villagers face tyrannical choices between the consequences of resistance and the consequences of compliance.

Overview of hydropower dams and green enclosures in the Cardamom Mountains

The Cardamom Mountains in Southwest Cambodia are a vast forested landscape, inhabited by a relatively small population of Indigenous and Khmer people. Since 2000, the area has been targeted by conservation groups due to the presence of globally significant biodiversity, like the endangered Asian elephant (*Elephas maximus*) and the critically endangered Siamese crocodile (*Crocodylus siamensis*), among other species (see Appleton et al. 2000). Conservation efforts have mostly involved partnerships between international NGOs and the Cambodian government. Early support came from the Global Environment Facility in 2001, which saw Flora and Fauna International (FFI) partnered with the Ministry of Environment (MoE), and Conservation International (CI) partnered with the Ministry of Agriculture, Forestry, and Fisheries to protect three contiguous forest areas of about 1 million hectares (Milne 2009; Paley 2015). Together, FFI and CI pursued rather conventional protected area management and community engagement activities.

Now, nearly 20 years on, the conservation landscape reflects contemporary currents of green neoliberalism and state territorialisation. While the NGOs CI and FFI remain engaged, the most prominent player at this stage is Wildlife Alliance – not a mainstream global conservation NGO, but a smaller organisation driven by the personal interest and commitment of an American environmental philanthropist.[1] One of the most recent initiatives led by Wildlife Alliance is Cambodia's largest forest carbon (REDD+) project, covering nearly 500,000 hectares, across the southern part of the mountain range. This project, implemented with the company Wildlife Works, will produce carbon credits in two years' time through "reduced emissions from deforestation and forest degradation" or REDD+.[2]

The extent of green enclosures across the mountains has increased since 2000, with new protected areas and biodiversity conservation corridors having been declared. Jurisdictional reforms in 2016 confirmed this, with the composition of protected areas now including: Central Cardamom Mountains National Park (NP) (401,313 ha.), Southern Cardamom Mountains NP (410,392 ha.), Phnom Aural Wildlife Sanctuary (WS) (253,750 ha.), Phnom Samkos WS (333,750 ha.), Tatai WS (144,275 ha.), and the Peam Krasop Wildlife Sanctuary (4,976 ha.), among others (see RGC 2017, Figure 16.1).

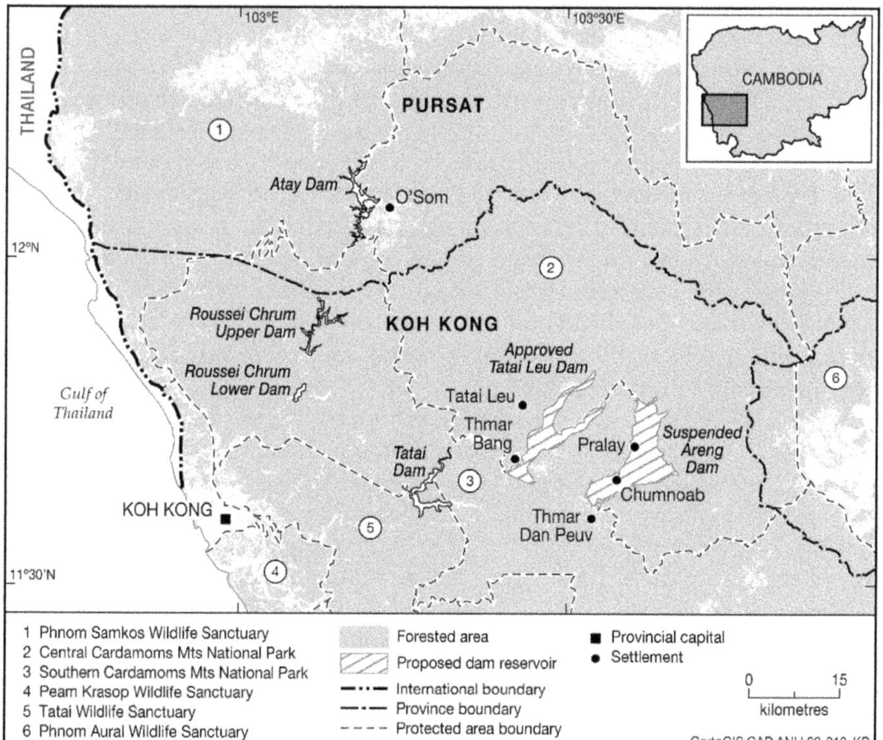

Figure 16.1 Dams and protected areas in the Cardamom Mountains[3]

Together this means that over *1.5 million hectares* have been enclosed for conservation purposes by the Cambodian government in the Cardamom Mountains alone, now to be managed under the MoE's Protected Area Law (RGC 2017). The landscape therefore represents a significant frontier for state-making and state control over natural resources, ostensibly for conservation and "green development" (Käkönen & Thuon 2019; Milne 2015).

Within the Cardamom Mountains conservation areas, a hydropower complex has also evolved since 2007, involving the construction of four dams, with another suspended in 2017, and another approved in 2020 (see Figure 16.1).[4] These dams are built by Chinese SOEs through Build Operate Transfer contracts, which effectively see arms of the Chinese government owning and selling electricity from the dams for the first 30–35 years of operation, before a handover to the Cambodian government towards the end of the dam's life (Sullivan 2015; Killeen 2012). Arrangements like these introduce a unique geopolitical dimension to the hydro-enclosures in the Cardamom Mountains, especially now that some of the investments are being publicly linked to China's ambitious Belt and Road Initiative (BRI).[5]

More problematic, however, is that the dams have given rise to resource speculation and illicit extraction, driven mainly by Cambodian elites and the ruling party. In the Cardamom Mountains, strong state control over forested areas, nominally for conservation purposes, produced a territorial monopoly ripe for exploitation (Milne 2015). Here, as dam construction got underway, the Cambodian government awarded contracts to local companies for the clearing and removal of trees from future dam reservoirs.[6] These contracts went to select tycoons and family members of Prime Minister Hun Sen, who then exploited the circumstances to conduct extensive illegal logging in the surrounding conservation areas. This logging focused upon high value luxury timber, mainly rosewood (*Dalbergia sp.*), which was laundered through the dam reservoir sites for later export to Vietnam and China. From 2009 to 2012, it is estimated that over US$220 million of rosewood was extracted from around the Atay dam, with a similar amount removed from the southern section of the range, around the Tatai dam (ibid.).[7]

The green-hydro complex in the Cardamom Mountains therefore led to the establishment of a regime of illicit extraction, which has been extremely lucrative for Cambodia's ruling elite – indeed, illegal logging rents have played a key role in contemporary state-making, building upon past post-conflict patterns (Le Billon 2000; Milne et al. 2015). We should therefore see the green enclosures and dams of the Cardamoms as linked interventions, led by an extractive authoritarian state.

International carbon markets have also entered the unusual "frontier constellation" that is found around the hydropower dams of the Cardamom Mountains (Käkönen & Thuon 2019). Here, the Kyoto Protocol's Clean Development Mechanism has been used at all of the dams to generate carbon credits – on the Atay dam, the Tatai dam, and the two Roussey Chrum dams. This mechanism introduces a layer of transnational technical governance that lends a veneer of legitimacy to the dams, subtly disguising or distracting from the dams' devastating environmental and social impacts. In the application of similar carbon market technologies elsewhere in Cambodia, this phenomenon has been characterised as "bureaucratic violence" (Milne & Mahanty 2019) – a term that describes the technocratic anti-politics of green economy processes in contested forested landscapes.

In light of these circumstances, it is tempting to portray the Cardamom Mountains as a political landscape, where the interests of state power, elite extraction, and transnational green governance combine. Yet, this perspective would not duly recognise that the forests of this region are inhabited and used by Indigenous villagers and others, making experiences *on the ground* highly contested and vitally important for the pursuit of social and environmental justice. To shed light on this, I now examine local perspectives on and responses to the Atay dam and the proposed Areng dam.

The Atay dam

In 2007, rumours circulated among NGO and government actors about a new dam that would be built in the Atay river. There was no public disclosure or consultation about the dam, to be located in O'Som commune, nestled between the Central Cardamoms NP and Phnom Samkos WS. Orders and approvals followed from the highest levels of government, and construction began in 2009. The China Datang Corporation was contracted to build the US$225 million 120-MW plant (Hydro Review 2012), which was officially launched in early 2014 by Hun Sen. Notably, the Datang Corporation is a state-owned power generator: a $25 billion dollar Fortune Global 500 company, whose president is keenly involved in the Belt and Road Initiative (Phnom Penh Post 2018). In promotional material, Datang's early work in Cambodia is celebrated:

> When Datang set out to build a hydropower dam and power grid in southwest Cambodia's tropical jungles, the challenges the engineers faced included not only the road-less hilly terrain and the debilitating weather, but also the wild beasts and gnawing boredom from weeks of isolation deep inside the forests.
>
> (ibid.)

This material depicts the jungle as a wild, uninhabited frontier, to be tamed by Chinese modernity. But Indigenous villagers of Chong ethnicity have lived in the area for centuries: displaced by the civil war, they returned to their homelands in the late 1990s, near to the famous Veal Veng wetland that is connected to the Krau river, a tributary of the Atay (Daltry et al. 2005). There are four old villages in O'Som commune, and at the time of the dam construction there was a total official population of about 1,000 people (Daltry et al. 2005; Seak et al. 2013).[8] Before the dam became operational, over 90% of O'Som residents were reported to rely upon forest products for their livelihood, alongside subsistence farming activities (ibid.). This forest dependence is reflected in an early community land use plan of the area which indicated a vast customary forest territory, including spirit forests and cardamom agro-forests, exceeding 13,000 hectares.[9]

But the dam soon unleashed new market forces and a wave of extraction driven by outsiders that local residents were poorly equipped to deal with. Indeed, the dam project was something that Indigenous villagers witnessed from the side-lines: the road was upgraded and Chinese workers and equipment arrived, while all construction activity was contained in a separate compound about a 20-minute motorbike ride from O'Som village. Tellingly, signage about the dam was in Chinese, not Khmer (Milne 2015). It was as if the villagers were not there. To be made invisible in this way is a form of dispossession and an act of "symbolic violence" (after Bourdieu 1979; see Grant & Le Billon 2019).

The response of the mainstream conservation NGOs engaged in O'Som was also symbolically violent, given that they did not voice concerns or advocate for local villagers' rights. The NGOs in this case (mainly CI and FFI) were careful to maintain their government relationships, so they did not question the dam or the ruling party's intentions. Rather, they saw the dam as an opportunity for "green development", or a way to harness the market value of the surrounding forest for its "watershed services" (Killeen 2012). This reasoning, for example, prompted FFI to launch a major European Union-funded Payments for Environmental Services (PES) scheme in the area, which aimed to levy a fee from the Datang Company to pay for forest conservation services provided by local villagers (Milne & Chervier 2014). But the lack of formal property rights for local villagers was a key stumbling block, which eventually meant that FFI's nascent PES scheme never came to fruition.

Meanwhile, dam construction proceeded apace, and the dam's "side effects" began to consume the surrounding forest and land. The most conspicuous problem was that of rosewood extraction, co-ordinated by the now infamous logging tycoon Try Pheap: this ran at fever pitch from 2009 to 2012, with many villagers, local officials, and park rangers sucked into the machinery of extraction (Milne 2015). Openly questioning the logging was not an option in O'Som, as dissenters were rapidly subject to persuasion from authorities, non-stop phone calls from middlemen, offers of bribes from those involved, and ultimately intimidation of those who were non-compliant.[10] It was a logging racket underpinned by violence and backed by state power, including armed forces and military police. Alongside the logging, the new road and soon-to-be-filled reservoir attracted land speculators from the lowlands. The land market arrived in full force, with significant land grabbing conducted by district and provincial officials, among others.

This atmosphere of violence, as well as the powerful connections of those who exploited the forests around Atay, are key reasons why there was no overt or organised local resistance. Other key reasons for the lack of mobilisation were: (i) this being the first of the Cardamom Mountains dams, and one of the earliest in Cambodia, many villagers and potential advocates were unprepared for the impending impacts, especially the logging; and (ii) while the Atay reservoir was going to submerge some villagers' farmland and shift agriculture zones, it was not going to submerge houses, so villagers were never considered for relocation or compensation by the authorities – they were simply ignored.

The conservation NGOs in the area, on the other hand, did engage local villagers – but they did so in an apolitical way, discouraging local resistance and anti-government advocacy. For example, with land speculation and logging gathering pace in O'Som over 2009–2010, during the dam construction phase, CI engaged villagers in a PES-like "conservation agreement". The agreement was aimed at protecting Siamese crocodiles and a small remaining patch of "cardamom forest" – a mere 600 hectares, compared with villagers' original claims for customary land of over 13,000 hectares. These highly bounded conservation goals seemed only to compound the sense of community demoralisation in the face of dispossession. Some villagers even asked: why should we collaborate with a conservation NGO whose park rangers are profiting from the rosewood?

These were the questions and circumstances that lured Chut Wutty back into the Cardamom Mountains. As a former solider and CI staff member, he knew the area well – he also knew the powerful individuals in government who were orchestrating the rosewood extraction in the area. Intent on exposing the corruption, Wutty had already made one trip with journalists to the Southern Cardamoms (see Boyle 2011). On the day of his assassination, Wutty was on another such trip, near to the Atay dam. For his efforts to give voice to the forest and its customary owners, Wutty paid the ultimate price. His murder sent shockwaves through Cambodia and the international community, finally putting O'Som on the map, albeit momentarily.

Justice for Wutty will probably never come under the current regime. Nor will villagers in O'Som ever regain their traditional way of life. This is a key aspect of the violence that prevails around the Atay dam, which is made even more painful by the silence of the conservation NGOs operating in the area, who witnessed these events unfolding. Indeed, in response to the Atay dam, foreign conservationists in O'Som chose to focus on the relocation of the globally significant population of Siamese crocodiles from the site when construction commenced in 2010 (e.g. see Newshub 2011), as if to reinforce that their priorities were biological rather than social or political.[11]

Finally, apart from these acts of symbolic and political violence, there was also physical violence built into and emerging from the dam itself – we might consider this infrastructural violence (see Li 2018). For example, in 2012, there was a partial collapse of the dam wall during construction, which caused several Chinese workers to be swept away, at least four of whom drowned (Hydro Review 2012). Furthermore, with the dam now complete, the Atay river has effectively died. I felt this in 2015, when I travelled downstream from the dam on the newly made Chinese road, looking for the site of Wutty's murder. Along the way, I stopped to visit a waterfall that I had once enjoyed in 2002 – back then, it was a torrent of crystal-clear water, teeming with life. This time, I witnessed a dry and perilous cliff. Looking down from the top, I saw only a stagnant and silent pond below. The violence was palpable.

The Areng dam

In contrast to experiences in O'Som, the story of the Areng dam is in many ways a tale of Indigenous people mobilising to protect their homelands. If built, the reservoir would displace over 1500 people, submerging 20,000 hectares of traditional villages, farmland, spirit forest, and village forest (CI 2007). But a unique combination of factors in and around the Areng Valley produced an unexpected outcome: an anti-dam resistance campaign of major significance for Cambodia.

There are four key factors that produced the resistance campaign. First, villagers in the Areng Valley saw or knew about what was happening in O'Som and at other "development" sites in Koh Kong,[12] so they had good reason to resist another dam and time to prepare. Second, there was more at stake in the Areng Valley, in terms of the dispossession of villagers from their residential, farming, and customary forest lands, and this fuelled local resistance. Third, a highly committed and charismatic foreign activist, Mr. Alex Gonzalez-Davidson, was based in the valley at the time of the dam's proposal. He was able to mobilise local people to resist and speak out, in a way that they otherwise would not have dared (Milne 2017; Rose-Jensen 2017). Fourth, and finally, was political timing: the Areng dam campaign ramped up during the 2013 Cambodian election and subsequent stand-off, when Hun Sen had to face off an existential challenge from the Cambodian National Rescue Party (CNRP).

As a product of this unusual confluence of factors, the Areng campaign was multi-sited and messy. It was not a consensus-based grassroots campaign that united all villagers against the dam. Rather, it relied on a small group of about 20–30 families in the valley who were prepared to stand in the firing line, alongside Alex and his Khmer collaborators. Importantly, Alex's collaborators – with whom he eventually went on to form the local NGO Mother Nature Cambodia – were highly motived volunteers, mainly from Phnom Penh. None of them planned to be activists necessarily, but through their passion for the area and its natural beauty, Alex and his collaborators effectively translated the struggle for the Areng Valley into an issue of relevance for young, urban Khmers, using Facebook and YouTube. This catalysed remarkable national and transnational solidarity against the dam, producing Cambodia's first domestic environmental movement.[13]

Not surprisingly, these developments soon attracted party political interests and dynamics. For example, the Areng campaign in its early stages was supported by the emerging CNRP, after the drama of the 2013 election. The ruling Cambodian People's Party then responded accordingly, with its usual combination of gossip, slander, and threats, all designed to intimidate and infect the campaign. Tensions soon escalated and resulted in a curious series of events which show Hun Sen's assertion of authority, on the one hand, and his apparent acquiescence to the people's campaign, on the other. What ensued, from 2014 to 2015, was as follows: the local campaigners' road-block, which aimed to stop Chinese trucks from entering the Areng Valley, was forcibly removed by the Cambodian military. Hun Sen then had Alex arrested and deported from Cambodia in dramatic fashion (see Tat et al. 2015). Subsequently, one of the resistance leaders, Mr. Ven Vorn, an Indigenous man from the Areng Valley, was arrested on spurious charges and sent to the provincial prison of Koh Kong for several months. With the campaign largely dismantled, Hun Sen then made a speech that suspended the dam on condition that local people "stopped talking about it" (Associated Press 2015). By 2017, plans for the dam were "suspended indefinitely", as new investments in coal-fired electricity were announced (Khuon & Zsombor 2017).[14]

These events have ultimately dampened and deferred the struggle against the Areng dam. But the ruling party has not let up: it has maintained concerted efforts to control the area, including its people and resources. For example, rumours have been used to intimidate people and to exclude so-called anti-dam resisters from political and economic life. For a remote Indigenous community, this has had profound social effects, including divisions within villages and families (Milne 2017). Such are the consequences of resistance in Cambodia: a suite of subtle and insidious dynamics, which together amount to a form of political violence. The Cambodian form of this violence relies on fear, the activation of past traumas, and the mobilisation of powerful social norms that expect villagers to respect local authorities as though they were uncles or grandfathers.[15]

Today, villagers are trying to move on with their lives, as more immediate threats emerge. Chinese-backed roads and transmission lines are now planned for the area, and land alienation driven by powerful outside interests is accelerating.[16] Furthermore, illegal timber extraction continues with the backing of armed forces, which has become a new focus for some of the original anti-dam activists (IWGIA 2019). As a result, the dynamics of violence in Areng have continued, even after the dam's apparent cancellation. Together, these circumstances add weight to the argument that the Areng dam proposal was never really about the dam. Some observers believe that the dam was never viable – a notion supported by the number of times the project changed hands, between different Chinese companies.[17] This suggests that the dam proposal was really a front for other extractive agendas. Indeed, the Areng dam might now be considered a "Damocles project" (see Kirchherr et al. 2018), or a threat that always hangs over local residents, like a form of "slow violence" (Ahmann 2018).

Conclusion

Like peeling an onion, there are layers of violence present in the contemporary landscape of the Cardamom Mountains. There is the violence of Cambodia's recent history and civil war, never far from people's memories (Zucker 2013); the destruction of the forest through new dams and roads, and ongoing appropriation of resources (Käkönen & Thuon 2019; Milne 2015); physical and symbolic violence against those who resist the incumbent regime, akin to that experienced elsewhere in Cambodia (Grant & Le Billon 2019); and structural violence exerted through Cambodian law, which alienates customary farmers and Indigenous people (Loughlin & Milne 2020).

In this violent environment, the two dam sites that I have discussed – Atay and Areng – illustrate both key dynamics and divergent outcomes of resistance in Cambodia. For example, local experiences of resistance are evidently shaped by unique contextual and site-specific factors, such as geography, resources, individuals, and community dynamics. Yet, common patterns of resistance and violence also appear. In both of the cases examined here, resistance was motivated by underlying struggles for control over forested land; the resistance was not necessarily local or grassroots, but national and transnational in scope. Furthermore, the modalities of violence in each case were similar, given that they were generated by the same state apparatus, adept at mobilising fear and intimidation.

Ultimately, the violence observed here underpins Cambodia's "unofficial regime of extraction" (Milne 2015). Core aspects of this involve the manipulation of green enclosures by authorities and the conduct of state-backed illegal logging. This system of extraction, known in Khmer as "making cakes without flour" (*twer num ot masao*), has been especially evident in the Cardamom Mountains. Yet, this tale is unique because it has occurred under the watch of a few big international conservation NGOs, which partner with the Cambodian government. These NGOs aim to save forests and biodiversity, but really, they act like a green icing on the government's cake: their apolitical presence subtly papers over the terrible effects of "green" hydropower for local Indigenous people and the environment. This signals a form of institutional hypocrisy and dissonance that must also be considered as violence.

Notes

1 The founder and CEO of Wildlife Alliance is Suwanna Gauntlet. She has worked in Cambodia since 2000.
2 See Wildlife Alliance's homepage: www.wildlifealliance.org/redd/ and Renard et al. (2020).
3 The location of the recently approved Tatai Leu dam is approximate, based upon the description provided here: "Investment in new 150-megawatt dam requested upstream on Tatai River" in *Construction & Property Magazine*, June 26th 2020, Phnom Penh. www.construction-property.com/a-new-150-megawatth-power-dam-project-requested-for-investment-on-upstream-tatairiver/.
4 The four constructed dams are Roussey Chrum Leu, Roussey Chrum Krom, Tatai and Atay. The suspended dam is Areng, and the recently approved dam is Tatai Leu (see Ry 2020 on this most recent dam).
5 For example, a US$200 million road upgrade to connect O'Som to Koh Kong and Pursat was announced in 2019 under the BRI (Narin & Soeung 2019). The 2015 Stung

Tatai dam is also tagged as a BRI investment by some, like the Reconnecting Asia database by the Centre for Strategic International Studies.

6 These contracts were meant to be for the removal of all vegetation from reservoir sites, including tree roots, but the contractors largely ignored this to focus on luxury timber removal from the wider area instead.

7 Rosewood extraction in Cambodia is strongly associated with elite power and state-making processes (Milne 2015), as in other countries like Madagascar (Remy 2017).

8 Daltry et al. (2005) state a population of 900 in 2004; Seak et al. (2013) report a population of 1059 in 2013.

9 J. Ironside, personal communication. Ironside facilitated the participatory land use planning in the area.

10 Source: Milne field-trips in 2010 and 2015, including key informant interviews in Phnom Penh in 2011.

11 Other international NGOs, like Global Witness, *did* condemn the murder of Wutty – a stance made possible by the fact that Global Witness is not trying to maintain a partnership with the Cambodian government.

12 For example, the Chinese SOE Union Development Group was involved in a controversial land grab in coastal Koh Kong at this time, affecting over 1000 villagers.

13 The Khmer diaspora was a key element of this; e.g. see short film by Khmer-American Kalyanee Mam for the *New York Times*: www.nytimes.com/2014/07/29/opinion/a-threat-to-cambodias-sacred-forests.html.

14 The Areng dam remains "suspended" in 2020; however, the nearby Tatai Leu dam was approved in late 2020. Secrecy shrouds this new dam, which will certainly impact Indigenous communities in Tatai Leu commune.

15 For example, local authorities refer to villagers as their "children and grandchildren" (*goan jao*).

16 See Khuon and Zsombor (2017) on the new transmission line. Regarding land alienation and roads in the valley, my source is a Cambodian research assistant who visited the valley in late 2019 to gather data.

17 Three different Chinese state-owned enterprises "held" the proposal for the Areng dam from 2006: China Southern Power Grid (2006); China Guodian (2011); and finally, Sinohydro (2012). See the Mother Nature Cambodia webpage (www.mothernature cambodia.org/saving-the-areng-valley.html).

References

Ahmann, C. (2018). It's exhausting to create an event out of nothing: Slow violence and the manipulation of time. *Current Anthropology*, *33*(1), 142–171.

Appleton, M., Bansok, R., & Daltry, J. (2000). *Biological survey of the Cardamom Mountains region, Southwest Cambodia*. Flora and Fauna International and Ministry of Environment. Cambridge and Phnom Penh.

Associated Press. (2015, February 24). Cambodia leader says work on mega-dam will not start until at least 2018. *The Guardian*. tinyurl.com/megadam2018

Beban, A., So, S., & Un, K. (2017). From force to legitimation: Rethinking land grabs in Cambodia. *Development and Change*, *48*(3), 590–612.

Bourdieu, P. (1979). Symbolic power. *Critique of Anthropology*, *4*(13–14), 77–85.

Boyle, D. (2011, December 21). Logging in the wild west. *The Phnom Penh Post*. www.phnompenhpost.com/national/logging-wild-west

CI. (2007). Summary of social and environmental impacts of the proposed Areng dam. *Conservation International*. www.rainforestinfo.org.au/cambodia/Areng%20dam%20-%20impact%20summary%20October%202007%20_2_.pdf

Daltry, J., Chheang, D., & Nhek, R. (2005, January). A pilot project to integrate crocodile conservation and livelihoods in Cambodia [Conference paper]. *17th Working Meeting of the IUCN/SSC Crocodile Specialist Group*. www.researchgate.net/publication/261109465

Grant, H., & Le Billon, P. (2019). Growing political: Violence, community forestry, and environmental defender subjectivity. *Society & Natural Resources*, *32*(7), 768–789.

Grant, H., & Le Billon, P. (2020). Unrooted responses: Addressing violence against environmental and land defenders. *Environment and Planning C*. https://doi.org/10.1177/2399654420941518

Hydro Review. (2012, March 12). Several missing after collapse at Cambodia's 120-MW Stung Atay hydropower plant. *Hydrovision International*. www.hydroreview.com/2012/12/03/several-missing-after-collapse-at-cambodias-120-mw-stung-atay-hydropower-plant/#gref

IWGIA. (2019). Indigenous peoples in Cambodia. *International Working Group for Indigenous Affairs (IWGIA)*. www.iwgia.org/en/cambodia/3420-iw2019-cambodia.html

Käkönen, M., & Thuon, T. (2019). Overlapping zones of exclusion: Carbon markets, corporate hydropower enclaves and timber extraction in Cambodia. *The Journal of Peasant Studies*, *46*(6), 1192–1218.

Khuon, N., & Zsombor, P. (2017, February 20). Government ditches hydropower dam for more coal power. *The Cambodia Daily*. https://english.cambodiadaily.com/news/government-ditches-hydropower-dam-for-more-coal-power-125402/

Killeen, T. (2012). *The Cardamom conundrum: Reconciling development and conservation in the Kingdom of Cambodia*. Singapore: NUS Press.

Kirchherr, J., Pomun, T., & Walton, M. J. (2018). Mapping the social impacts of "Damocles projects": The case of Thailand's (as yet Unbuilt) Kaeng Suea Ten Dam. *Journal of International Development*, *30*, 474–492.

Le Billon, P. (2000). The political ecology of transition in Cambodia 1989–1999: War, peace and forest exploitation. *Development and Change*, *31*, 785–805.

Li, T. (2018). After the land grab: Infrastructural violence and the "Mafia system" in Indonesia's oil palm plantation zones. *Geoforum*, *96*, 328–337.

Loughlin, N., & Milne, S. (2020). After the grab? Land control and regime survival in Cambodia since 2012. *Journal of Contemporary Asia*. doi:10.1080/00472336.2020.1740295

Martin, M. (1997). *Les Khmers Daeum, "Khmers de l'origine": Société montagnarde et exploitation de la forêt, de l'écologie à l'histoire*. Paris: Presses de l'Ecole française d'Extrême-Orient.

Milne, S. (2009). *Global ideas, local realities: The political ecology of payments for biodiversity conservation services in Cambodia*. PhD thesis. University of Cambridge, Cambridge, UK.

Milne, S. (2015). Cambodia's unofficial regime of extraction: Illicit logging in the shadow of transnational governance and investment. *Critical Asian Studies*, *47*(2), 200–228.

Milne, S. (2017). On the perils of resistance: Local politics and environmental struggle in Cambodia. *International Institute of Asian Studies: The Newsletter*, *78*(Autumn), 32–33.

Milne, S., & Chervier, C. (2014). A review of Payments for Environmental Services (PES) experiences in Cambodia. Working paper 154. Bogor: Centre for International Forestry Research.

Milne, S., & Mahanty, S. (2019). Value and bureaucratic violence in the green economy. *Geoforum*, *98*, 133–143.

Milne, S., Pak, K., & Sullivan, M. (2015). Shackled to nature? The post-conflict state and its symbiotic relationship with natural resources. In S. Milne & S. Manhanty (Eds.), *Conservation and development in Cambodia: Exploring frontiers of change in nature, state and society* (pp. 28–50). Abingdon, UK: Routledge.

Narin, S., & Soeung, S. (2019, April 23). Ahead of second BRI forum, Cambodia sees benefit of massive Chinese project. *Voice of America*. www.voacambodia.com/a/ahead-of-second-bri-forum-cambodia-sees-benefit-of-massive-chinese-project/4887886.html

Newshub. (2011, April 14). Crocodiles rescued from Cambodia dam destruction. *Newshub Archive*. www.newshub.co.nz/environmentsci/crocodiles-rescued-from-cambodia-dam-destruction-2011041414

Paley, R. (2015). Managing protected areas in Cambodia: The challenge for conservation bureaucracies in a hostile governance environment. In S. Milne & S. Mahanty (Eds.), *Conservation and development in Cambodia* (pp. 159–177). Abingdon, UK: Routledge.

Peluso, N., & Watts, M. (Eds.). (2001). *Violent environments*. Ithaca, NY: Cornell University Press.

Phnom Penh Post. (2018, December 20). Kingdom wants more China energy investment. Originally run in the *China Daily*. www.phnompenhpost.com/business/kingdom-wants-more-china-energy-investment

Remy, O. (2017). Rosewood democracy. In A. Williams & P. Le Billon (Eds.), *Corruption, natural resources, and development* (pp. 142–153). Cheltenham: Edward Elgar.

Renard, Q., Nhem, S., Leng, C., Silvermann, J., & Lee, D. (2020). Cambodia: Building a nested system to protect remaining forests. *Ecosystem Marketplace: Forest Trends*. www.ecosystemmarketplace.com/articles/cambodia-embarks-on-building-a-nested-system-to-protect-remaining-forests/

RGC. (2017). *National protected area strategic management plan*. Royal Government of Cambodia Ministry of Environment. https://redd.unfccc.int/uploads/54_2_cambodia_nat_protected_area_strategic_plan_eng_27_jul_2017.pdf

Rose-Jensen, S. (2017). "Everything we do is democracy": Women and youth in land rights social mobilization in Cambodia. *Journal of Mason Graduate Research*, *5*(1), 1–16.

Ry, S. (2020, October 26). Tatai Loeu river hydropower dam project approved *Khmer Times*. www.khmertimeskh.com/50776651/tatai-loeu-river-hydropower-dam-project-approved/

Schoenberger, L., & Beban, A. (2018). "They turn us into criminals": Embodiments of fear in Cambodian land grabbing. *Annals of the Association of American Geographers*, *108*(5), 1338–1353.

Seak, S., Phat, C., & San, V. (2013). Importance of forest ecosystem services for the livelihood activities of a rural community in O'Som commune, Pursat province, Cambodia. *Natural resource governance in Cambodia*. Research Papers Vol. 1. pp. 5–22. Phnom Penh: Department of Natural Resource Management and Development, Royal University of Phnom Penh.

Sullivan, M. (2015). Contested development and environment: Chinese-backed hydropower and infrastructure projects in Cambodia. In S. Milne & S. Mahanty (Eds.), *Conservation and development in Cambodia* (pp. 120–138). Abingdon, UK: Routledge.

Tat, O., May, T., & Pye, D. (2015, February 23). Defiant activist deported. *The Phnom Penh Post*. www.phnompenhpost.com/defiant-activist-deported

Temper, L., del Bene, D., & Martinez-Alier, J. (2015). Mapping the frontiers and front lines of global environmental justice: The EJAtlas. *Journal of Political Ecology*, *22*(1), 255–278.

Zucker, E. (2013). *Forests of struggle: Moralities of remembrance in upland Cambodia*. Honolulu: University of Hawaii Press.

17 Pacifying autonomous land defenders in Oaxaca, Mexico

Human rights groups as social warfare mechanisms

Alexander Dunlap and Martín Correa Arce

The terms 'wind park', 'wind farm' and 'utility-scale wind energy' do not capture the existent reality behind wind energy development. We might even say that these words subtly preform public relations by referring to parks, farms and public amenities to describe a type of power plant or private factory. Wind energy development, as a method of 'green extractivism', actually spreads *wind factories*, or *wind factory zones*, to capture the vital force of winds. Extraction is understood as pulling, drawing out and harnessing (usually with special effort, skill or force) various minerals, hydrocarbons or vital energy from wind, sun, hydrological and so-called human resources (Dunlap & Jakobsen 2020). While windmill technology emerges from ancient civilizations (Pasqualetti et al. 2004), industrial wind turbines are a relatively new type of power plant assemblage regarded as a 'clean', 'green' and 'renewable' energy source. It is often popular in environmental policy – especially in terms of combating the ecological crisis, and by extension climate change – although this prestige is questionable at best, and completely unjustified at worst.

Agrarian change, conflict and related real or anticipated negative socio-ecological impacts from wind factories are increasing (Avila 2018; Backhouse & Lehmann 2019; Franquesa 2018; Lawrence 2014; Siamanta 2019; Zografos & Martínez-Alier 2009). Revisiting the Isthmus of Tehuantepec region, known locally as the Istmo (see Figure 17.1), this chapter examines the struggle against the new Electricité de France (EDF) wind factory called Gunaa Sicarú (meaning 'Beautiful Woman' in Zapotec), situated in Unión Hidalgo, Oaxaca, Mexico. The Isthmus region is recognized by the International Finance Corporation (IFC 2014: 1) as "home to some of the best wind resources on earth", and by the mid-2000s the region experienced a 'wind rush' – rapid and intensive wind factory development – that lasted until recently (Dunlap 2019a).[1] The EDF wind factory comprises 96 wind turbines that can generate 252 megawatts (MW) and is located on the communal land between Unión Hidalgo and La Ventosa (see Figure 17.1). This megaproject became official on 29 June 2017, when President Enrique Peña Nieto approved the factory without consultation through the Energy Regulatory Commission (CRE), making it the 29th wind factory in the region (Manzo 2019). This wind factory was immediately contested by Zapotec farmers (*campesinos*) and fishers (*Comuneros*).

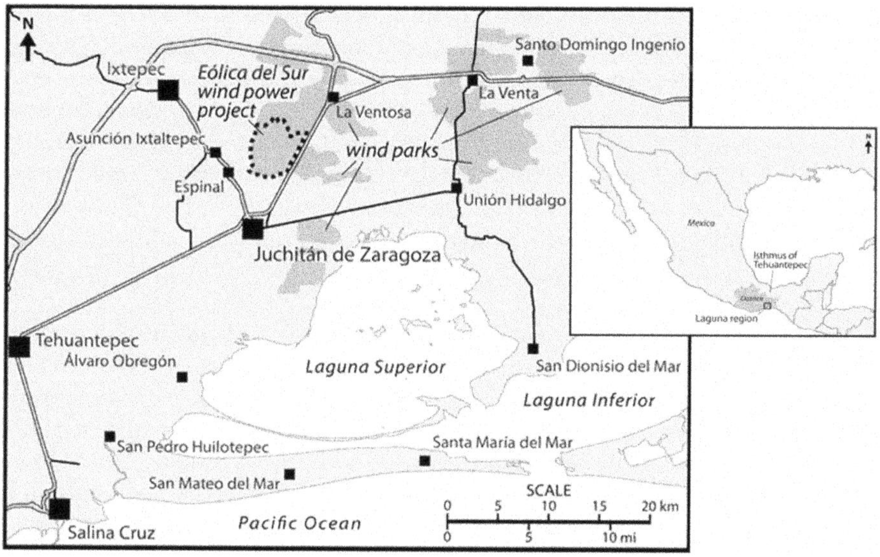

Figure 17.1 Map of the Isthmus of Tehuantepec and Mexico, Oaxaca

Source: Carl Sack.

In the meantime, on 7 September 2017, the Isthmus region shook with an 8.2 magnitude earthquake – the second largest in Mexico's history – resulting in more than 37 deaths; about half of Juchitán's 14,000 buildings suffered severe structural damage (Dunlap 2020a). The earthquake (combined with poor urban planning and negligent relief efforts) caused serious psychological and emotional trauma, radical insecurity and displacement, provoking construction delays for the EDF wind factory.

While the construction of Gunaa Sicarú is incomplete, the threat of agrarian – and corresponding socio-ecological – change has spawned resistance in Unión Hidalgo. In August 2018, EDF renewed its efforts to build the wind factory. Drawing on the political ecology of counterinsurgency and applying social war theory, this chapter examines the political ecology of wind factory development. Responding to Marta Conde and Philippe Le Billon's (2017: 693) call for additional research into "the repression of resistance" by extraction companies and "the micro-politics and psychological dimension of conflict escalation", this investigation delves into the (colonial) micro-politics of wind factory *socio-ecological pacification* in Unión Hidalgo. Socio-ecological pacification indicates how both landscapes and people are simultaneously subjected to processes of domestication and control for extractive development. The Isthmus retains a history of struggle for regional autonomy by the Zapotec, Ikoot, Zoques and other native nations (see Dunlap 2019a; Lucio 2016: Manzo 2011), which perform cultural revitalization projects and territorial land defense against extractive megaprojects. Building on previous research in the Isthmus (Altamirano-Jiménez 2017; Dunlap 2017, 2019a; Howe 2014; Lucio 2016; Nahmad et al. 2014; Oceransky 2011), specifically on political violence (Dunlap 2018a, 2019a), the chapter details the micro-political processes involved in the socio-ecological pacification required for wind factory development. While documenting a spectrum of coercive techniques, this chapter argues that (human rights) non-governmental organizations (NGOs) and 'activist' identities are an underestimated weapon of social warfare, playing on existing weaknesses of 'Leftist' (hierarchical) organization and disrupting processes of resistance by fragmenting, isolating and, ultimately, pacifying autonomous land defenders.

This research builds on long-term engagements with land defenders from the Isthmus region that began in December 2014 (see Dunlap 2019a). This complemented recent fieldwork conducted in December 2018 and from December 2019 to March 2020. Fieldwork resulted in 41 semi-structured recorded interviews with 35 different people, comprising roughly 26 hours of audio. The semi-structured interviews had a male to female 2:1 ratio and came from various towns throughout the Isthmus, Oaxaca and Mexico City. The people interviewed were land defenders (farmers, fishers, teachers, herbalists, lawyers, laborers, professional activists, etc.), human rights NGOs, electricity grid administrators and the regional director of EDF. There were also upwards of 30 informal interviews conducted in public transportation and social events. Interviews were triangulated with secondary research material: books, public relations material, blogs, newspapers and academic articles. Because of the level of political violence in this region, names of research participants were altered or omitted.

The chapter proceeds by discussing colonial theory in relation to social warfare and the philanthropy sector, preparing readers for a critical analysis of an NGO. Then a brief background to wind factory development in the Isthmus is laid out, before outlining the specific situation and dynamics in Unión Hidalgo. This leads to discussions about money, *Sicarios* (hitmen) and NGOs-*Comunero* relationships to understand processes of social-ecological pacification. These issues of money and, to a lesser degree, *Sicario* violence, are well known. Instead, space is prioritized to discuss the complicated NGO dynamic that formed with the *Comuneros* in Unión Hidalgo. The chapter concludes by reflecting on 'the problem of organization' and the construction of good/bad consultations in relation to colonial practices, supporting (slow) ecocidal–genocidal dynamics in the name of climate change mitigation.

Consuming territory: (neo)colonization and social warfare

Megaproject development in the Isthmus, as mentioned, has been described as 'neocolonization', 'internal colonization' and colonization generally. We contend that 'internal colonization' is legalistic and nation state-centric, as it demarcates artificial boundaries in a fluid and increasingly convoluted process of transnational capital accumulation. The line between 'internal' and 'external' actors becomes redundant, especially – but not limited to – regional collaboration that is further blurred by (trans)national investment, influence and regulation. Colonization or (neo)colonization[2] honor the continuity, organizational trajectory and ethos underpinning the statist project (Dunlap 2018c). The 'democratic' state propagates a liberal value system, centered on individualism, rationalism, egalitarianism and ameliorism/progressivism, that "is territorialized onto the nation-state" (Gilbert 2009: 199). Colonization in the Isthmus, and elsewhere, is, however, distinctly infrastructural, retaining various and simultaneous modalities of colonization/statism, adaptation and resistance (Mignolo 2005). Colonization entails variegated and progressive forms of ecocide and ethnocide (Short 2016), whereby 'infrastructural colonization' enacts a specific modality (Dunlap 2020c). Wind factory development, alongside other megaprojects, contributes to this trajectory by necessitating "various degrees of destruction and/or disciplinary transformation of plants, animals, water and people, altering existing land relationships, creating new prohibitions and denying the free and qualitative aspects of medicinal herbs" (Dunlap 2018b: 565). Megaprojects are territorial weapons that, to various intensities, dispossess populations and 'roll out' an apparatus of spatial, economic and psychosocial management. The Miguel Alemán and Cerro de Oro dams, among others in Southern Mexico (Manzo 2011), are examples of infrastructural warfare against Indigenous populations. Infrastructural colonization spreads social war predicated on the ideology of modernity, discourses of progress and the fabrication of desires/aspirations[3] of populations – near and far – to justify the socio-ecological repercussions. Social warfare recognizes that *invasion is infrastructural.*

The embrace, negotiation and rejection of infrastructural colonization is multifaceted (Borras et al. 2012; Hall et al. 2015). Land control and social war theory allow us to dissect the technologies of infrastructural colonization. Social war is a colonial theory (Gardenyes 2012; Trocchi 2011) emerging from the Roman Social War (91–89 BC). The Roman Republic learned the indispensability of political concessions and political–military 'hold' techniques for maintaining internal stability (Foucault & Ewald 2003; Gardenyes 2012; Trocchi 2011). Social war discourse, embodying Foucault and Ewald's (2003: 15) "politics as a continuation of war by other means", views the colonial/state apparatus and its politics (see Loadenthal 2017: 170), economy, divisions of labor and hierarchical orderings as a complex apparatus of subjugation (Anonymous 2001; Dunlap 2014, 2019b; Gardenyes 2012; Trocchi 2011;). Related to anarchist political ecology (see Brock 2020), social warfare dissects the 'hard' coercion and 'soft' social technologies of pacification emblematic of counterinsurgency (see Dunlap 2014, 2019b), while highlighting the social and/or psycho-geographical aspects as essential to colonial–statist intervention.

Social war is rooted in "a way of seeing the world" (Gardenyes 2012: 7), which depends on creating "an imagined geography that begins the regimentation of space" and influences political terrain as well as "the imaginations, desires, and possibilities of people" (Dunlap 2014: 57–8). Emphasizing affinity over identity, social war, as opposed to class war, recognizes "the enemy is not a class but a point of view, [a] subjectivity" that cuts across class, ethnicities, (non)genders[4] and implies specific ecological relationships. Social war, following Joseph Gardenyes (2012: 11), is "a struggle against the structures of power that colonize us and train us to view the world from the perspective of the needs of power itself" (see also Bonanno 1998). In this view, the root of colonization is "the logic of control in and of itself" (Gardenyes 2012: 7), of which megaprojects are a flagrant expression. Social war acknowledges a psycho-social war embedded in differential processes of primitive accumulation or accumulation by dispossession[5] to capture the 'hearts' and 'minds' of (traditionally native) populations to regiment them as 'citizens', 'proletariats', 'bourgeoisie' or, generally, claimed subjects of states and colonial empires. Liberalism emerges as not only a colonial export, but also a technology of social warfare and counterinsurgency.

The section 'Indirect Methods for Countering Insurgencies' of the *Insurgencies and Countering Insurgencies* Field Manual (FM 3–24 2014) advocates the 'generational engagement' approach. This approach is designed "to educate and empower the population to participate in legal methods of political discourse and dissent". It is applicable "in both high threat" areas and "situations where an insurgency is at its infancy and combat is less intense". Keeping in mind the 'preparation period' and the targeting of non-violent social movements (see Brock & Dunlap 2018; Dunlap 2014), the counterinsurgency field manual advises to "undertake this method as soon as possible to affect the next generation" of people by building "on a foundation of education, empowerment, and participation" (FM 3–24: 10–2). The generational engagement diagram (see Figure 17.2) is an official representation of the liberal model as a social control system, celebrating a *counterinsurgency colonial model* to stabilize territories. NGO structures frequently project and emerge from this liberal vision. Echoing development discourse (see Moe & Müller 2017), this social warfare diagram openly advocates pacification by political means, employing 'voting', 'education', 'town meetings', 'youth programs', 'empowerment', 'participation' and, elsewhere, 'sustainable development' as devices of pacification to integrate members of a target population. Said differently, (representative) democracy, civil society groups and education articulate a foundational method of social war, illuminating the strategies and tactics to engineer social pacification. NGOs emerge as an important mechanism promoting this model, which we later discuss in the context of Unión Hidalgo.

NGOs and social warfare

NGOs are mechanisms of social warfare and (neo)colonization. Emerging in the United States at the turn of the 20th century, the NGO culture emanates from philanthropic foundations (e.g. Carnegie, Ford and Rockefeller). Philanthropy was designed to establish capitalist ideological hegemony, subdue

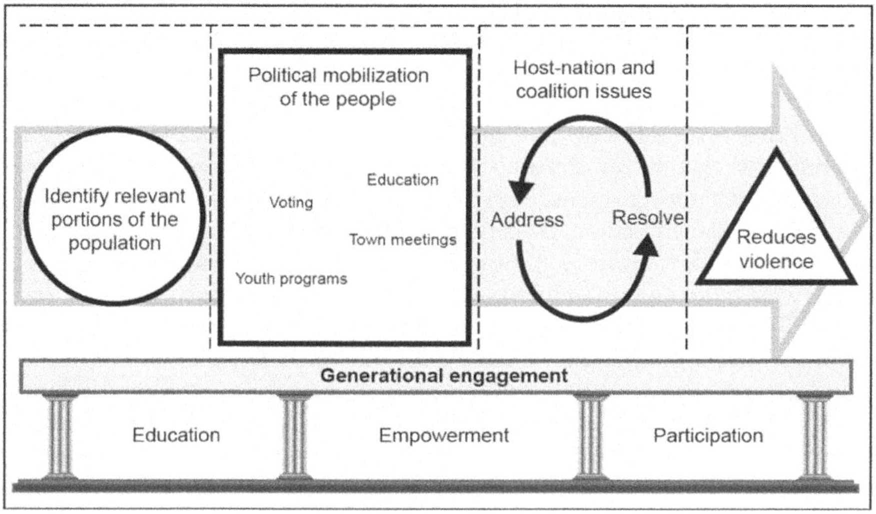

Figure 17.2 Generational engagement model

Source: FM 3–24.

class antagonisms and advance US foreign policy interests (Berman 1983; Esteva & Prakash 1998). This process has further morphed and diversified with the onslaught of neoliberalism. Structural adjustment programs intensified the spread of NGOs, which James Petras (1997: 11, 14) contends "became the 'community face' of neoliberalism" that created "a political world where the appearance of solidarity and social action cloaks a conservative conformity with the international and national structure of power" (see also Esteva & Prakash 1998). NGOs, Petras (1997: 12–15) continues, are "a new type of cultural and economic colonialism and dependency" that "co-opts the language of the left", promotes "non-confrontational politics", creates "competition between communities for scarce resources" and, overall, creates a system of "self-exploitation of the poor" (see also Petras & Veltmeyer 2001). Raúl Zibechi (2012: 274) agrees, arguing that NGOs "went from playing an oppositional role to collaborating with the state, specializing in consultation, mediating in social processes, and managing or promoting people's local participation, but without questioning the macroeconomic politics".

Generational engagement and 'local turn' in counterinsurgency perform the same task as neoliberalism. This 'local turn', Moe and Müller (2017: 14) contend, has penetrated even further the logics of "counterinsurgency into the politics, governance arrangements and life worlds of the local". Neoliberal governance and counterinsurgency strategies simultaneously 'roll out' governance frameworks instrumental for "global liberal order making" that produce "individual subjects capable of self-governing" themselves (Moe & Müller 2017: 19).

NGOs, civil society groups and their programs of 'empowerment', 'participation', and 'sustainable development' are instrumental governance technologies. NGOs, Aziz Choudry and Dip Kapoor (2013: 5–7) explain, "that frame their demands in liberal social democratic traditions tend to demand a humanized form of capitalism and a retooled state, which lead them towards "managing and structuring dissent" by "channeling this into organizational structures and processes that do not threaten underlying power relations". The Mexican federal and state governments have a history of coordinating efforts in Oaxaca that employ armed action, psychological propaganda and "the allocation of social funds with Preventive Actions that are aimed at promoting new agreements with trade unions and political and social organizations" (Arronte et al. 2000: 75). 'Soft' counterinsurgency initiatives in Oaxaca tend to include social scientists mapping contested terrains, environmental programs and corporate social responsibility initiatives to "deactivate the social movements" (Dunlap 2018a: 645). Walker et al. (2008: 539) acknowledge that NGOs dispersing technical assistance in Oaxaca is "a major vehicle for cleansing 'civil society' of its oppositional political possibilities, rescripting it as the social realm in which communities are improved through human capital acquisition". It should be recognized, however, that these interventions are frequently rejected, negotiated or subverted to peoples' needs and that NGOs can take various forms.

There are, however, reoccurring and general attributes of NGOs. "In order to commodify struggle, it must first be objectified", explains the Indigenous Action Media's (IAM 2014: 1–6) pamphlet, *Accomplices Not Allies*. This pamphlet voicing an Indigenous perspective offers three general caricatures of NGOs.[6] The first, 'Salvation aka Missionary Work & Self Therapy', speaks to "romantic notions of oppressed folks" that people tokenize and wish to 'help', fostering mutually unhealthy relationship in the process. Second, 'Exploitation & Co-optation' speaks to the individual and institutional profiteering of liberation struggles, which simultaneously imposes liberal (reformist) agendas in structurally patronizing ways. Third, 'Gatekeepers' "seek power over, not with, others". Gatekeepers are NGOs, or persons within them, that create dependency by controlling and/or withholding information, resources, connections and support (see also Choudry & Kapoor 2013). NGOs can be instrumental to political struggle and are frequently appreciated by opposition groups. NGOs, however, operate along a spectrum of enacting intentional corporate subversion or navigating solidarity in land struggles through a liberal politics of strengthening 'rights', 'due process' and 'inclusion' within statist capitalist systems. These overlooked issues deserve recognition and scrutiny in environmental conflicts, which are examined in the last section.

Istmeño wind factory struggles

Ardent resistance has characterized wind factory development in the Isthmus, primarily coming from Zapotec, Ikoot and Zoque native populations struggling to defend their land, sea, dignity and livelihoods from megaproject development (Altamirano-Jiménez 2017; Dunlap 2019a; Lucio 2016; Oceransky 2011). There are two broad types of resistance, explains Saturnino Borras et al. (2012: 413): struggle against exploitation – or terms of incorporation – and dispossession (see also Hall et al. 2015). While struggles over (deceptive) terms of incorporation began in 2007 with the first wind factories in the northern coastal Isthmus (Lucio 2016; Oceransky 2011), it was not until 2011–2016 (after the northern Isthmus developmental experience), when wind factories arrived in fishing communities, that Indigenous resistance and violent repression, characterized wind factory development in the southern regions (Dunlap 2018a, 2019b). "The Mexican state has always been very cautious about imposing extractivist projects in the Isthmus, due to the fear of social reactions", as a human rights defender explains, yet "wind energy extraction was promoted and defended by the Left" in the Isthmus. The Leftist heroes and defenders of 'the people' became agents 'promoting transnational capital',[7] and as a daughter of a COCEI[8] militant explains, "ended up being land grabbers and cacique intermediaries for the companies, justifying themselves by saying they bring more development to the region".[9]

The repression dispensed by state forces, private security and extra-judicial actors became particularly acute in 2013, only intensifying once a 'drug war'-style of violence came to the region (see Correa-Cabrera 2017; Paley 2014; Zavala 2018). 'Wild Tiger' describes his experience with wind factory development as "murderous energy (*energia asesina*)", because "from the moment that energy arrives, it comes with death threats to Indigenous people". People living within their ecosystems – farming, fishing and participating in reciprocal spiritual–material relationships – do not find wind energy development 'green' or 'clean'. The real and anticipated destructive effects of wind factory development on lands and social bonds triggered militant land defense. Corporate and governmental efforts to protect investments against popular resistance in the region, especially from 2011 to 2015, resulted in violent repression, numerous deaths[10] and the documentation of employing pluralist and culturally sensitive counterinsurgency protocols (Dunlap 2018a, 2019a, 2020b). Experiences of conflict, counterinsurgency and violent repression – characteristics of fossil fuel+ development projects – transform *wind energy to murderous energy*.

Manufacturing wind factory development: money, *Sicarios* and NGOs

Despite wind factory popularity with the government, politicians, landowners and landless workers, the agrarian change towards dependency, the loss of control over land and the failure to provide (noticeable) collective benefits has made the EDF wind factory unpopular. The socio-ecological pacification necessary for the Gunaa Sicarú development has necessitated three interlinked process of *distributing money*, *dispensing Sicarios* and *NGO intervention*. These three processes manufacture social–ecological pacification and enforce wind factory 'social license' to operate. The distribution of money, gunmen and human right NGOs are three central mechanisms of facilitating extractive wind factory development. The intensity of violence surrounding wind factory development in the Isthmus remains under-acknowledged by researchers,[11] instead prioritizing moderate positions and voices, deserving an expanded discussion on the role that money and *Sicarios* play in the Isthmus. While this dynamic is mentioned, the central focus of this chapter is to examine the underestimated role of human rights groups in confronting the arrival of the new EDF wind park.

Human rights groups, however, exist against the backdrop of money and *Sicarios*. Development funds play an extremely divisive role in rural communities. The strategic distribution of money to pacify resistance challenging wind factory development, as previously discussed (Dunlap 2018a), bears resemblance to the counterinsurgency Integrated Monetary Shaping Operations (IMSO) protocol that employs social development as a method to permit wind energy extraction. While overlapping and (relatively) synchronized, the analytical distinction between 'development money' – land contract, employment and socio-infrastructural programs – and 'repressive money' – intimidation, physical attacks and killings – is important. "That strategy of offering money or big rewards for going over to their side as opposed to being killed has been successful at dismantling resistance in Oaxaca", explains 'Wild Tiger'. Aspirations for 'development money' can lead towards 'repressive money' that employs people to commit violence against land defenders and other unions.

This leads to *Sicarios*, or hitmen. The levels of violence have skyrocketed in Unión Hidalgo and the Isthmus in general (Santos 2018). Guchachi counts five murders in Unión Hidalgo (UH) since the arrival of wind factories, while George claims there have been '15 to 25 murders' in the region since 2009, which 'is something that did not use to happen here'. There have always been armed groups and mafias in the region (Dunlap 2019a; Lucio 2016), yet the rise of *Sicarios* in the region is new and relates to three overlapping factors: repressing Indigenous autonomy, union conflicts and cartel struggles. Repression of Indigenous autonomy secures wind factory and corresponding infrastructural investments. Wind factory development in Unión Hidalgo created a scramble for company money, causing unions to proliferate and compete over construction contracts. Construction unions need 'to protect themselves and to threaten others [unions] or those who oppose the project' to secure contracts. This competition between unions also entails a confrontation with cartels and criminal organizations. Finally, violence emanating from cartel territorial struggles and industry activities, according to land defenders, has become a serious issue since 2015. The relationship between large-scale capital investments into the region and the escalation of violence resonates with experiences in other parts of Mexico (Correa-Cabrera 2017; Paley 2014; Zavala 2018). The arrival and distribution of money and *Sicarios* remain central noticeable issues, surrounding the more subtle and complicated issue of human rights NGOs.

Human rights NGOs

Until recently, President Andres Manuel Lopez Obrador (AMLO) employed consultations as the preeminent mechanism to implement his ambitious megaprojects and industrial corridors (AMLO 2018). This includes the Isthmus Inter-Oceanic Corridor, which seeks to intensify extractive development in the Isthmus (SIPAZ 2020). Within this political terrain, human rights NGOs have the potential to function as mechanisms of pacification by *shifting rebellious groups from total rejection to negotiation*. This shift is reinforced by legal impositions, (statist) threats of coercive force and widening existing social divisions (Dunlap 2018a, 2019a, 2020b). "No one is going to finance you to oppose a megaproject – nobody", explains a local human rights defender, "you are going to get financing if you engage in long-term cultural promotion for community development, gender empowerment and disaster reconstruction". The September 2017 earthquake invited NGOs with such programs to the region. Wind and mining companies distributed humanitarian assistance plastered with 'company logos', which many felt was 'taking advantage of publicity space'. Proud that the Isthmus is 'listed as a problematic region for NGOs', Natalie asks: "Why would we work with NGOs when their dynamic tend towards creating dependency based on their own agenda? In turn, that agenda is based on their own moral beliefs on how Indigenous populations should behave". The nonprofit ProDESC (the Economic, Social, and Cultural Rights Project) working in UH deserves further attention.

Since 2007, ProDESC has been working with the *Comuneros* in UH (ProDESC 2019b). In 2013, ProDESC submitted judicial proceedings to nullify DEMEX land contracts, yet, as they described on their website, "a judgement has not been issued recognizing the human rights violations committed by the company against community members, nor the cancellation of the contracts" (ProDESC 2019b). Founded by Alejandra Anceita, winner of the Martin Ennals Prize (known as the Nobel Prize for Human Rights), ProDESC (2019a) describes their mission to "defend and promote an integral DESC [Economic, Social and Cultural Rights] perspective of disadvantaged groups with respect to the full enjoyment of their rights". While their work has been highly appreciated by some *Comuneros*, others have voiced concern, inside and outside the *Comunero* Assembly.

"ProDESC is following the same line as the rest of the NGOs in [AMLO's] the Fourth Transformation, which is to go straight for the consultations", explains Broma, "they are not against the [wind] projects, they are fighting for consultation processes that line up strictly to ILO 169 (see Dunlap 2019a). They say they have achieved results, but the only thing they have accomplished is restarting the consultation process". Remembering the Juchitán consultation experience (see Dunlap 2019a), another *Comunero*, Gueu, claims that the *Comunero* Board acts "as if a divine light will awaken the hearts of the town's people during the consultation deliberation phase", but in reality "all the governments and international organisms are behaving as if the project will be constructed, and they [*Comunero* Board] take that for granted". The unfolding dynamic is complicated, resting on the relationship between the *Comunero* Board and ProDESC.

Two issues emerge from this process. The first relates to *Comunero* participation in the consultation. When consultation planning and implementation was clearly faulty, or 'simulated', the *Comunero* assembly agreed to reject the EDF wind factory consultation. Yet, many still participated. Some dissident *Comuneros* explain: "we have all pointed out its theatrics, I then fail to understand why the fuck some *Comuneros* are participating?" The second, more substantive concern is about filing an agrarian court injunction. The agrarian injunction would prohibit megaproject development on communal land, reserving this space for agrarian activity. Filing an Isthmus agrarian court injunction has been in the works since 2013, if not longer, yet Carlos González – a respected Nahua National Indigenous Congress (CNI) lawyer specializing in agrarian law – prepared an agrarian injunction for the *Comuneros*. The *Comunero* Board, working with ProDESC since 2007, wanted the lawyers to collaborate. González believes that, even if carried out under ILO 169, consultations are surrendering Indigenous territories to the Mexican State (Camacho 2019). The agrarian injunction challenges ProDESC's civil injunction requesting an FPIC consultation. Crafting an agrarian injunction comes from decades of research and experience. ProDESC's organizational and legal work presumably did not inspire collaboration. It reached a point where the two lawyers 'are not talking to each other' and 'neither wants to'. Bluntly summarizing the situation, Broma explains:

> Speaking honestly, having a consultation process start all over again due to a court injunction is a big media propaganda move. What happens? The agrarian court injunction instead cuts the process at the roots. It would make it possible to say: 'Why the heck do we have to fight for consultations when they should not even step into our territories in the first place.' If a state grants you consultation rights, he is immediately violating your self-determination. Consultations are in total contradiction and conflict, because it is forcing a decision upon you. But because these guys [ProDESC] make a living out of long legal processes, it makes them very uncomfortable to see a court injunction that attacks the root of the problem and prevents a longer legal process. Thus, the agrarian court injunction leaves them without a job.

The *Comuneros* agreed in September to file this agrarian injunction, yet they wanted to have ProDESC's comments. The CNI lawyer was indirectly forced to share the injunction, and ProDESC submitted an opinion at the *Comuneros*' request.

ProDESC's comments and the agrarian injunction were the center of the 26 January 2020 *Comunero* Assembly. Another resistance group, who initially brought ProDESC to Unión Hidalgo, berated the *Comuneros* and ProDESC for not filing the injunction, as it concerns all of the Isthmus' communal lands. ProDESC's representative replied that based on their 'expertise', they found some 'weaknesses' and 'mistakes' in the agrarian injunction. ProDESC's recommendations, however, were not read to the assembly and the *Comuneros'* lead representative "did not remember with detail its content", nor did they bring ProDESC's injunction opinion.[12] This provoked outrage in the assembly. Referring to this assembly, Broma claims: "So evidently ProDESC is halting the [agrarian] court injunction, because its success would end their long-term gold mine". ProDESC claims, according to the *Comunero* Board, that it is "wrong to have land written down as communal" because existing land titles "legally prove they are the landowners". Broma says these claims are "absurd", stressing that they are not agrarian lawyers and have a vested interest for drawing out the consultation process. In La Ventosa, an agrarian court injunction was recently filed against a Riverside Resources mine. The injunction was successful, yet quickly resulted in severe acts of intimidation. Heavily armed men, with modified bullet-proof trucks, went door to door offering 80,000 pesos to injunction signatories. The implication was: take the money or disappear. "The state and companies are sensing the check-mate and they have been toughening up their legal defense", explains Broma, who assisted with the mine injunction. The agrarian injunction strategy is demonstrating some forms of success, resulting in immediate extra-judicial repression (Parada 2019). This has forced *Comuneros* in La Ventosa to revoke their signatures or leave town. Legal success, according to Broma, is causing "judges to demand documents that certify the Indigenous identity of every signatory of the agrarian injunction" as it threatens to terminate all megaproject development on communal land.

The *Comunero* Assembly is not organized along horizontal lines, with a *Comunero* Board exhibiting vanguardist tendencies. Assembly organization creates controls on information; meanwhile, an informal hierarchy emerges between landed and landless *Comuneros*. This allows the *Comunero* Board to concentrate power, becoming the principal liaison with ProDESC. A mutually reinforcing 'gate keeper' dynamic forms between the assembly vanguard and ProDESC. "Undertaking paternalistic attitudes, some *Comuneros* make the decision on how to manage information in a way that will be to the liking of the [ProDESC] lawyer", explains Mixtu, continuing, the "*Comuneros* want to feel that they have a bond with the [ProDESC] lawyers to feel secure and they think that by promoting an idea that the lawyers will feel comfortable with", it can be accomplished. This behavior empowers ProDESC influence and decision making within the assembly. Asking a ProDESC representative working in UH (repeatedly) about the agrarian injunction in a phone interview, they replied:

> Yes, we wrote them because the assembly of *Comuneros* asked us to. That is why we sent them directly to them and you are not anyone to know about that topic. *I do not even see the relevance of that topic.*

Mixtu argues that "even if *Comuneros* do not agree with the ProDESC dynamic, they never speak out because they fear being marginalized or kicked out from the committee".

There are micro-politics that ProDESC engages with, not uncommon among NGOs. First, although they have not been charging legal fees to the *Comunero* Board, cases like UH serve to create a portfolio favorable to grant writing. ProDESC has been expanding as an organization, repeatedly hiring new staff.[13] This case serves as 'a gold mine' to present themselves as not only advancing the liberal politics of integration, in line with AMLO's consultation strategy – which land defenders have identified as 'a discourse of defeat' and 'legal dispossession' (see also Camacho 2019) – but also fashion themselves as fighting for social justice. "They spread fear with their comments", explains Mixtu. "I believe they distort the information that they present". Others mention how ProDESC representatives employ an upper-class lawyer aesthetic, using technical terms and flaunting their credentials as 'experts' to convince the *Comunero* Board to allow them to guide the process.

Second, ProDESC provides resources and a sense of (legal) security by creating struggle representatives. Human rights groups create marketable representatives to showcase their personality on their websites or in pithy promotional videos. While raising awareness about an issue, it simultaneously operates within a market-based logic. The struggle is transformed into social and grant capital, establishing organizational purpose and fighting for improved consultation procedures as well as (more equitable) participation in wind factory development. In terms of developmental reform, they are a showcase success: distributing resources and creating new opportunities for (select) *Comuneros*. In the struggle for Indigenous autonomy and land reclamation, however, they are solidifying capitalist relationships, pacifying resistance and eliminating socio-ecological alternatives to development. The NGOization of the UH struggle can reconcile development, but not Zapotec autonomy and communal land.

The manufacturing of struggle representatives also has deeper political and psycho-social implications. *Representative making*, we can say, mirrors a micro-political colonial strategy to 'appoint leaders to horizontal societies', because, as Peter Gelderloos (2017: 19, 22) tells us, a "society needs to be accustomed to having leaders for a foreign power to effectively be able to appoint puppet rulers". Representative making – in ideally horizontal – Indigenous collectives replicates the colonial–statist model on the local level (see Dunlap 2018b). The pre-existing (informal) hierarchical nature of the *Comunero* council made them susceptible to the manufacturing of leaders, as this process employs allure and creates material possibilities for those willing to be professionalized as Indigenous or struggle representatives. Describing the persistence of the ProDESC–Assembly dynamic, Mixtu explains

> I think one of their main strategies [for ProDESC] to remain in the assembly, aside from not charging and absorbing all of the judicial costs, is that they formed moral links with people inside the committee. People they

> eat with, or that they pay visits to. They name them communitarian land defenders and they launch them as a 'poster boy' for the media and they will take them to workshops in Oaxaca or Mexico City, they will even take them to summits and international events. It might seem silly, but people like it [being made into human right celebrities]. They like going to Europe, they love being interviewed in UN meetings and seeing themselves on video being land defenders. You see, so they weave this net of trinkets that wrap people up and make them feel morally indebt with them. So obviously those *compañeros* will defend them.

The micro-politics of land defender identity formation, celebration and professionalization are potent weapons of social warfare. Greta Thunberg, and other 'climate youth leaders', are another large-scale example of this process (see Morningstar 2019), which relates to the objectification and spectacularization of socio-ecological struggle. The subjectivities of land defenders are bureaucratized, conditioning them to less overtly violent methods of struggle, micro-colonial meeting structures, political norms and working habits that alter struggle priorities and, by extension, their outcomes. The NGOization of struggle and the spectacularization of land defenders remain a potent weapon in social war for pacifying land defenders and progressively add to a trajectory of ecocidal and genocidal social change.

Conclusion

Wind factory development in the Isthmus has materialized significant, negative socio-ecological impacts. Outlining the impacts and controversies related to wind factory development in Unión Hidalgo, this chapter demonstrates that (human rights) NGO are important mechanisms of infrastructural colonization. Money creates important footholds among political elites, landowners and construction unions that enact political violence in favor of extractive 'green' development. The (post)developmental aspirations of Indigenous land defenders and *Comuneros* are diametrically opposed and contest the entire legal regime of permitting land titling and privatization that intensifies and strengthens the (neo)colonial apparatus. Coercive repression and legal pressures make the legal help from NGOs and social movement lawyers indispensable, yet internal disagreement emerges over information sharing and the legal route established. NGO influence intertwines with *Comunero* organizational dynamics to create a dangerous level of social control. Money, *Sicarios* and NGOs, in the end, are crucial to social engineering political acquiescence into people and, consequently, gaining the 'social license' of EDF's Gunaa Sicarú project.

Acknowledgement

This research was conducted with Martín Correa Arce and could not have been possible without the confidence and care provided by the Zapotec land defenders in the region. Let this chapter enable discussion and self-reflection in the furtherance of Zapotec and Ikoot comunalidad, and corresponding visions of (post)development.

Notes

1 The AMLO presidency has de-prioritized wind energy in the Isthmus, preferring other methods of extractivism for the moment.
2 The bracketed 'neo' pays homage to organizational development and computational shifts since the 19th century.
3 The recent 'rural aspirations' focus is important (Bennike et al. 2020: 2, 10), yet the historical–political neglect in the construction or influence of aspiration is oddly apolitical.
4 See *Baedan 1: Journal of Queer Nihilism.*
5 This is the original or continuous dispossession of public and private lands; the redistribution of state resources; the manipulation of crises to advance economic privatization objectives; and the general advancement of financial mechanisms.
6 The text also details insights into NGO 'ally' personalities, 'Parachuters' and 'Navigators', which also include insightful hostility towards academics.
7 Ibid.
8 The Coalition of Workers, Peasants, and Students of the Isthmus.
9 See also Altamirano-Jiménez (2017) & Dunlap (2019a).
10 Documented calculations based on Dunlap (2020b) calculates eight deaths, numerous instances of wounding and countless physical attacks. There has been a dramatic rise in murder since 2015 (Santos 2018), which blurs issues related to cartels, politicians and megaproject development.
11 Lucio (2016) is a notable exception.
12 A woman next to him offered to go get the file he was supposed to bring, and he declined her offer.
13 See: https://twitter.com/ProDESC/status/1229435340198334465. On 25 February 2020, an ex-ProDESC member working in Unión sent notice of their voluntary resignation to the *Comuneros*, noting that they felt their activity was compromising their ethics. This has happened before.

References

Altamirano-Jiménez, I. (2017). "The sea is our bread": Interrupting green neoliberalism in Mexico. *Marine Policy*, *80*, 28–34.

AMLO. (2018). *Programas y proyectos de AMLO obtienen aprobación ciudadana superior al 90% en la Consulta Nacional Programas Prioritarios*. https://lopezobrador.org.mx/temas/consulta-ciudadana/

Anonymous. (2001 [1998]). *At daggers drawn with the existent, it's defenders and false critics*. London: Elephant Editions.

Arronte, E. L., Soto, G. E. C., & Lewis, T. P. (2000). *Always near, always far: The armed forces in Mexico*. San Francisco, CA: Global Exchange.

Avila, S. (2018). Environmental justice and the expanding geography of wind power conflicts. *Sustainability Science*, *13*(3), 599–616.

Backhouse, M., & Lehmann, R. (2019). New 'renewable' frontiers: Contested palm oil plantations and wind energy projects in Brazil and Mexico. *Journal of Land Use Science*, *15*(3), 1–16.

Bennike, R. B., Rasmussen, M. B., & Nielsen, K. B. (2020). Agrarian crossroads: Rural aspirations and capitalist transformation. *Canadian Journal of Development Studies*, 1–17.

Berman, E. (1983). *The ideology of philanthropy*. New York: State University of New York Press.

Bonanno, A. M. (1998). *The anarchist tension*. London: Elephant Editions.
Borras, S., Kay, C., Gómez, S., & Wilkinson, J. (2012). Land grabbing and global capitalist accumulation: Key features in Latin America. *Canadian Journal of Development Studies*, *33*(4), 402–416.
Brock, A. (2020). 'Frack off': Towards an anarchist political ecology critique of corporate and state responses to anti-fracking resistance in the UK. *Political Geography*, *82*, 1–15.
Brock, A., & Dunlap, A. (2018). Normalising corporate counterinsurgency. *Geography*, *62*(1), 33–47.
Camacho, Z. (2019). *La lucha indígena de hoy, puede ser la última: Carlos González*. www.contralinea.com.mx/archivo-revista/2019/05/16/la-lucha-indigena-de-hoy-puede-ser-la-ultima-carlos-gonzalez/
Choudry, A., & Kapoor, D. (2013). *NGOization: Complicity, contradictions and prospects*. London: Zed Books.
Conde, M., & Le Billon, P. (2017). Why do some communities resist mining projects while others do not? *The Extractive Industries and Society*, *4*(3), 681–697.
Correa-Cabrera, G. (2017). *Los Zetas Inc*. Austin: University of Texas Press.
Dunlap, A. (2014). Permanent war: Grids, boomerangs, and counterinsurgency. *Anarchist Studies*, *22*(2), 55.
Dunlap, A. (2017). 'The town is surrounded': From climate concerns to life under wind turbines in La Ventosa, Mexico. *Human Geography*, *10*(2), 16–36.
Dunlap, A. (2018a). Counterinsurgency for wind energy: The Bíi Hioxo wind park in Juchitán, Mexico. *The Journal of Peasant Studies*, *45*(3), 630–652.
Dunlap, A. (2018b). The 'solution' is now the 'problem': Wind energy, colonization and the 'genocide-ecocide nexus' in the Isthmus of Tehuantepec, Oaxaca. *The International Journal of Human Rights*, *42*(4), 550–573.
Dunlap, A. (2018c, May 10). End the "green" delusions: Industrial-scale renewable energy is fossil fuel+. *Verso*. www.versobooks.com/blogs/3797-end-the-green-delusions-industrial-scale-renewable-energy-is-fossil-fuel
Dunlap, A. (2019a). *Renewing destruction: Wind energy development, conflict and resistance in a Latin American context*. Lanham, MD: Rowman & Littlefield.
Dunlap, A. (2019b). 'Agro si, mina NO!' The Tia Maria copper mine, state terrorism and social war by every means in the Tambo Valley, Peru. *Political Geography*, *71*, 10–25.
Dunlap, A. (2020a, February 15). Disaster breeds disaster in Oaxaca. *Toward Freedom*. https://towardfreedom.org/story/disaster-breeds-disaster-in-oaxaca/
Dunlap, A. (2020b). Wind, coal, and copper: The politics ofland grabbing, counterinsurgency, and the social engineering of extraction. *Globalizations*, *17*(4), 1–22.
Dunlap, A. (2020c). Bureaucratic land grabbing for infrastructural colonization: Renewable energy, L'Amassada and resistance in Southern France. *Human Geography*, *13*(2), 1–22.
Dunlap, A., & Jakobsen, J. (2020). *The violent technologies of extraction*. London: Palgrave Macmillan.
Esteva, G., & Prakash, M. S. (1998). *Grassroots postmodernism*. London: Zed Books.
FM3-24. (2014). *Insurgencies and countering insurgencies*. http://fas.org/irp/doddir/army/fm3-24.pdf
Foucault, M., & Ewald, F. (2003). *"Society must be defended": Lectures at the Collège de France, 1975–1976*, Vol. 1. London: Palgrave Macmillan.
Franquesa, J. (2018). *Power struggles: Dignity, value, and the renewable energy frontier in Spain*. Bloomington, IN: Indiana University Press.
Gardenyes, J. (2012). Guerra Social: Tension antisocial. *The Anarchist Library*. https://josepgardenyes.files.wordpress.com/2012/10/tension-antisocial.pdf

Gelderloos, P. (2013). *The failure of nonviolence: From the Arab spring to occupy*. Seattle, WA: Left Bank Books.

Gelderloos, P. (2017). *Worshiping power: An anarchist view of early state formation*. Chico, CA: AK Press.

Gilbert, E. (2009). Liberalism. In R. Kitchi & N. Thrift (Eds.), *International encyclopedia of human geography* (pp. 195–206). Amsterdam: Elsevier.

Hall, R., Edelman, M., Borras Jr, S. M., Scoones, I., White, B., & Wolford, W. (2015). Resistance, acquiescence or incorporation? An introduction to land grabbing and political reactions 'from below'. *Journal of Peasant Studies*, *42*(3–4), 467–488.

Howe, C. (2014). Anthropocenic ecoauthority. *Anthropological Quarterly*, *87*(2), 381–404.

IAM. (2014). Accomplices not allies: Abolishing the ally industrial complex. *Indigenous Action Media*. www.indigenousaction.org/accomplices-not-allies-abolishing-the-ally-industrial-complex/

IFC. (2014). Investments for a windy harvest. *International Finance Corporation*. www.ifc.org/wps/wcm/connect/a0f55458-988a-4756-8ebd-f456235bc644/IFC_CTF_Mexico.pdf?MOD=AJPERES&CVID=kCCelk9

Lawrence, R. (2014). Internal colonization and Indigenous resource sovereignty. *Environment and Planning D*, *32*(6), 1036–1053.

Loadenthal, M. (2017). *The politics of the attack*. Manchester: Manchester University Press.

Lucio, C. F. (2016). *Conflictos socioambientales, derechos humanos y movimiento indígena en el Istmo de Tehuantepec*. Zacatecas: Universidad Autónoma de Zacatecas.

Manzo, C. (2011). *Comunalidad, resistencia indígena y neocolonialismo en el Istmo de Tehuantepec, siglos XVI-XXI*. Mexico: Ce-Acatl.

Manzo, D. (2019). *Energía limpia y contratos sucios en eólicas de Oaxaca*. http://imparcialoaxaca.mx/oaxaca/382866/energia-limpia-y-contratos-sucios-en-eolicas-de-oaxaca/

Mignolo, W. D. (2005). *The idea of Latin America*. Hoboken, NJ: John Wiley & Sons.

Moe, L. W., & Müller, M. M. (2017). *Reconfiguring intervention*. London: Palgrave Macmillan.

Morningstar, C. (2019). The manufacturing of Greta Thunberg for consent has been written in six acts. *The Wrong Kind of Green*. www.wrongkindofgreen.org/2019/01/17/the-manufacturing-of-greta-thunberg-for-consent-the-political-economy-of-the-non-profit-industrial-complex/

Nahmad, S., Nahón, A., & Langlé, R. (2014). *La visión de los actores sociales frente al los proyectos eólicos en el Istmo de Tehuantepec*. Mexico: Consejo Nacional para Ciencia y Tecnología.

Oceransky, S. (2011). Fighting the enclosure of wind. In K. Abramsky (Ed.), *Sparking a worldwide energy revolution* (pp. 505–522). Chico, CA: AK Press.

Paley, D. (2014). *Drug war capitalism*. Chico, CA: AK Press.

Parada, G. V. (2019, May 31). Arma frente La Ventosa contra la minería. *NVI Noticias*. www.nvinoticias.com/nota/117464/arma-frente-la-ventosa-contra-la-mineria

Pasqualetti, M. J., Righter, R., & Gipe, P. (2004). Wind energy, history of. *Encyclopedia of Energy*, *6*, 419–433.

Petras, J. (1997). Imperialism and NGOs in Latin America. *Monthly Review*, *49*(7), 10–27.

Petras, J., & Veltmeyer, H. (2001). *Globalization unmasked*. Delhi: Madhyam Books.

ProDESC. (2019a). *¿Quiénes somos?* https://prodesc.org.mx/inicio/quienes-somos/

ProDESC. (2019b). *Comunidad Agraria de Unión Hidalgo*. https://prodesc.org.mx/acompanamiento-de-casos/union-hidalgo/

Santos, F. (2018, August 7). El horror está presente en el Istmo. *El Imparcial*. https://imparcialoaxaca.mx/policiaca/201619/el-horror-esta-presente-en-el-istmo/

Short, D. D. (2016). *Redefining genocide: Settler colonialism, social death and ecocide*. London: Zed Books.

Siamanta, Z. C. (2019). Wind parks in post-crisis Greece: Neoliberalisation vis-à-vis green grabbing. *Environment and Planning E*, *2*(2), 274–303.

SIPAZ. (2020, March 6). *Corredor Transístmico, un proyecto no tan novedoso del nuevo gobierno*. www.sipaz.org/enfoque-corredor-transistmico-un-proyecto-no-tan-novedoso-del-nuevo-gobierno/

Trocchi, A. (2011). For the insurrection to succeed, we must first destroy ourselves. In A. Vradis & D. Dalakoglou (Eds.), *Revolt and crisis in Greece* (pp. 299–328). Chico, CA: AK Press.

Walker, M., Roberts, S. M., Jones III, J. P., & Fröhling, O. (2008). Neoliberal development through technical assistance: Constructing communities of entrepreneurial subjects in Oaxaca, Mexico. *Geoforum*, *39*(1), 527–542.

Zavala, O. (2018). *Los cárteles no existen: Narcotráfico y cultura en México*. Barcelona: Malpaso Ediciones.

Zibechi, R. (2012). *Territories in resistance*. Chico, CA: AK Press.

Zografos, C., & Martínez-Alier, J. (2009). The politics of landscape value: A case study of wind farm conflict in rural Catalonia. *Environment and Planning: A*, *41*(1), 1726–1744.

18 Land defenders, infrastructural violence and environmental coloniality

Resisting a wastewater treatment plant in East Nablus

Jeanne Perrier

In 1997, an initial feasibility study called for the establishment of two large-scale wastewater treatment plants (WWTPs) for the city of Nablus: one in the west and one in the east. The first one was built in 2013 and is already feeding agricultural reuse of treated wastewater (Perrier 2020; Trottier & Perrier 2018). The second has a more sinuous and laborious history. Some villagers from Azmut, Deir Al Hattab, and Salem oppose the project and constitute the driving force behind the protest against it.

At first glance, this project had all the makings of a successful one, since it had already been carried out on the western side of Nablus and in other Palestinian governorates. Moreover, health concerns transformed this project into an inevitable solution to improve sanitation in the eastern region. According to the agricultural and water strategies produced by the Palestinian Authority (PA), wastewater treatment should aim at achieving "efficient" water resource management and improving the living environment of the inhabitants. It therefore seems to be an obligatory step, but one that is struggling to assert itself as such on the ground.

Different discourses are clashing over the WWTP project in East Nablus. The PA, the Palestinian Water Authority (PWA), the Municipality of Nablus and the donor see it as a "development" opportunity for the eastern part of Nablus and a technical solution to the sanitary and environmental problems. In contrast, some villagers are opposed as they perceive it as yet another confiscation of their land and a new landscape degradation for their region. Each side uses arguments that the other does not seem to hear, creating a gap between the narratives concerning this project. The contestation of this WWTP project cannot be reduced to a cultural problem. The socio-political context, issues of power and issues of territorialization play a role in the contestation.

It creates tensions and generates different types of violence, mainly infrastructural and epistemic. This chapter focuses on infrastructural violence in the planning process of a development project, bridging literatures on infrastructural violence and land defenders. It allows for the exploration of practices of resistance and negotiations deployed by land defenders, as well as practices of repression used by the PA and its allies.

The first section puts forward the theoretical framework of this work. The second section briefly presents the project and the context of land dispossession by the Israeli occupation in which it takes place. Palestinian villagers are once again facing land dispossession but this time from Palestinian institutions. This contextualization provides the necessary tools to analyze the arguments of land defenders in the third section. I explore the tensions and resistance strategies generated by infrastructural violence and by governmental practices of repression, amounting to signs of coloniality. In the last section, I demonstrate the epistemic violence at stake inside the overall infrastructural violence and how land defenders use it as a negotiation tool.

Infrastructural violence and land defenders

Conceptualizing infrastructural violence allows us to highlight the role of non-humans in the production of injustice. Rodgers and O'Neill (2012) coined the term to express the fact that infrastructure is not simply a product of violence but is often an instrument of it. It not only reflects but also contributes to strengthening the dominant social order and thus produces violence. Rodgers and O'Neill (2012: 406–407) distinguish between two types of infrastructural violence: "active" violence, which refers to infrastructures whose very purpose is violence, and "passive" violence, which is the result of limitations and omissions related to the infrastructure. A WWTP may fall into the second category because it generates processes of dispossession and exclusion.

A growing body of social science work is concerned with infrastructure as objects embedded in social and power relations and playing a role in the reconfiguration of societies. Rodgers and O'Neill (2012) along with Harvey et al. (2017) bring together different analyses of the complexities, fragility, and forms of governance of infrastructure in different fields (e.g. development, urban, environmental, digital). Li (2018) analyzes plantations as "routinely" violent infrastructure in Indonesia, highlighting their disastrous long-term impacts on social life and the environment. Dunlap (2020) focuses on infrastructural violence brought about by renewable energy in France, while Nolan et al. (2020) study the various forms of violence a power station inflicts on a fishing community in Ghana.

Water infrastructure is the subject of much research, which is mainly focused on the analysis of urban water supply networks as tools of governance, territorialization, and domination. Some analyze how domestic water networks become a tool for governing urban populations (Coelho 2004; Desai 2018; Jensen 2017) and reveal racist practices, as in Mumbai (Anand 2012). Others analyze certain sociotechnical objects, such as dams (Kaika 2006), water meters (Loftus 2006), or wells (Barnes 2012, 2017).

This chapter hopes to bridge the literature on infrastructural violence and land defenders by adding a third type of violence to the two identified by Rodgers and O'Neill (2012): the violence related to the conflict over infrastructural project, here the implementation of a wastewater treatment plant. It reflects on the violence caused by infrastructures even before their construction and helps explore the resistance practices of land defenders, as well as the politics of repression against them. In this chapter, I draw on Zoomers' (2018) reflections about the exclusion of local communities from participation in the design of development projects imposed from "the outside". Dunlap (2020) also offers an interesting theoretical tool by calling out the implementation of an energy transformer substation in a French village as an "infrastructural colonization". It brings into perspective the view of those directly affected, even before the construction of a project, and the strong resentment towards the state. This chapter builds on this perspective in a complex political context, between Israeli colonization and Palestinian state-building objectives. Thus, land defenders from the Palestinian villages explored here face practices of coloniality, as the PA uses tactics and strategies of repression and attrition very similar to those used by the Israeli occupation.

Abundant literature documents the direct, structural, and cultural violence resulting from the Israeli occupation of Palestinian territories and affecting agriculture or water management (Selby 2013; Trottier 1999, 2007a; Weinthal & Sowers 2019; Zeitoun 2008). Some authors (Fustec 2014; McKee 2020; Trottier 2007b; Trottier et al. 2019a; Trottier & Perrier 2017) have also encouraged the analysis of intra-Palestinian violence mechanisms in water management to highlight inequalities in access to water and the structuring and recomposition of power relations through discourses and development projects.

Mobilizing the concept of infrastructural violence allows us to adopt a bottom-up approach, shedding light on the criticisms made of the PA and certain development projects. Entry through infrastructural violence also provides an understanding of the PA's difficulties in becoming spatially but also sustainably anchored. Indeed, infrastructure is a political tool for long-term territorialization since it is established in a space for a relatively long period of time. This is particularly the case for wastewater treatment plants, whose lifetime extends into decades or more.

This chapter considers large sewage treatment plant infrastructures as potential vectors of violence. They do not adapt to the environment but rather expect the environment, and its people, to adapt to these massive infrastructures. This allows us to reflect on the dynamics of violence in sustainable development projects, which are often seen as necessarily having a positive impact on the environment and surrounding populations while they produce what Dunlap (2017) calls "sustainable violence". This chapter analyzes the effects of these infrastructures imposed from above in inhabited and lived spaces. Using the lens of infrastructural violence allows for the highlighting of land defenders' practices and resistance. Thus, this chapter aims to explore the struggle of Palestinian villagers against a state- and donor-led development project and the repression practices to silence their movement. It articulates the infrastructural violence in peri-urban areas, where marginalization and inclusion affect discourses and practices of land defenders.

Implementing a WWTP in a context of Israeli infrastructural violence and pre-existing land defenders' resistance

The wastewater treatment plant project in East Nablus is co-financed by the European Union (EU) and the German government's development bank (KfW). It includes the construction of the WWTP, main distribution networks, some domestic connections, pre-treatment of industrial effluents, operational assistance, implementation of a wastewater reuse pilot project, and project supervision (EU 2013).[1] The WWTP was supposed to be ready by 2010, with an extension planned for 2012. The administrative obstacles imposed by the Israeli occupation considerably slowed down the progress of the WWTP construction in Nablus and in the West Bank in general (Selby 2013; Trottier et al. 2019b; World Bank 2009). The contestation of the project by the inhabitants of the eastern villages from 2014 onwards accentuated the delay. This section first briefly presents the project and the criteria selected to choose the site construction. Then it explores the context of land dispossession by the Israeli occupation in which the project is embedded.

Building a conventional WWTP: the selection of the construction site on technical but approximate criteria

The future station should collect wastewater from the eastern part of the city of Nablus and six villages: Azmut, Deir Al Hattab, Salem, Beit Fourik, Rujeib, and Kufr Qalil (Figure 18.1). In these six villages, the inhabitants have individual pits to store wastewater. Some of the sewage infiltrates, polluting the groundwater, and the rest is pumped by private trucks, which dump the sewage at different points in the *wadi* Al Sajour. Different solutions and coverage are considered for constructing sewage systems in the different villages (KfW 2019: 87).

In 1990, the Municipality of Nablus had already expropriated a plot of 7 hectares in anticipation of the construction of the WWTP located at the entrance to the three villages of Azmut, Deir Al Hattab, and Salem (site n° 1 on Figure 18.2). Despite this, the EIA conducted in 2013[2] proposes to compare seven potential sites (Figure 18.2) based on technical criteria and potential perimeters for the reuse of treated water in agriculture. The seven sites were classified according to nine criteria, mainly technical and economic.[3]

The choice of site n° 1 in 2013 is essentially based on economic criteria. No compensation is needed for land expropriation. The proximity of the site to the flow of wastewater flowing into the *wadi* makes it possible to convey it by gravity to the plant without a pumping system. Proximity to dwellings is one of the disadvantages of the site, listed in the document, but no criteria take this into account in the final rating table. Moreover, these criteria do not in any way address the socio-political contexts in which these different parcels of land are located.

In the spring of 2018, some rumors of relocating the project circulated within the Municipality of Nablus. At the beginning of 2019, the officialization of the

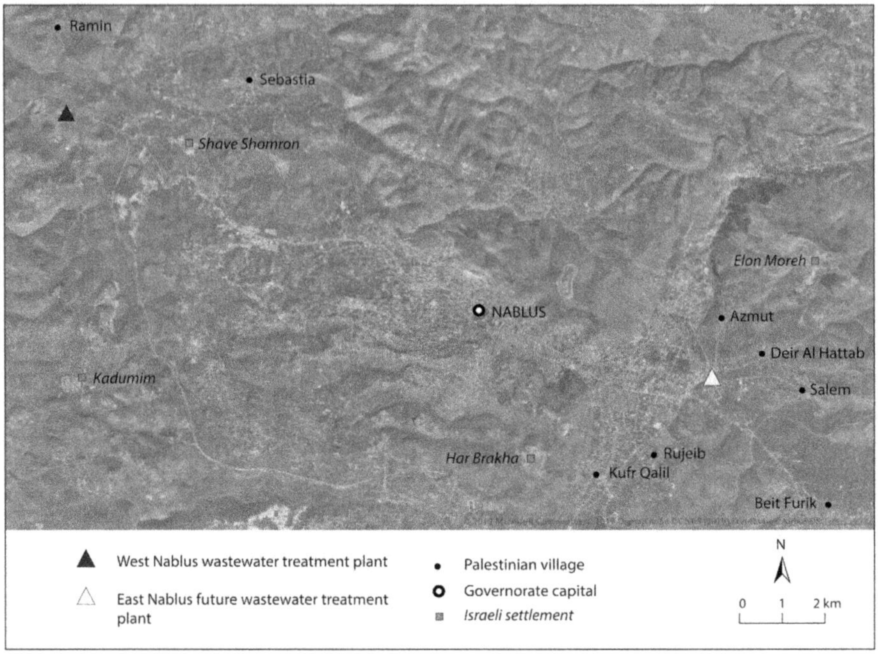

Figure 18.1 Situation map of the wastewater treatment plant in Nablus

relocation arrived: the WWTP project would finally take place on another site, a few hundred meters away from the first that had caused so many conflicts and negotiations (Figure 18.3). Reasons for changing the location oscillated between technical arguments and the desire to unlock the situation. The relocation implies a renewed interest in the land question. A new wave of expropriation has been underway since the beginning of 2019, conferring ownership of new land to the Municipality of Nablus.[4]

The project's blind spot: a context of land dispossession by Israeli colonization

The WWTP project in East Nablus takes place in a historical context of oppression and land violence in the villages of Azmut, Deir Al Hattab, and Salem. Presenting the stages and tools of Israeli colonization on the lands of these three villages sheds light on the land prejudice already suffered by the villages. Since the 1980s, they have lost access to about 1,500 hectares of land,[5] not to mention the difficulties of access and use of land located in Area C since the Oslo Agreements.[6] It has prevented any extension of the villages to the east and has largely constrained the use of these lands. Villagers opposing the WWTP use this argument to contest its construction on the few remaining lands.

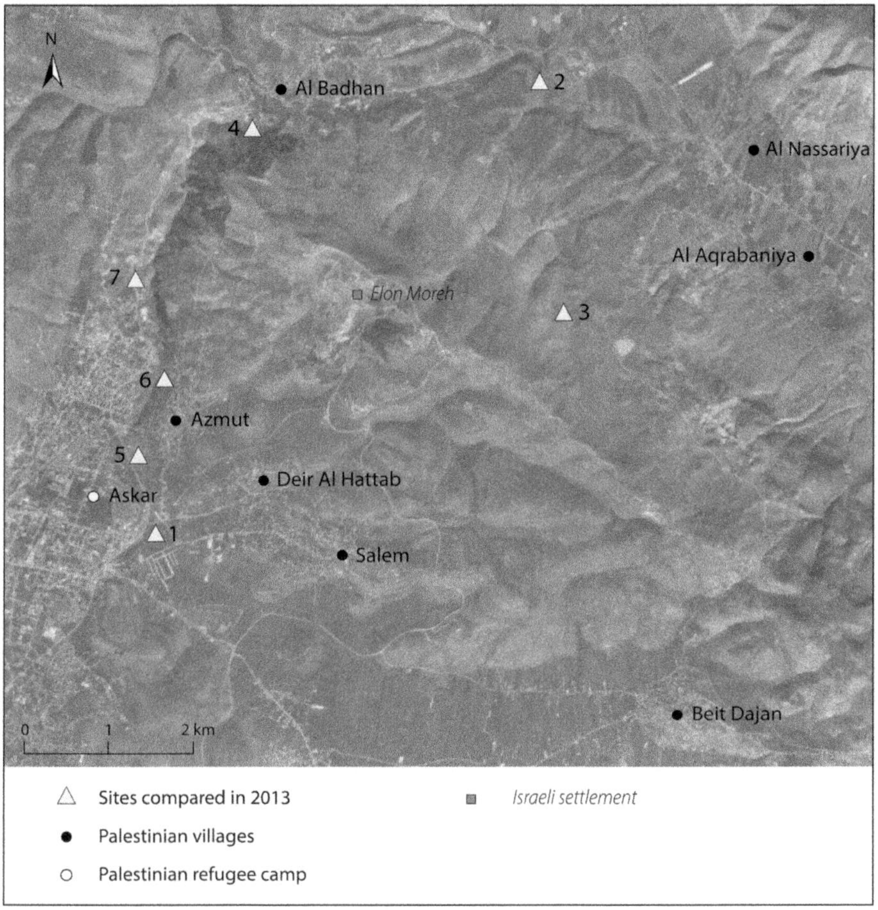

Figure 18.2 Potential sites for the project in 2013

Land as a tool of infrastructural violence of Israeli colonization

The establishment of the Elon Moreh settlement on the lands of Azmut and Deir Al Hattab represents a turning point in the history of Israeli colonization from a legal and land point of view.[7] It was mainly established on state land registered as such before 1967,[8] and developed through both confiscations by the Israeli army for "military reasons" and use of the Jordanian law of 1953 on the expropriation of land for public interest (Gazit 2003; Lustick 1981).[9]

The Israeli government's new land confiscation policy takes advantage of complex Ottoman land legislation, and a new interpretation of it, while reversing the burden of proof, which now falls on Palestinian landowners.[10] The Israeli government also used other legal tools to secure the settlement and

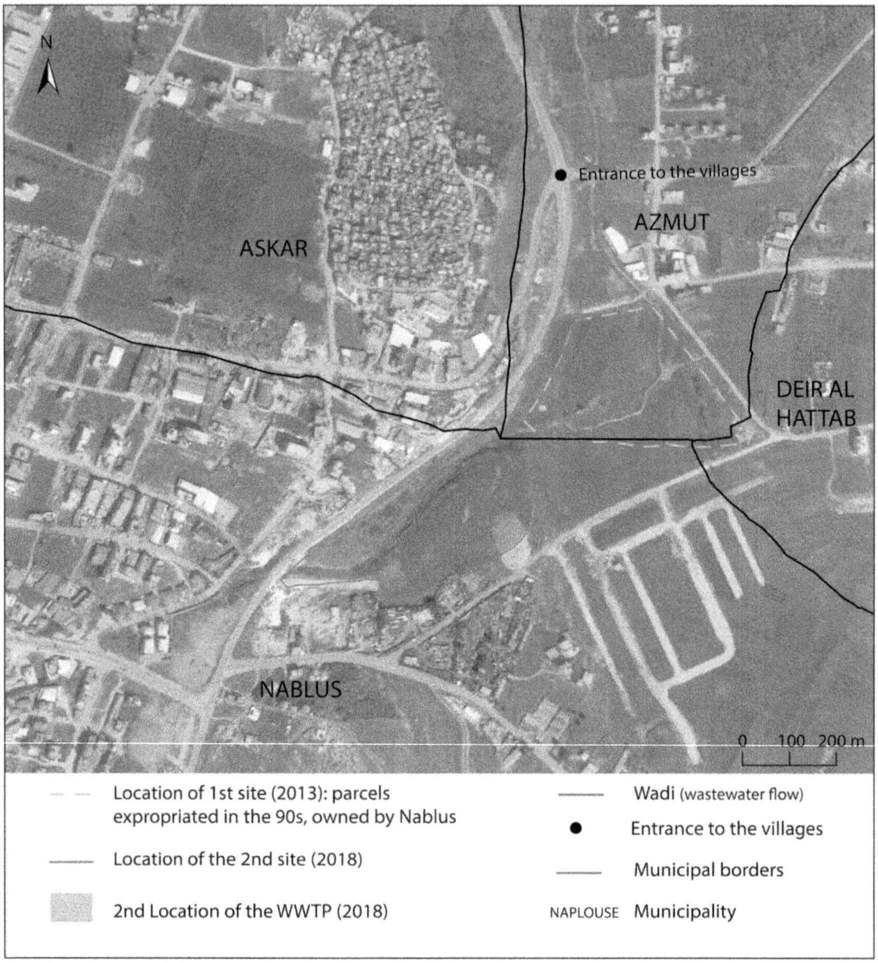

Figure 18.3 Location of the future wastewater treatment plant on the new expropriated lands in 2019

allow access to it. Two years after its establishment, the Israeli government created a nature reserve on the land around Elon Moreh, covering nearly 1,500 hectares (Etkes 2015). Establishing a nature reserve on these lands is supposed to protect nature by preventing any construction or cultivation. However, this process has, above all, considerably restricted access to the land by Palestinian residents and landowners. Dispossessing and expropriating by imposing a natural reserve or park is a frequent policy used by Israel against Palestinians (Braverman 2019b). Finally, some land was also confiscated by military order to build a road to access the settlement of Elon Moreh. These military requisitions continue to this day. As in other parts of the West Bank, the Israeli colonization policies generate direct, structural, and infrastructural violence. The territorial division into three areas, A, B, and C, following the Oslo Agreements, has reinforced the infrastructural violence by locating the majority of the land of the three villages in area C, acting as a buffer zone between the settlement of Elon Moreh and the Palestinian residential areas.

Dynamics of dispossession through usage restrictions

The regular conflicts with the colony accentuate the structural and infrastructural violence suffered by the three villages since the 1980s. The colony of Elon Moreh appropriated the main spring, *Ein al kbira* (the big spring), which alone supplied the three villages with domestic water. In 1983, they were connected to the supply system of Mekorot, the Israeli national water company, but continued to use the spring for domestic and agricultural purposes as well as for animal watering. The transformation of the spring into a swimming pool by the settlers (Braverman 2019a) considerably dried up the spring below, *Ein al sghira* (the small spring), which was connected to the spring upstream by a system of pipes to carry the water. In 2007, the Elon Moreh colony dumped its wastewater on the lands of Deir Al Hattab and Azmut because of a technical problem in the colony's WWTP. This situation continued for several years, causing the death of many olive trees and decreasing the yield of others.[11] Finally, the Israeli infrastructures, military or civilian, have considerably, if not totally, restricted the access of the inhabitants to their lands located in area C.[12]

The various project documents for East Nablus WWTP do not mention the Israeli occupation, except for the administrative issues of permit applications. They portray the project as a technical one and ignore the colonial context in which the WWTP is to be built. This may seem paradoxical because the PA regularly builds its political resistance discourses around the Israeli occupation and its consequences in terms of resource dispossession. Here, since the project is in area B, there is little direct interference with the occupation as compared to projects in area C. As a result, this dimension has disappeared from project documents and institutional discourses, in a way silencing the violent experiences of the inhabitants of the three Palestinian villages.

The numerous past and present conflicts between the three Palestinian villages and the settlement of Elon Moreh demonstrate the persistence of infrastructural violence. Going back over these entanglements allows us to understand the argument developed by the villagers and the parallel drawn between Israeli expropriation policies and the ones envisaged by the PA and the Municipality of Nablus for the WWTP project. Making this parallel does not mean comparing two unequal governmental authorities, as one is occupying the other. However, it helps in understanding the narratives and resistance practices deployed by villagers opposed to the project and how such tensions around land arose. Land defenders from these Palestinian villages already resist Israeli occupation by demonstrating every Friday, marching on their lands.

Thus, resistance techniques have been in place for a long time and will serve to defend their lands from the construction of the future WWTP.

The discourses of contestation: land concerns, feelings of landscape degradation, and coloniality

The choice of site is the major point of opposition from the villagers. They recognize the collective interest of having a WWTP, but they refuse to give up their own lands for the benefit of others, and especially for Nablus' interests. However, their struggle also encompasses social and political aspects, transforming it into what Sébastien (2017) would call "enlightened resistance". To understand the blockages around the realization of this project, it is necessary to explore the three main arguments deployed by the inhabitants and the narratives they construct. First, for landowners, the starting point of the dispute dates back to the expropriations carried out at the end of the 1990s. Second, at the village level, the installation of a WWTP represents yet another sign of the landscape degradation in the eastern region of Nablus and the "racism" of the city against its eastern rural surroundings. Finally, the inhabitants accuse the PA of mimicking Israeli practices with regard to land expropriations and enforcement practices of the project, which I analyze as signs of coloniality.

Expropriations as a starting point for contestation

There are many conflicting stories about the process of land expropriation. For the expropriated owners, it is an injustice. For the Municipality of Nablus, it is simply an expropriation for the public interest, as has been done elsewhere for development projects (Perrier 2020). The issue of expropriation adds to the list of land dispossessions suffered by the inhabitants of these villages and accentuates the infrastructural violence in this area.

At the end of the 1990s, a presidential decree, signed by Yasser Arafat, then-president of the Palestine Liberation Organization (PLO) and the PA, ratified the process of expropriation of 7 hectares on the lands of Azmut. Several employees of the Municipality of Nablus confirmed that the expropriation decree named the WWTP project as the reason for expropriation.

There are differing accounts of how the expropriation process unfolded. The Water and Sewerage Department of the Municipality of Nablus states that no problems occurred at the time of the expropriation and that the municipality compensated the owners as required by law.[13] However, one of the expropriated owners denounced the relentlessness of the Municipality of Nablus. According to his account, the latter and its former mayor had already tried to buy the land in 1994–1995, to no avail. This failure prompted the municipality to use a presidential decree to expropriate the land. According to the owner, he resisted and refused to sell his land on several occasions. He experienced the expropriation as a coercive maneuver.[14] The director of the Water and Sanitation Department at PWA knows about opposition and discontent but believes the issue lies in the amount of compensation received, considered below market prices by the owners.[15]

Despite the prohibition in the decree on using expropriated land, the parcels in question continue to be cultivated. Most of the former owners cultivated rain-fed wheat on these plots. These agricultural activities, which are a priori illegal, show the reluctance of landowners to accept the expropriation of their land. Moreover, the laissez-faire attitude of state institutions towards these activities illustrates both a lack of interest in the expropriated area, as long as the WWTP project was not implemented, and a chronic difficulty in implementing the decisions adopted.

With the establishment of the settlement of Elon Moreh and the gradual confiscation of more and more land by the Israeli army, the possibilities of expansion for the three villages are diminishing. The western part is the only space available for the construction of new homes, new industries, or the continuation of agricultural activities. On 3 May 2015, the mayors of the villages of Salem, Deir Al Hattab, and Azmut wrote a letter to the Governor of Nablus, detailing their arguments against the construction of the WWTP on the chosen site in 2013. The first argument they mention refers to the issue of the future expansion of the villages. They write: "the sewage treatment plant is at the entrance of the three villages and the construction possibilities for the expansion of the three villages are only in the western zone, which is intended for the plant".[16] In addition, since the first feasibility studies were carried out in 1997, the constructed areas of the three villages have expanded. The new EIA carried out in 2019 notes these changes.

Landscape degradation arguments: aesthetic, olfactory, and health concerns

Opponents of the WWTP project see it as further evidence of Nablus' disinterest in the eastern part of the governorate. During the demonstrations held in January 2016, the inhabitants held up a banner that read: "The Eastern region is not a garbage dump". It reveals the way they perceive the different actions of the Nablus Municipality carried out in this region. In the same letter of 3 May 2015, quoted previously, the mayors of the three villages explain that their area is in great need of development projects to improve the living conditions of the inhabitants. However, they explain that it remains a "very ignored" area. They add that several projects, such as a sports field, a hospital, and a branch of Al-Najah University in Nablus, should have been built in this area, but were eventually established elsewhere.

These criticisms of abandonment reflect a feeling of exclusion and marginalization that induces a certain distrust among the inhabitants towards the Municipality of Nablus and the governorate. These discourses, whether the disinterest is real or not, affect the interactions between the different actors involved. The feelings of exclusion and neglect come up in the interviews conducted, in the documents written by the village mayors, and in the analysis of the infrastructural landscape of the area.

Different infrastructures surround them and reinforce their feelings of suffocation and isolation. Opposite Azmut, on the other side of the *wadi*, is an open-air waste transfer station: the Al Serafi station, which receives all types of waste. Sometimes the waste is burned, leaving toxic fumes for the surrounding villages.[17] In addition, some residents dislike the presence of a refugee camp near their villages. The Askar refugee camp runs along the *wadi* and is located opposite Azmut. Established in 1950, it expanded in 1964 to accommodate population growth. Relations between Askar and the three neighboring villages are tense. Finally, the Israeli settlement of Elon Moreh and the conflicts with it further illustrate the landscape degradation for its inhabitants.

For the Municipality of Nablus, the low development of this area is, on the contrary, an argument in favor of the construction of the WWTP in East Nablus. The EIA carried out in August 2019 mentions a fairly common natural environment in terms of biodiversity, already highly affected by "anthropogenic activities", and a non-industrial landscape destined to change as the Municipality of Nablus plans to set up an industrial and commercial zone (KfW 2019: 98). This representation produced by the Municipality of Nablus runs counter to the one built by the inhabitants.

The problem of solid waste treatment adds up to that of the untreated wastewater flowing into the *wadi*, along the road, and near the houses and fields of the three villages. In the summer, when rainfall is scarce, the *wadi* carries only wastewater coming from Nablus city and the surrounding villages. Wastewater is discharged into this *wadi* by tanker trucks a few hundred meters upstream from the entrance to the three villages. The flow of this wastewater represents a health issue, to which the construction of a WWTP could put an end. The three villages consider this project necessary to improve the sanitary situation, but contest the location chosen for the construction.

Some opponents of the project fear landscape degradation with the construction of a WWTP. The proximity of the future WWTP to the existing houses would imply a degradation of the living environment of the closest populations. The boundaries of the constructed areas have changed since 1997 due to population growth. As a result, the distance originally planned between the WWTP and the first settlements has decreased significantly.

Proximity to the future WWTP raises concerns about odors and the potential dangers of daily inhalation of gases. Some villagers fear an increase in cancer cases because of the proximity of the WWTP.[18] In October 2017, a report conducted by an independent Palestinian consultant reinforces the fears related to air pollution. It was commissioned by a pharmaceutical company located in Salem and by the three villages concerned. The Municipality of Nablus contests its impartiality, particularly because the consultancy was paid for by the pharmaceutical company, which opposes the construction of the WWTP. Finally, during the demonstrations organized in 2016 and 2017 on the chosen site for the WWTP, the residents clearly expressed their worries, comparing the plant to a source of epidemic. This chapter does not aim at knowing if these fears are justified or not. The fact inhabitants formulated them illustrates their concerns and helps identify their arguments. However, they remain ignored and discredited by the PA and the Nablus Municipality.

Resistance practices against signs of coloniality from the PA

To demonstrate their discontent, the villages and their allies used different methods of protest: demonstrations, legal action, and sabotage. These tools do not have the same objective. The demonstrations and actions of deterioration are the concrete expression of their refusal to accept the WWTP on their lands. The legal recourse has focused more on challenging the expropriation of land at the end of the 1990s. These contestation mechanisms have resulted in authoritarian responses from the PA. This confrontation between village demonstrators and the PA is sometimes similar to, and uses certain methods of, confrontation with the Israeli occupation.

The contestation by the three villages took different forms between 2014 and 2017. The evolution of resistance practices, from official letters to demonstrations, illustrates the various forms of infrastructural violence against defenders. In the first two years, village representatives expressed their opposition through meetings and official letters. Villagers, and even their local representatives, were excluded from consultation processes during planning. During 2016, the project remained at a standstill, mainly due to the municipal electoral calendar. Municipal elections were scheduled for 2016 but were eventually postponed to the spring of 2017 by the PA. This appears to have delayed the court decision and also influenced the publication of the ministerial decree ordering the implementation of the project by the police force.

Demonstrations resumed in November 2017 following an attempt by the Municipality of Nablus to start the project by creating facts-on-the-ground on the expropriated plots. This step amounted to a violence of material imposition for land defenders. Following the demonstration, Palestinian police arrested several residents of Deir Al Hattab. The Municipality of Nablus justified these arrests by the damage caused at the site of the WWTP during the demonstration. For the village council and the residents of Deir Al Hattab, these arrests were reprisals for the actions against the construction of the WWTP.

The protest mechanisms used in East Nablus are reminiscent of those mobilized against the Israeli occupation. In their struggle against the WWTP, the inhabitants of the villages use the same tools as the inhabitants of other Palestinian villages protesting against the Israeli separation wall. In the latter, inhabitants organize weekly demonstrations every Friday after the collective prayer. These demonstrations are one of the symbols of the perseverance of resistance to the Israeli occupation. Reusing this tool of resistance against a development project carried by local government institutions is a strong symbol of opposition. It also demonstrates the importance of this issue for the demonstrators. These resistance practices used by land defenders illustrate the violence of what they see as "infrastructural colonization", as Dunlap (2020) frames it, referring to a second colonization brought by their own Palestinian governmental representatives. Thus, land defenders mobilize resistance strategies they are already used to, but against practices of coloniality by their own government, supported by donor funds.

The repressive mechanisms envisaged and used by the PA also refer to certain tools used by the Israeli occupation. The repression put in place by the PA, through arrests, threats, and blackmail, is all the more contested and experienced as a betrayal because it is carried out by Palestinian institutions. One land defender from Deir Al Hattab also explains the strategies of attrition used by the PA and the Municipality of Nablus to take down the resistance movement, as well as attempts to divide the movement. For instance, village councils were invited separately to the negotiation table, allowing other parties to pit one council against another. Israeli military forces regularly use this strategy against Palestinians. It highlights the complexity for the PA to realize itself as a state, caught between the will to impose its decisions despite local opposition and the political risk of being compared to the Israeli occupation.

This case study sheds light on the complexity of the relations between the villages concerned and the Municipality of Nablus, a sign of a mode of government in peril. Land defenders feel marginalized by the city because of a lack of basic infrastructure, while they also refuse the form of inclusion promoted by the Nablus Municipality and resulting in infrastructural violence associated with the industrial complex and wastewater infrastructures that mostly benefit the urban population.

Epistemic violence under green development and its counterpart

Since the project is scientifically approved and supposed to improve the quality of life for inhabitants, why are some villagers "stubbornly" opposing it? Asked only from the point of view of scientific and technical arguments, this question closes the door to a broader understanding of the problem. It reproduces the asymmetry constructed by scientists and their allies between "knowledge" and "belief", making any other argument illegitimate (Latour 1987).

Faced with the arguments put forward by the contestations, institutional actors, such as the PWA, the Municipality of Nablus, and donors, adopted a strategy of *irrationalizing* the local voice. They devalue local speech by presenting it as irrational thinking, incapable of understanding and accepting scientific arguments. On the other hand, land defenders also play on their reputation of "stubborn" and "closed-minded" people[19] to support their resistance movement.

The rational hegemony of scientific discourse

According to the defenders of the project, the scientific argument cannot be questioned and is superior to any other argument. For them, the project represents only a technical solution to an equally technical problem of pollution and mismanagement of water resources. The very nature of the project encourages this discourse because pollution is a threat to the environment and populations, which accentuates the necessity of the construction of the WWTP.

The institutional actors also mobilize the rhetoric of state-building to convince the villages to accept this project as a step towards a future Palestinian state. The establishment of a centralized WWTP is presented as a tool for state-building because it requires significant management capacity that must be coordinated at the regional or even national level. Moreover, the reuse project envisaged in East Nablus echoes the arguments mobilized in West Nablus, arguing that irrigated agriculture will strengthen the Palestinian economy (Perrier 2020). Thus, by drawing this parallel between the construction of the WWTP and state construction, the PA and the Municipality of Nablus further isolate local protest.

The dialogue is compromised because the different parties to this conflict deploy arguments that respond to their own understandings of the project and representations of the environment, while ignoring those of the opposing parties. Analyzing the different types of discourse deployed by each of the parties, and those that they ignore, reveals the "dialogue of the deaf" that has been going on for several years around this project and the epistemic violence at play against the arguments of the contestation.

Some mediators have attempted to establish common ground between the different parties. The final objective of this consultation was essentially to convince the villages, thus postulating from the outset a refusal to engage in dialogue and a willingness to impose the dominant narrative of the PA, the Municipality of Nablus, and the donor. In fact, it underlined the marginalization of the villages and the lack of legitimacy of the PA in trying to juggle different actors, local and international. It also noted a lack of communication and dialogue on the part of the Nablus Municipality at the beginning of the project. The three mayors of the villages already deplored the lack of field visits by the consultants and the lack of information to which they had access. This criticism had been voiced two years earlier by the village representatives, but the consultancy finally allowed it to be taken into account. This demonstrates the epistemic violence at work since the villagers' words become intelligible when they are reported by an "expert".

Representatives of the three villages repeatedly proposed alternative locations for the project further downstream in the Al Sajour valley. The Municipality of Nablus and the donor refused to consider these proposals as it would have entailed additional financial cost.[20] However, political reasoning may be at stake, as some land defenders put it, because Nablus Municipality refuses to construct the project outside its municipal border, as it would lose its control over it.

Portraying local arguments as irrational

The way in which each party represents the other creates a hierarchy of discourse that makes dialogue difficult. According to Latour (1987: 478), accusations of irrationality appear as soon as the paths of different parties cross, revealing different worlds made up of different elements that make sense for each party. By portraying local people as irrational, the Nablus Municipality and its allies legitimize their own discourse and present it as indisputable. This discourse is part of what Latour calls "the interior", that is, the world of scientists, which constructs a demarcation between the "knowledge" produced internally by science and the "beliefs" existing externally that are therefore considered "non-scientific". The fact that the land defenders maintain their opposition despite these technical explanations is seen as proof of their irrationality.

The PA and the Municipality of Nablus are deploying what I call a strategy of *irrationalizing* of the local voices. First, this strategy homogenizes the protesting group through character traits with negative connotations. Second, it explains their irrationality by external social, cultural, and psychological factors that prevent them from accessing scientific "knowledge" (Latour 1987). Land defenders are also portrayed as impressionable people who change their minds very easily. This description devalues the villagers' decision-making process. It also implies that the decision-making process itself is based on fragile arguments that come from informal and private discussion. The villages of Salem, Deir Al Hattab, and Azmut are well aware of their reputation as stubborn and closed-minded people and are playing along, hoping to derail the project or obtain the concessions they feel necessary.

The construction of such representation and reputation is nourished by the cleavage between the urban bourgeoisie and the rural population. This cleavage was already found under the Ottoman Empire (Doumani 1995) and continued in the 20th century (Robinson 1997) during the formation and transformation of the urban elite and its relations with the peasantry. In East Nablus, the urban/rural divide predates the WWTP project and represents an important fissure fueled by development projects.

Conclusion: infrastructural violence and environmental coloniality

This chapter demonstrated that a wastewater treatment plant project can be a source of infrastructural violence as early as its design phase. Public interest legitimates the construction of such a project, as it reduces pollution and health hazards. However, it mostly serves Palestinian urban areas, here Nablus city, while being located in peri-urban Palestinian villages already suffering from Israeli infrastructural violence and what land defenders see as landscape degradation and their marginalization as a peri-urban industrial area. Moreover, land defenders protest against a third form of violence from Palestinian governmental institutions implementing practices of repression and coercion, mimicking Israeli colonial practices. Thus, while protesting against a development project, land defenders also fight against practices of coloniality from their own governmental institutions.

Coloniality also reflects in the way Palestinian institutions apprehend the environment, clashing with the view of land defenders. Palestinian national water and planning strategies build on colonial concepts, such as "water efficiency", which silence local knowledge and practices on environment sustainability (Perrier 2020). Epistemic violence has a long history as a tool of power in colonial contexts, justifying several projects deemed to "improve" colonized environments (Broich 2013; Davis 2016; Davis & Burke 2011; Mitchell 2002). In the case study presented here, epistemic violence represents a form of violence against land defenders as part of the overall infrastructural violence. It also adds up to practices of coloniality used by the PA to discredit land defenders.

Finally, this chapter highlights the usefulness of thinking in terms of coloniality to understand how the PA appropriated means of repression from its own experience as a colonized institution and people. Perrier (2020) introduces the term "environmental coloniality", defined as the strong influence of colonial representations of environment on post-colonial regimes as to how they perceive and manage it. Even though the PA shall not be considered as a post-colonial regime because of the ongoing Israeli occupation, thinking in terms of coloniality implies considering modes of appropriation from colonial knowledge and practices. Expropriation led by the PA and supported by donors in the example of the WWTP in East Nablus clearly demonstrates the penetration of the colonial imaginary of land management.

Notes

1 I analyzed the feasibility study carried out in 1997 and the Environmental Impact Assessment (EIA) of August 2019 for the eastern WWTP. The first document details the projects for the east and west of Nablus.

2 I obtained an Arabic version of the EIA and translated the criteria into English.

3 The first two criteria concern the ownership status of the plots under study to evaluate the financial cost of a possible transfer of ownership. Expropriating private land, for example, implies foreseeing the cost of the financial compensation that will be allocated to the owner(s). Site n° 1, owned by the Municipality of Nablus and already expropriated in the 1990s, and site n° 3, owned by the PA, scored the maximum. The other sites, classified as private land, scored 0, showing that land belonging to a public institution is a major asset for the project. No further details are provided on the land issue and the tensions, actual or potential, around it. The fifth and sixth criteria determine the geographical and topographical characteristics to be considered in the choice of site. These include the possibility of transferring wastewater by gravity to the future WWTP, without the need for pumping stations, and the ease of access to the site. Land that is bumpy or located in mountainous areas does not correspond or would require additional work, and therefore a significant additional financial cost.

4 Perrier (2020) details the selection for the new site and the new discourse portraying the previous chosen site as technically not suitable anymore. She also details the new expropriation wave that began in 2018.

5 This estimate comes from the report of the Israeli NGO Kerem Navot (Etkes 2015): "Elon Moreh was established in 1980, and its jurisdictional area is approximately 1,278 dunams. Approximately two years after its establishment, 24,226 dunams west of the settlement were declared as the Mt. Kabir Nature Reserve". I triangulated this estimate

using QGIS by calculating the vector layer areas of the Elon Moreh settlement and the Har Kabir Nature Reserve. I obtained an estimate equivalent to that of Kerem Navot.

6 Zone C contains 41% of Azmut's land, 48% of Deir Al Hattab's land, and 73% of Salem's land.

7 For more detail on the Court's reasoning, see Justice Landau (1980).

8 Not all state land in the West Bank has the same status: some has been "registered" and some has been "declared". This difference is reflected in the changes to Israeli Military Order 59 of 1969 (for a compilation and comments of Israeli military orders, see Rabah & Fairweather 1995). "Registered" land refers to land that was recognized as state land prior to the Israeli occupation of the West Bank in 1967, and is thus registered as such in the land register by the Ottoman authorities or British or Jordanian agents. Military Order 59 defines state land as land belonging to a hostile state or to an organization linked to that hostile state. Since the State of Israel has been the occupying power in charge of the Palestinian occupied territories since 1967, state land designated as such before 1967 reverts to it. This is the case for part of the land on which the Israeli government established the settlement of Elon Moreh. However, Military Order 364 of 1969, amending Order 59, broadened the original definition by establishing a presumption of statehood. Thus, state land includes any land whose owner is unable to prove ownership before the Israeli military committee (Rabah & Fairweather 1995). This has allowed the Israeli government to declare large portions of land as state land, relying on the complexity of appeal procedures to prevent and discourage challenges by potential Palestinian owners. In 1984, Military Order 59 was further amended by Military Order 1091, which made it possible to declare state land after 1967. This amendment thus made it possible to avoid a second setback after the judgment of Elon Moreh. These various amendments allow the Israeli government to extend the surface of state land.

9 This information comes from various NGO reports ("A Guide to Housing, Land and Property Law in Area C of the West Bank" 2012; Aloni 2016). Lustick (1981) also mentions the establishment of the colony of Elon Moreh on state land, without specifying whether this is a status granted before or after 1967. We were able to triangulate this information with an interactive map produced by the Economic Cooperation Foundation (ECF), where the legend differentiates between "state land registered" and "state land declared". The map is available online: https://ecf.org.il/ (accessed on July 18, 2020).

10 According to Article 78 of the Ottoman Land Code, a farmer could obtain a title to *miri* land provided that he had cultivated the land for at least ten years. The Ottoman Land Code of 1858 defined five categories of land status, including *miri* land, which is owned by the state and includes cultivated land (with usufruct right for users) or land within 2.5 km of the boundary of the residential area (Bisharat 1994; for more information on these categories of land tenure, see Granott 1952; Shehadeh 1988). The British and then Jordanian authorities adopted the doctrine of "*reasonable* cultivation" to deal with the case of *miri* land of poor quality and topology: an individual could acquire title if he had cultivated the land, for at least ten years, even sporadically, according to the attributes of the land (A Guide to Housing, Land and Property Law in Area C of the West Bank 2012). Instead of applying the doctrine of "reasonable culture", Israeli authorities considered that land that has not been cultivated for more than three years becomes state land, unless the individual has a title to the land. The Jordanian authorities had begun a cadastral census in 1953, which was completed for only half of the land in the West Bank. This left some Palestinian landowners without a Jordanian title deed, which was considered as irrefutable proof of ownership for the Israeli administration (Lustick 1981).

11 Interview conducted on 08/02/2017 in Deir Al Hattab. This information is consistent with that reported by the Israeli NGOs B'Tselem (Hareuveni 2009) and Rabbis for Human Rights (https://rhr.org.il/eng/2017/10/palestinian-olive-trees-killed-sewage-run-off-elon-moreh-settlement/.

12 See for instance road 555, built on expropriated private Palestinian land for Israeli settlers only, which makes access to the land north of this road almost impossible. Similarly,

the establishment of an Israeli outpost, Skali's Farm, inside the nature reserve, where no construction is allowed under Israeli law, prevents Palestinian residents from accessing the land to the southeast of the settlement.

13 Interview conducted on 20/02/2017 in Nablus.

14 Interview conducted on 08/02/2017 in Azmut.

15 Interview conducted on 01/06/2017 in Ramallah.

16 Letter in Arabic dated 3/05/2015 and accessed during fieldwork. Translation into English made by the author.

17 Stamatopoulou-Robbins (2019) examines waste management in the West Bank and the issues related to the lack of a national strategy by the PA. Other research focuses on the health consequences of toxic e-waste burning in the West Bank (Davis & Garb 2019).

18 Interviews conducted on 08/02/2017 in Azmut and 14/11/2017 in Deir Al Hattab.

19 Members of the PWA and the Nablus Municipality used these adjectives several times to qualify the inhabitants of these villages during the interview conducted for this study.

20 Interviews conducted in July and August 2017 at the Municipality of Nablus.

References

Aloni, A. (2016, December). Expel and exploit: The Israeli practice of taking over rural Palestinian land. *B'Tselem*. www.btselem.org/download/201612_expel_and_exploit_eng.pdf

Anand, N. (2012). Municipal disconnect: On abject water and its urban infrastructures. *Ethnography*, *13*(4), 487–509. https://doi.org/10.1177/1466138111435743

Barnes, J. (2012). Pumping possibility: Agricultural expansion through desert reclamation in Egypt. *Social Studies of Science*, *42*(4), 517–538.

Barnes, J. (2017). States of maintenance: Power, politics, and Egypt's irrigation infrastructure. *Environment and Planning D: Society and Space*, *35*(1), 146–164.

Bisharat, G. E. (1994). Land, law, and legitimacy in Israel and the Occupied Territories. *The American University Law Review*, *43*(467), 467–561.

Braverman, I. (2019a). Silent springs: The nature of water and Israel's military occupation. *Environment and Planning E: Nature and Space*. https://doi.org/10.1177/2514848619857722

Braverman, I. (2019b). Nof kdumim: Remaking the ancient landscape in East Jerusalem's national parks. *Environment and Planning E: Nature and Space*. https://doi.org/10.1177/2514848619889594

Broich, J. (2013). British water policy in Mandate Palestine: Environmental orientalism and social transformation. *Environment and History*, *19*(3), 255–281.

Coelho, K. (2004). *Of engineers, rationalities, and rule: An ethnography of neoliberal reform in an urban water utility in South India*. Tucson, AZ: The University of Arizona.

Davis, D. K. (2016). *The arid lands: History, power, knowledge*. Cambridge, MA: MIT Press.

Davis, D. K., & Burke, E. (Eds.). (2011). *Environmental imaginaries of the Middle East and North Africa*. Athens, OH: Ohio University Press.

Davis, J.-M., & Garb, Y. (2019). A strong spatial association between e-waste burn sites and childhood lymphoma in the West Bank, Palestine. *International Journal of Cancer*, *144*(3), 470–475.

Desai, R. (2018). Urban planning, water provisioning and infrastructural violence at public housing resettlement sites in Ahmedabad, India. *Water Alternatives*, *11*(1), 86–105.

Doumani, B. (1995). *Rediscovering Palestine: Merchants and peasants in Jabal Nalbus, 1700–1900*. Berkeley: University of California Press.

Dunlap, A. (2017). Wind energy: Toward a "sustainable violence" in Oaxaca. *NACLA Report on the Americas*, *49*(4), 483–488.

Dunlap, A. (2020). Bureaucratic land grabbing for infrastructural colonization: Renewable energy, L'Amassada, and resistance in southern France. *Human Geography*, *13*(2), 109–126.

Etkes, D. (2015). *A locked garden: Declaration of closed areas in the West Bank*. Jerusalem: Karem Navot. https://f35bf8a1-b11c-4b7a-ba04-05c1ffae0108.filesusr.com/ugd/cdb1a7_5d1ee4627ac84dca83419aebf4fad17d.pdf

EU (2013). *Commission decision on a special measure (Part III): Action fiche for Palestine*. Brussels: European Union.

Fustec, K. (2014). *Processus multi-échelles, enjeux environnementaux et construction étatique. Le cas de l'Autorité palestinienne, des politiques de gestion de l'eau et du changement climatique*. Montpellier: Université Montpellier 3.

Gazit, S. (2003). *Trapped fools: Thirty years of Israeli policy in the territories*. London: Frank Cass.

Granott, A. (1952). *The land system in Palestine: History and structure*. London: Eyre & Spottiswoode.

NRC (2012). *A guide to housing, land and property law in area C of the West Bank*. Oslo: Norwegian Refugee Council.

Hareuveni, E. (2009, June). Foul play: Neglect of wastewater treatment in the West Bank. *B'Tselem*. www.btselem.org/sites/default/files/sites/default/files2/200906_foul_play_eng.pdf

Harvey, P., Jensen, C. B., & Morita, A. (2017). Introduction: Infrastructural complications. In P. Harvey, C. B. Jensen, & A. Morita (Eds.), *Infrastructures and social complexity: A companion* (1st ed., p. 42). Abingdon, UK: Routledge.

Jensen, C. B. (2017). Pipe dreams: Sewage infrastructure and activity trails in Phnom Penh. *Ethnos*, *82*(4), 627–647.

Justice Landau. (1980). Israel: Supreme court judgment with regard to the Elon Moreh settlement in the occupied West Bank. *International Legal Materials*, *19*(1), 148–178.

Kaika, M. (2006). Dams as symbols of modernization: The urbanization of nature between geographical imagination and materiality. *Annals of the Association of American Geographers*, *96*(2), 276–301.

KfW (2019). *Palestine–wastewater management Nablus East: Revised environmental impact assessment (draft version)*. Frankfurt: KfW, EU, and Nablus Municipality.

Latour, B. (1987). *La science en action. Introduction à la sociologie des sciences*. Paris: La Découverte.

Li, T. M. (2018). After the land grab: Infrastructural violence and the "Mafia system" in Indonesia's oil palm plantation zones. *Geoforum*, *96*, 328–337.

Loftus, A. (2006). Reification and the dictatorship of the water meter. *Antipode*, *38*(5), 1023–1045.

Lustick, I. (1981). Israel and the West Bank after Elon Moreh : The mechanics of de facto annexation. *Middle East Journal*, *35*(4), 557–577.

McKee, E. (2020). Divergent visions: Intersectional water advocacy in Palestine. *Environment and Planning E: Nature and Space*, 251484862090938.

Mitchell, T. (2002). *Rule of experts: Egypt, techno-politics, modernity*. Berkeley: University of California Press.

Nolan, C., Goodman, M. K., & Menga, F. (2020). In the shadows of power: The infrastructural violence of thermal power generation in Ghana's coastal commodity frontier. *Journal of Political Ecology*, *27*(1), 775–794.

Perrier, J. (2020). *Quelle gouvernance des eaux pour quelle construction étatique dans les territoires palestiniens? L'étude des constellations hydropolitiques des eaux douces et usées*. Montpellier: Université Paul Valéry, Montpellier 3.

Rabah, J., & Fairweather, N. (1995). *Israeli military orders in the occupied Palestinian West Bank, 1967–1992* (2nd ed.). Jerusalem: Jerusalem Media & Communication Centre.

Robinson, G. E. (1997). The growing authoritarianism of the Arafat regime. *Survival*, *39*(2), 42–56.

Rodgers, D., & O'Neill, B. (2012). Infrastructural violence: Introduction to the special issue. *Ethnography*, *13*(4), 401–412.

Sébastien, L. (2017). From NIMBY to enlightened resistance: A framework proposal to decrypt land-use disputes based on a landfill opposition case in France. *Local Environment*, *22*(4), 461–477.

Selby, J. (2013). Cooperation, domination and colonisation: The Israeli-Palestinian joint water committee. *Water Alternatives*, *6*(1), 1–24.

Shehadeh, R. (1988). *Occupier's law: Israel and the West Bank* (2nd ed.). Beyrouth: Institute for Palestine Studies.

Stamatopoulou-Robbins, S. (2019). *Waste siege: The life of infrastructure in Palestine*. Palo Alto, CA: Stanford University Press.

Trottier, J. (1999). *Hydropolitics in the West Bank and Gaza strip*. Jerusalem: PASSIA.

Trottier, J. (2007a). A wall, water and power: The Israeli "separation fence". *Review of International Studies*, *33*(1), 105–127.

Trottier, J. (2007b). *Eau, Pouvoir et Société* [Thèse d'Habilitation à diriger des recherches]. Bordeaux: Université Montesquieu-Bordeaux.

Trottier, J., Leblond, N., & Garb, Y. (2019a). The political role of date palm trees in the Jordan Valley: The transformation of Palestinian land and water tenure in agriculture made invisible by epistemic violence. *Environment and Planning E: Nature and Space*, *3*(1), 114–140.

Trottier, J., & Perrier, J. (2017). Challenging the coproduction of virtual water and Palestinian agriculture. *Geoforum*, *87*, 85–94.

Trottier, J., & Perrier, J. (2018). Water driven Palestinian agricultural frontiers: The global ramifications of transforming local irrigation. *Journal of Political Ecology*, *25*(1), 292–311.

Trottier, J., Rondier, A., & Perrier, J. (2019b). Palestinians and donors playing with fire: 25 years of water projects in the West Bank. *International Journal of Water Resources Development*. https://doi.org/10.1080/07900627.2019.1617679

Weinthal, E., & Sowers, J. (2019). Targeting infrastructure and livelihoods in the West Bank and Gaza. *International Affairs*, *95*(2), 319–340.

World Bank. (2009). *Assessment of restrictions on Palestinian water sector development*. No 47657-GZ, Washington, DC.

Zeitoun, M. (2008). *Power and water in the Middle East: The hidden politics of the Palestinian-Israeli water conflict*. I.B. Tauris. http://choicereviews.org/review/10.5860/CHOICE.46-1153

Zoomers, A. (2018). Plantations are everywhere! Between infrastructural violence and inclusive development. *Geoforum*, *96*, 341–344.

19 Defending territory from the extraction and conservation nexus

Philippe Le Billon

Extraction and conservation seem to be polar opposites, yet spaces of resource extraction and wildlife conservation are increasing intertwined and threaten the territories of many Indigenous and traditional agrarian communities. Extraction–conservation partnerships connect a wide array of organizations, objects, narratives, and practices, including extractive companies, conservation organizations, local communities, government authorities, geological formations, ecosystems, protected areas, carbon sinks, and endangered species lists. Extraction and conservation partnerships have a broad set of costs and benefits (Adams 2017) and create unevenly distributed potential and realized values (Enns et al. 2019). Building on a growing literature investigating the articulation of extraction and conservation (Norris 2017; Symons 2018; Sonter et al. 2018), this chapter focuses on the exclusionary terrain of their (joint) operations and the challenges that they pose to environmental and land defenders seeking to protect their communal lands and associated traditional livelihoods.

From a conceptual perspective, I draw on political ecology, including inputs from 'new materialism', 'post-developmentalism', and epistemic 'decolonization' efforts (Schulz 2017), to discuss relations between extraction and conservation and implications for the defense of territories. I focus on the ways extractive corporations and conservation organizations co-create spaces of double exception to secure their activities and derive new values out of their 'harnessing of nature'. Conceptually, the main argument is that despite progressive discourses, extraction and conservation alliances often seek to rid the land of people resisting extractivism and neoliberal conservation. Empirically, I seek to show how 'biodiversity offsets' pretending to environmentally compensate for the impacts of extraction can result in further violence against local communities and ecosystems. Following this introduction, I briefly review relations between extraction and conservation and the threats that these represent for Indigenous and local community lives and territories.

Extraction and conservation nexus

Extraction and (neo)extractivism are now well-established concepts within studies of exploitation and subjectification associated with 'development'

(Acosta 2013; Junka-Aikio & Cortes-Severino 2017; Veltmeyer 2016). Mostly identified in the form of mining or oil and gas activities, extraction is more generally understood as the extirpation of select 'natural' materials for human purposes, including export-oriented commodity trade and industrial production. Here, I extend this conception to include not only other 'resources' (Douglas & Alie 2014; Van Vliet et al. 2016), but also 'land' itself – in the sense of loss of control and access to a 'territory of life' fundamental to Indigenous and traditional agrarian communities (Correia 2019; Li 2014; see also Milanez, this volume). From this perspective, conservation can constitute one of the gravest forms of extraction, as the establishment of 'protected areas' generally results in territorial losses for local communities.

Many studies of extraction come from, or relate to, political ecology approaches, especially as a result of the commodity boom initiated in the early 2000s (Bridge 2004), exposing uneven power relations in resource control and extractive governance (Peluso 1992), the ambivalent attitudes to and effects of extraction (Bebbington et al. 2018; Bridge 2008), the various forms of violence associated with extraction (Navas et al. 2018; Nixon 2011; Watts 2001), and the numerous struggles against extractivism (Condé & Le Billon 2017; Temper et al. 2015). Some studies have also focused on the claims of sustainability and contributions to climate change mitigation and biodiversity conservation that extractive companies have advanced, notably to counter their negative public image and reduce resistance to their activities (Boon 2019; Dahl & Fløttum 2019). Here, I join other studies engaging with the extraction and conservation nexus (see Büscher & Davidov 2016; Enns et al. 2019; Menton & Paul, this volume).

Connecting extraction and conservation

Despite the many distinctions and contradictions between extraction and conservation, both also share many similarities, including common dialectical imaginaries and materialities reshaping landscapes for their own purposes (Büscher & Davidov 2013). As Norris (2017) points out, those sectors seek to commodify nature and calculate the value of these commodities (proven mineral resources and protected areas marketed for tourism) in order to persuade landowners to accept these new land use regimes or surrender their lands through a mix of promises, compensation, and coercion (Adams 2017; Watts 2000). Both extraction and conservation seek, at times together, to bring about new imaginaries of places and re-territorialize them through regimes of exception that legitimize particular forms of inclusion/exclusion for local communities.

As discussed here, extraction and conservation share in this regard some common enmities, notably towards local communities, land uses, and livelihoods that are seen as incompatible with large-scale extraction and mainstream forms of conservation. As a result, growing affinities between extraction and conservation have resulted in a flurry of initiatives on both sides. One of the main critiques of these affinities is that they are in effect *sustaining extraction* through green-washing, engagement in conservation, or post-extraction rewilding; with conservation organizations legitimating extraction industries in a world that seems increasingly critical of their actions and without halting the consumption of the goods they help to produce, nor considering some of the negative socio-environmental impacts of these partnerships. The logics of conservation and extraction, then, are in many ways coming together.

Extractive footprints and conservation

Extraction and conservation entertain multiple and complex relations. Many extractive activities are taking place in conservation-valued areas, such as with mining and oil extraction in the Western Amazon (Finer et al. 2008; Sonter et al. 2017). About 86% of industrial mines for key metals are located in areas of high or intermediate plant diversity (Murguía et al. 2016), and about a third are either inside or within 10 km of a protected area (see Durán et al. 2013). This threatening proximity, or even blatant overlap, is pushing some major conservation organizations to work with extractive companies and their financiers, often based on the argument that 'minerals will still need to be mined, even in a fully renewable age. Any extractive activity must [therefore] be conducted in an environmentally and socially responsible way, causing the least possible damage' (WWF 2019). Studies of relations between extraction and conservation vary in number according to the different extractive sectors involved and the prominence of the affected habitats for the conservation community, as most studies come from the conservation literature and many focus on reducing site-specific impacts through a mix of guidelines and restrictions. Such impacts have generally pitted local communities and allied environmental organizations against extractive projects (see Temper et al. 2015). In turn, many extractive companies have confronted resistance of their projects through a mix of incentives and coercion, generally with the active support of host country authorities and local elites (Condé & Le Billon 2017). The result has often been escalation of conflicts over extractive activities, leading in turn to grave human rights abuses, including killing (Middeldorp & Le Billon 2019; Le Billon & Lujala 2020). Given the high stakes for all parties and the risks of confrontational approaches, many major Western extractive companies and conservation NGOs have come to closely collaborate for the purpose of advancing extraction *and* conservation.

Growing affinities between extraction and conservation

Conservation organizations (COs) constitute the 'professional' embodiment of civil society involvement in mainstream biodiversity protection. As such, and unlike local and often more 'radical' environmental groups, COs have been at the forefront of conservation partnerships with extractive industries. An often already corporate-friendly branch of the broader family of environmental non-governmental organizations (ENGOs), COs are frequently funded by high-earning individuals and their foundations (Holmes 2012; Spence 1999; Tsing 2005), build their programs on the legacies of (extractive) colonial conservation practices, and do not hesitate to use a 'firm' (and deadly) hand with local populations resisting relocation as well as ownership and user rights to their lands as a result of 'fortress conservation' (Duffy 2014). Some conservation projects have extended or even directly benefited from the dispossession of traditional agrarian communities and Indigenous people by extractive activities, such as in the case of gold mining in Yosemite National Park (Spence 1999). Conservation can be instrumented as a 'green alibi' to displace communities and more easily access mineral resources, as reported in the case of diamonds for the Basarwa in Botswana (Boonzaier 2011; Mosweunyane 2017). Many COs have long welcomed funds from corporations, but initially mostly through the personal fortunes of the owners of these corporations (Chapin 2004). Direct funding from extractive corporations rose from the mid-1990s onward, with often different purposes in mind. Rather than green philanthropy perhaps assuaging guilt for old deeds or realizing dreams of 're-wilding' places, such direct corporate funding was pragmatic, profit-driven, and 'future oriented'. Often, such funding sought to secure future environmental license for extractive activities and 'produce value out of nature' in novel ways for both extractive corporations and conservation organizations (Apostolopoulou et al. 2019; Dempsey 2016). Taking the form of public relations campaigns, hired consultancy, operational sub-contracting, and strategic partnerships, the 'new' relations have been much more instrumental in enrolling conservation within on-going and future extractive activities. Bringing their name, expertise, and networks to the table, COs have been able to facilitate the conservation claims and projects of extractive companies. At their most basic, these relations have allowed extractive companies to populate their sustainability reports with images of charismatic wildlife, ecofriendly 'natives', and the green logos of conservation. At their most elaborate, extraction–conservation partnerships have created and run vast protected areas or led to the co-design of corporate strategic plans on future climate policies.

This move by major conservation NGOs to seek corporate support found an echo in the mid-1990s among a number of large Western mining companies that started to commit to changing their environmental practices, and engaged for this purpose with select conservation organizations (Adams 2017), including through funding, research partnerships, and conservation offset schemes. In short, there can be some close affinity between extractive companies needing green expertise and credentials, and conservation organizations needing access and finance. Among ENGOs, Conservation International has been the most exposed to critiques, as a result of its partnership with companies such as BHP, BP, Cargill, Chevron, Exxon, Monsanto, Rio Tinto, and Shell (Choudry 2003). Asked in an industry blog how she would respond to critiques of a $50 million partnership with Australian mining giant BHP – a Top 20 Global Carbon Major (Taylor & Watts 2019) with 177 million tons of embedded

CO_2e/year from its Australian coal mines (Moss 2019) – CI's senior director for responsible mining and energy responded that 'the world needs the expertise, agility and funding the private sector brings to the table to tackle and scale conservation solutions' (Evans 2019). Funding can also be part of counterbalancing specific negative media coverage of an extractive company, as suggested by Hamann and Kapelus (2004) in the case of First Quantum in Zambia, which announced conservation funding shortly after facing criticisms for low tax payments.

Explaining extraction and conservation partnerships

Mobilizing insights from critiques of 'neoliberal nature', Enns et al. (2019: 969) have argued that not only that 'extraction and conservation activities increasingly occur in the same spaces and make use of similar logics, strategies and technologies', but also that 'biodiversity conservation is increasingly being carried out through partnerships between extractive and conservation actors in pursuit of shared or complementary interests' (see also Büscher & Davidov 2013; Norris 2017; Seagle 2012). Several factors or motives explain these partnerships.

The first explanation is the neoliberalization of much of conservation, which 'shift[ed] the focus from how nature is used in and through the expansion of capitalism, to how nature is conserved in and through the expansion of capitalism' (Büscher et al. 2012: 4). This neoliberalization, in turn, provided the extractive sector, facing a double crisis of legitimacy and exhaustibility, with a spatial fix opening up new lands and reserves and an ecological fix flipping conservationists from opponents to partners through market solutions (Enns et al. 2019). Conservation can also render land around extractive sites 'investable' (Le Billon & Sommerville 2017), notably through eco-tourism (Büscher & Davidov 2016). Without conservation, local residents may be seen by extraction as a threatening 'surplus population' in need of costly pacification. Through their partnership, conservation and extraction can thereby render the land 'investable' not only for extraction by reassuring investors through increasing the green credentials of the project and lowering reputational risk, but also for conservation as environmental organizations benefit from a source of funding and some of the extractive infrastructure – such as roads and local air strips – to develop eco-tourism (Smith 2013). This new source of value, in turn, can be used to the advantage of extractive companies to buy in local communities through new livelihoods and 'co-management' as well as keep job-seekers and migrants at bay through stricter human settlement rules.

Enmity and the extraction–conservation nexus

Concerns relating to the common (rather than reciprocal) enmity associated with affinities between extraction and conservation are particularly acute for Indigenous and traditional agrarian communities facing a criminalization of their land uses and livelihoods – including artisanal mining, hunting, and shifting agriculture – echoing colonization processes that have been well documented within the political ecology literature (e.g. Neuman 2004; Walker 1998). At least seven major conservation activities associated with extractive companies can be identified, from the commissioning of conservation research to the creation of new protected areas (see Table 19.1). While some of these activities can have positive impacts on some communities or least some members of local communities, such as easier access to health services and additional sources of livelihoods reducing out-migration (Smith 2013), there can also be negative ones. To sum up, by fixing some of the crises of extraction, conservation enables its reproduction or, in other words, helps to 'sustain extraction'. While conservation initiatives may help to counterbalance some of the 'bads' of extraction, the resulting extraction–conservation also extends its own impacts onto local communities and the world at large.

Conclusion

Crisis conservation and virulent extraction have produced seemingly odd partners. Conservation is supposed to fend off extraction, but neoliberal logics have led many conservation organizations to embrace extractive corporations eager to revamp their image as defenders of the environment. Beyond opportunistic funding and green-washing, affinities between conservation and extraction also rest on a readiness to dispossess local communities in the name of a 'greater

Table 19.1 Impacts of the extraction–conservation nexus

Activities within the extraction–conservation nexus	*Potential negative impacts on local agrarian communities and land defenders*
Commissioning conservation research	Shift in understandings of value for 'nature' and potential land use for area; extractive bioprospecting
Raising community awareness about conservation initiatives	Shaming of traditional practices and closure of alternative 'sustainabilities' to those allowed by conservation
Initiating captive breeding and rewilding programmes	Increased risk of damage from wildlife
Establishing community-based conservancies and resource management programs	Increased inequalities, including uneven distribution of revenues increasing intra-community inequalities and tensions
Training and equipping park rangers	Increased fines and human rights abuses
Building new security infrastructure in and around protected areas	Further loss of access to land and resources
Establishing new protected areas	Restriction of access to land and resources, displacement, loss of individual and community assets

cause', whether de-humanized understandings of biodiversity or false promises of economic development (see Nixon, this volume). In effect, the extraction–conservation nexus sustains large-scale extraction through community dispossession. Environmental and land defenders seeking to oppose these logics face a wide array of practices, from enrollment into new extractive and conservation roles to brutal forms of eviction. This poses several challenges for defenders.

The first one is to grasp the seemingly paradoxical convergence of extraction and conservation and their growing mutual dependence, and thus the fact that some supposed 'allies' within the environmental/conservation NGOs may actually be on the opposite side. As discussed earlier, the neoliberalization and growth of extractive and conservation activities mean that extraction increasingly needs to partner with conservation to access land and show environmental credentials while, in turn, conservation projects benefit from the extraction's funding, land-base, and ruling elite support. These relations are not only pragmatic; they also reflect shared discursive imaginaries and material practices regarding (future) valuations of nature. In this regard, some defenders may be at risk of being influenced by these imaginaries, especially as some of their own identities and values become mobilized by extractive companies and conservation projects to reshape their own corporate image. The extraction–conservation nexus is thus not only a marriage of convenience, but also a mutually reinforcing system of value production (and destruction) that can come to permeate local communities. The more persuasive these imaginaries become, the more likely this system can displace and overrule existing alternatives – such as independent conservation organizations openly criticizing and opposing extractive corporations, or local communities able to resist rather than being dispossessed and selectively integrated through extraction/conservation social responsibility programs.

A second challenge is the need to identify and counter the spaces created by the extraction and conservation nexus. Extraction–conservation partnerships take shape through different types of spaces of 'double exception', including extractive areas within protected areas ('degazetted PAs'); protected areas within or around extractive areas to legitimize and/or secure extractive activities ('integrated or adjoining PAs'); and zones of extraction being offset through protected areas ('biodiversity offsetting PAs'). Each of these presents specific risks for local communities and land defenders, as they vary in their temporality (e.g. whether conservation or extraction comes first) and their spatiality (e.g. whether exclusionary conservation measures will take place in the immediate vicinity of a mine, or – as in the case of biodiversity offsets – in a different and often distant area that will affect an unsuspecting local community). While seeking allies and policy reforms, land defenders thus also need to be wary of the indirect effects of their struggles, notably in terms of the ways environmental concerns can be instrumentalized to push excluding forms of conservation measures on local or more distant communities.

The third challenge is for land defenders to be able to represent a credible threat or alternative to extraction and conservation projects in order to gain influence in decision making. In this respect, counter-narratives by defenders are crucial to deflect those so frequently mobilized by extraction and conservation proponents (e.g. anti-development; backwards; foreign-funded; poachers). Many defenders, from this perspective, can articulate their land rights, traditional livelihoods, and defense of agrarian cultures and landscapes in terms of viable socio-environmental developmental paths – including in terms of Community Conservation Areas (see Shaw, this volume) – therefore providing a credible competing imaginary, demonstrating other practices of 'sustainability', and consolidating a broad community of purpose able to transcend (often racialized) critiques of environmental and land defenders as 'selfish and ideologically driven local interests'.

Bibliography

Acosta, A. (2013). Extractivism and neoextractivism: Two sides of the same curse. In M. Lang, L. Fernando, & N. Buxton (Eds.), *Beyond development: Alternative visions from Latin America* (pp. 61–86). Amsterdam: Transnational Institute.

Adams, W. (2017). Sleeping with the enemy? Biodiversity conservation, corporations and the green economy. *Journal of Political Ecology*, *24*, 243–257. http://jpe.library.arizona.edu/Volume24/Volume_24.html

Apostolopoulou, E., Greco, E., & Adams, W. (2019). Biodiversity offsetting and the production of 'equivalent natures': A Marxist critique. *ACME: An International E-Journal for Critical Geographies*, *17*(3), 861–892.

Bebbington, A. J., Bebbington, D. H., Sauls, L. A., Rogan, J., Agrawal, S., Gamboa, C., Imhof, A., Johnson, K., Rosa, H., Royo, A., Toumbourou, T., & Verdum, R. (2018). Resource extraction and infrastructure threaten forest cover and community rights. *Proceedings of the National Academy of Sciences*, *115*(52), 13164–13173.

Boon, M. (2019). A climate of change? The oil industry and decarbonization in historical perspective. *Business History Review*, *93*(1), 101–125.

Boonzaier, E. (2011). An 'historic victory' for the Basarwa in Botswana?: Reading the evidence. *Anthropology Southern Africa*, *34*(3–4), 96–103.

Bridge, G. (2004). Mapping the bonanza: Geographies of mining investment in an era of neoliberal reform. *The Professional Geographer*, *56*(3), 406–421.

Bridge, G. (2008). Global production networks and the extractive sector: Governing resource-based development. *Journal of Economic Geography*, *8*(3), 389–419.

Büscher, B., & Davidov, V. (Eds.). (2013). *The ecotourism-extraction nexus: Political economies and rural realities of (un) comfortable bedfellows*. Abingdon, UK: Routledge.

Büscher, B., & Davidov, V. (2016). Environmentally induced displacements in the ecotourism-extraction nexus. *Area*, *48*(2), 161–167.

Büscher, B., Sullivan, S., Neves, K., Igoe, J., & Brockington, D. (2012). Towards a synthesized critique of neoliberal biodiversity conservation. *Capitalism, Nature Socialism*, *23*, 4–30.

Chapin, M. (2004). *A challenge to conservationists*. Washington, DC: World Watch Institute.

Choudry, A. (2003, October 1). Conservation International: Privatizing nature, plundering biodiversity. *Seedling*. Barcelona: GRAIN.

Condé, M., & Le Billon, P. (2017). Why do some communities resist mining projects while others do not? *The Extractive Industries and Society*, *4*(3), 681–697.

Correia, J. E. (2019). Unsettling territory: Indigenous mobilizations, the territorial turn, and the limits of land rights in the Paraguay-Brazil borderlands. *Journal of Latin American Geography*, *18*(1), 11–37.

Dahl, T., & Fløttum, K. (2019). Climate change as a corporate strategy issue. *Corporate Communications: An International Journal, 24*(3), 499–514.

Dempsey, J. (2016). *Enterprising nature: Economics, markets, and finance in global biodiversity politics*. Hoboken, NJ: John Wiley & Sons.

Douglas, L. R., & Alie, K. (2014). High-value natural resources: Linking wildlife conservation to international conflict, insecurity, and development concerns. *Biological Conservation, 171*, 270–277.

Duffy, R. (2014). Waging a war to save biodiversity: The rise of militarized conservation. *International Affairs, 90*(4), 819–834.

Durán, A. P., Rauch, J., & Gaston, K. J. (2013). Global spatial coincidence between protected areas and metal mining activities. *Biological Conservation, 160*, 272–278.

Enns, C., Bersaglio, B., & Sneyd, A. (2019). Fixing extraction through conservation: On crises, fixes and the production of shared value and threat. *Environment and Planning E: Nature and Space, 2*(4), 967–988.

Evans, S. (2019, December 4). Q&A: The inside track on BHP & conservation international's unlikely partnership. *Mining Technology*. www.mining-technology.com/features/the-inside-track-on-bhp-conservation-internationals-unlikely-partnership/

Finer, M., Jenkins, C. N., Pimm, S. L., Keane, B., & Ross, C. (2008). Oil and gas projects in the western Amazon: Threats to wilderness, biodiversity, and indigenous peoples. *PloS One, 3*(8), e2932.

Hamann, R., & Kapelus, P. (2004). Corporate social responsibility in mining in Southern Africa: Fair accountability or just greenwash? *Development, 47*(3), 85–92.

Holmes, G. (2012). Biodiversity for billionaires: Capitalism, conservation and the role of philanthropy in saving/selling nature. *Development and Change, 43*(1), 185–203.

Junka-Aikio, L., & Cortes-Severino, C. (2017). Cultural studies of extraction. *Cultural Studies, 3*(2–3), 175–184.

Li, T. M. (2014). What is land? Assembling a resource for global investment. *Transactions of the Institute of British Geographers, 39*(4), 589–602.

Le Billon, P., & Lujala, P. (2020). Environmental and land defenders: Global patterns and determinants of repression. *Global Environmental Change, 65*, 102163.

Le Billon, P., & Sommerville, M. (2017). Landing capital and assembling 'investable land' in the extractive and agricultural sectors. *Geoforum, 82*, 212–224.

Middeldorp, N., & Le Billon, P. (2019). Deadly environmental governance: Authoritarianism, eco-populism, and the repression of environmental and land defenders. *Annals of the American Association of Geographers, 109*(2), 324–337.

Moss, J. (2019). *Australia's carbon majors report 2019*. Practical Justice Initiative. UNSW Sydney.

Mosweunyane, D. (2017). Brutal development agenda by political panjandrums in Botswana: How CKGR evictions massacred the native citizens (Basarwa) through HIV/AIDS. *African Educational Research Journal, 5*(1), 75–90.

Murguía, D. I., Bringezu, S., & Schaldach, R. (2016). Global direct pressures on biodiversity by large-scale metal mining: Spatial distribution and implications for conservation. *Journal of Environmental Management, 180*, 409–420.

Navas, G., Mingorria, S., & Aguilar-González, B. (2018). Violence in environmental conflicts: The need for a multidimensional approach. *Sustainability Science, 13*(3), 649–660.

Neumann, R. P. (2004). Moral and discursive geographies in the war for biodiversity in Africa. *Political Geography, 23*(7), 813–837.

Nixon, R. (2011). *Slow violence and the environmentalism of the poor*. Cambridge, MA: Harvard University Press.

Norris, T. B. (2017). Shared social license: Mining and conservation in the Peruvian Andes. *Antipode, 49*(3), 721–741.

Peluso, N. L. (1992). *Rich forests, poor people: Resource control and resistance in Java*. Berkeley: University of California Press.

Schulz, K. A. (2017). Decolonizing political ecology: Ontology, technology and 'critical' enchantment. *Journal of Political Ecology, 24*(1), 125–143.

Seagle, C. (2012). Inverting the impacts: Mining, conservation and sustainability claims near the Rio Tinto/QMM ilmenite mine in Southeast Madagascar. *Journal of Peasant Studies, 39*(2), 447–477.

Smith, T. J. (2013). Crude desires and 'green' initiatives: Indigenous development and oil extraction in Amazonian Ecuador. In B. Büscher & V. Davidov (Eds.), *The ecotourism-extraction nexus* (pp. 169–190). Abingdon, UK: Routledge.

Sonter, L. J., Ali, S. H., & Watson, J. E. (2018). Mining and biodiversity: Key issues and research needs in conservation science. *Proceedings of the Royal Society B, 285*(1892), 20181926.

Sonter, L. J., Herrera, D., Barrett, D. J., Galford, G. L., Moran, C. J., & Soares-Filho, B. S. (2017). Mining drives extensive deforestation in the Brazilian Amazon. *Nature Communications, 8*(1), 1013.

Spence, M. D. (1999). *Dispossessing the wilderness: Indian removal and the making of the national parks*. Oxford: Oxford University Press.

Symons, K. (2018). The tangled politics of conservation and resource extraction in Mozambique's green economy. *Journal of Political Ecology, 25*(1), 488–507.

Taylor, M., & Watts, J. (2019, October 9). Revealed: The 20 firms behind a third of all carbon emissions. *The Guardian*. www.theguardian.com/environment/2019/oct/09/revealed-20-firms-third-carbon-emissions

Temper, L., Del Bene, D., & Martinez-Alier, J. (2015). Mapping the frontiers and front lines of global environmental justice: The EJAtlas. *Journal of Political Ecology, 22*(1), 255–278.

Tsing, A. L. (2005). *Friction: An ethnography of global connection*. Princeton, NJ: Princeton University Press.

Van Vliet, N., Cornelis, D., Beck, H., Lindsey, P., Nasi, R., LeBel, S., Moreno, J., Fragoso, J., & Jori, F. (2016). Meat from the wild: Extractive uses of wildlife and alternatives for sustainability. In R. Mateo, B. Arroyo, & J. Garcia (Eds.), *Current trends in wildlife research* (pp. 225–265). Cham: Springer.

Veltmeyer, H. (2016). Extractive capital, the state and the resistance in Latin America. *Sociology and Anthropology, 4*(8), 774–784.

Walker, P. (1998). Politics of nature: An overview of political ecology. *Capitalism Nature Socialism, 9*(1), 131–144.

Watts, M. J. (2000). Contested communities, malignant markets, and gilded governance: Justice, resource extraction, and conservation in the Tropics. In C. Zerner (Ed.), *Plants, people, and justice: The politics of nature conservation* (pp. 21–51). New York: Columbia University Press.

Watts, M. J. (2001). Petro-violence: Community, extraction, and political ecology of a mythic commodity. In N. Peluso & M. Watts (Eds.), *Violent environments* (pp. 189–212). Ithaca, NY: Cornell University Press.

WWF. (2019). Responsible oil, gas and mining. *WWF-UK*. www.wwf.org.uk/what-we-do/area-of-work/responsible-oil-gas-and-mining

20 BINGOs and environmental defenders

NGO complicity in atmospheres of violence and the possibilities for decolonial solidarity with defenders

Mary Menton and Paul R. Gilbert

As highlighted in other chapters in this book, environmental and land defenders face myriad forms of violence. It is this very context, of unfolding 'slow violence', threats of violence, forced displacement, criminalization, and even direct physical violence, that we focus this chapter. It is here that international actors are often complicit in violence against defenders and that there are many opportunities for decolonial solidarity. While national trends are important, most cases of violence against defenders are localized and related to conflicts over a particular mine, an area of forest, or a specific piece of land, for example. As such, it is important to look at atmospheres of violence operating at the local and subnational levels, but also to understand these violent environments as the product of specific corporations' involvement in particular extractive enterprises. In this chapter, it is not our intention to 'throw stones from the side-lines' and occupy the 'comfortably radical' role of a priori critiquing partnerships between extractive industry corporations and environmental BINGOs. Rather, we draw attention to the ways in which environmental BINGOs – through their own efforts to scale up conservation, and through partnership with extractive industry firms that help to fund and facilitate the scaling up of conservation through offsetting – are complicit in a 'necropolitical ecology' whereby some environmental defenders are systematically exposed to more risk of harm, violence, and death. We argue that partnerships between BINGOs and extractive industry firms can lead to a 'greenwashing of violence' that is perpetrated by extractive companies. At the same time, attempts BINGOs make to maintain a 'Thin Green Line' and militarize conservation creates a dangerous double standard by valuing some lives while cultivating silence about the complicity of park guards in the deaths of others.

The remainder of this chapter proceeds by first outlining our analytical approach to 'necropolitical ecology' and the 'Thin Green Line'. We then introduce in more detail the BINGOs (big international non-governmental organizations) and GONGO (government organized non-governmental organization) that are the focus of our analysis here: CI, TNC, WWF (BINGOs), and IUCN (GONGO). We have chosen these organizations because of their size, influence, partnerships with large extractive firms (see also Le Billon

2021), and recent deployment (or in some cases co-optation) of the language of 'environmental defenders'. The main body of the chapter focuses firstly on the complicity of some BINGOs in 'green violence' perpetrated towards defenders, and secondly on the partnerships maintained by environmental BINGOs and GONGOs with extractive industry firms associated with contributing to atmospheres of violence and shaping necropolitical ecologies. We conclude by reflecting on the possibilities for moving towards cultivating decolonial forms of solidarity with 'environmental defenders'.

Necropolitical ecology and the Thin Green Line

The term 'necropolitics' was introduced by Mbembe (2003) as a way to understand how certain sovereign authorities exercise the right to expose people (in colonies, plantations, and militarized zones) to death, or the risk of death. Building on this concept, Banerjee (2008) introduced the notion of 'necrocapitalism', or a system of organizing and managing economic activity that subordinates life to the power of death, all in the pursuit of capital accumulation and growth. Banerjee (2018) notes numerous cases in which companies (including those in high-profile partnerships with environmental BINGOs) have used a combination of corporate social responsibility (CSR) and philanthropy, and the co-optation of police violence to quash opposition, to deal with conflicts between local communities and corporations seeking to extract natural resources from their lands. More recently, political ecologists have taken up Mbembe's work on necropolitics, noting its resonance with colonial 'shoot on sight' responses to poaching (Cavanagh & Himmelfarb 2015: 59). Notably, two of the organizations discussed here (WWF and IUCN) had among their founders members of the Society for the Protection of Wild Fauna of Empire, an organization explicitly concerned with criminalizing 'poaching' along racialized lines while maintaining license for colonists to hunt for 'sport'. Here, as with more contemporary shoot on sight responses to poaching, sovereign authority is clearly exercised to expose certain people – often subsistence hunters criminalized as 'poachers' – to the risk of death.

In this chapter, we examine the 'necropolitical ecology' of the spaces in which environmental BINGOs conduct conservation activities, as well as the extractive zones in which their corporate partners operate. In both cases, we identify a reproduction of "deathly spaces, where certain people are more systematically assured of exposure to greater risk of bodily harm and death" (Margulies 2019: 152). We do not, therefore, make claims about the *intent* of either environmental BINGOs or their corporate partners, but instead highlight their complicity in creating atmospheres of violence: zones where certain deaths are more likely and only certain deaths are mourned. In particular, we stress the significance of the notion of the 'Thin Green Line' in policing the deathly spaces of conservation and the role that corporate partnerships play in greenwashing violence by providing kudos to corporations who are linked to atmospheres of violence.

Methodology and focus

This chapter builds on an analysis of the way the 'Big Three' conservation organizations – WWF, TNC, and CI – and one GONGO that is often identified as an NGO (IUCN) position themselves in relation to 'environmental defenders'. We focus equally on the complicity of environmental BINGOs in 'green violence' associated with militarized conservation, and on BINGO partnerships with extractive industry corporations, some of which are implicated in producing atmospheres of violence and of violence against defenders. The focus is on extractive industry corporations in particular for several reasons. Firstly, extractive industries positioned themselves as 'first movers' in the turn to corporate social responsibility and partnerships with NGOs from the turn of the 21st century, and have developed sustained and longstanding partnerships with environmental BINGOs (Adams 2017; Gilbert 2015; Rajak 2011; also Rainey et al. 2015). Secondly, several of the extractive industry corporations with whom environmental BINGOs maintain significant partnerships have been identified by the Climate Accountability Institute (2019; Licker et al. 2019) as among the top 20 fossil fuel emitters collectively responsible for 35% of emissions (1965–2017) and 51% of ocean acidification (1965–2015).[1] This raises important questions about the degree to which environmental BINGOs can achieve conservation gains by working with 'nature's enemies' (see later in this chapter). Finally, we focus on partnerships with extractive industries because extractives are among the industries from which environmental defenders are at the most risk (Global Witness 2018).

Based on 303 attempts that the Business & Human Rights Resource Centre (2015) made to contact UK-listed corporations over allegations of human rights abuses, 47% of approaches pertained to extractive industry firms, of which 54% pertained to environmental abuses and 35% to land rights and displacement. Likewise, the *Corporations and Human Rights Database* trial for Latin America has drawn on Business & Human Rights Resource Centre data to examine 1,308 allegations for 916 firms (2000–2014), of which natural resource firms account for 36% of allegations (Bernal-Bermudez & Olsen 2016). While our analysis of environmental BINGOs' complicity in green violence draws on press coverage and responses from BINGOs themselves, our analysis of BINGOs' complicity with extractive industry firms subject to allegations of human rights abuse follows Bernal-Bermudez and Olsen (2016) by drawing on the Business & Human Rights Resource Centre database of allegations and company responses. We also draw on a review of the academic and grey literature.

Three of the four organizations studied in this article have not only entered into partnerships with extractive industry firms, but also explicitly engaged with discourse on environmental defenders in recent years. In 2018, IUCN worked with UN Environment and Global Witness to promote the UN Environmental Rights Initiative, and IUCN Netherlands works with defenders in five countries to provide safety, support, and monitoring as well as contributing to advocacy at the UN. It is perhaps unsurprising that the IUCN has engaged most fully with environmental defenders, given its constitution as a GONGO which counts among its members states, state agencies, and NGOs – as well as corporations – even if Indigenous Peoples' Organizations were only recognized as a membership category in 2016. The Nature Conservancy (2016) has framed its partnerships with Indigenous people in terms of relations with 'nature's first defenders', based on a (perhaps belated) recognition that upholding forest community rights is associated with lower deforestation rates. 'Defenders' here, however, are figured as protectors *of* nature for humanity as a whole, rather than those subjected to atmospheres of violence for defending their land, livelihoods, and environment. This framing of defenders as protectors *of* a global nature forms part of the 'Thin Green Line' discourse which we examine in more detail later.

WWF has engaged more broadly with environmental defender discourse than the other BINGOs/GONGOs. Representatives of their Governance Practice and SDG Hub, Ganapin and Osieyo (2019), draw attention to Frontline Defenders' figures (Frontline Defenders 2014) in order to argue for the significance of SDG 16 (on promoting peaceful and inclusive societies, providing access to justice, and building accountable institutions) for building the good governance needed for 'protecting and restoring nature'. While this use of the defenders' discourse most clearly echoes the now-dominant human rights-based framing of environmental defenders, at other times WWF representatives frame former 'illegal loggers' turned anti-poaching 'game scouts' as 'forest defenders' (Skinner et al. 2018). Yet, as we show in the next section on green violence, WWF in particular has demonstrated a rather different response when it comes to the deaths of eco-guards and park rangers, as compared to those who die or are assaulted *at the hands of* allegedly WWF-backed park guards. We argue that this reflects a form of necropolitical ecology, whereby the lives of those who died in the service of 'protecting nature for all humanity' are valued more than those who die at the hands of militarized conservation forces. By upholding a 'Thin Green Line' between park guards and those who die at the hands of park guards, environmental BINGOs distort figures on deaths of environmental defenders, render certain forms of slow violence invisible, and reproduce colonial forms of nature conservation that undermine prospects for decolonial solidarity with defenders. As we discuss in the subsequent section, prospects for decolonial solidarity are further undermined by violent partnerships with extractive industry corporations.

Green violence

Büscher and Ramutsindela (2016) contest the ethics behind 'green violence', or violence that is carried out to protect nature. As Fletcher (2018) finds, many states distinguish between green violence they deem legitimate (e.g. violence linked to anti-poaching) and 'illegitimate' green violence (e.g. environmentally oriented political violence in the form of sabotage by organizations like Earth First). Much of the literature around green violence focuses on the militarization of conservation – the use of armed park guards to patrol national parks and combat poaching. It is important to note, however, that green violence also comes in the form of forced displacement of local people from protected areas (Ybarra 2018; Lunstrum & Ybarra 2018), restrictions on access to natural resources upon which local communities depend, criminalization of traditional hunting and harvesting practices, and symbolic and discursive violence. In some cases, the violence comes in the form of the use of conservation as a means of territorialization (Bluwstein & Lund 2018; see Ndoinyo, this volume). The militarization of conservation represents a 'war for biodiversity' (Duffy 2014) or 'green wars' (Marijnen & Verweijen 2016). More recently, this has also transformed into a 'war by conservation' through alignment of conservation NGOs with global security projects that promote a 'poachers-as-terrorists' narrative (Duffy 2016). Many environmental NGOs and BINGOs frame conservation and anti-poaching efforts as a war, with Conservation International's film *Hotspots* calling it 'the mother of all wars' (see Büscher & Fletcher 2018).

In 2019, *BuzzFeed* and the *Kathmandu Post* published a series exposing the apparent involvement of WWF-trained and -funded park guards in the torture and killing of local residents and Indigenous people in Nepal, Cameroon, and CAR (Baker & Warren 2019). Specifically, WWF was said to have celebrated the acquittal of guards connected to the death and torture of a Nepalese man (Shikharam) by Chitwan National Park guards, making unsubstantiated allegations that the victim was a 'crime convicted individual', and either hired or handed awards to park guards and soldiers accused of this and other offenses. In Cameroon, WWF was accused of suppressing findings that villagers in a proposed park site feared abuse by forest rangers, and elsewhere WWF-supported eco-guards were accused of numerous human rights abuses – of which WWF appears to have been aware at least one year prior to the *BuzzFeed* exposé (Baker & Warren 2019a, 2019b, 2019c). Indeed, public reports of links between WWF and park rangers accused of human rights abuses pre-date the *BuzzFeed* report by some years (e.g. Corry 2015), as do allegations that WWF had been non-responsive when presented with evidence of abuses carried out by rangers and eco-guards (Survival International n.d.).

When WWF announced a review (see WWF 2019) to be carried out by former UN Commissioner for Human Rights Navi Pillay, Indigenous rights organizations expressed concern about the limitation of the review to the *BuzzFeed/Kathmandu Post* allegations when "these issues are endemic rather than isolated", and many abuses had been "reported to WWF previously" (Counsell et al. 2019). The report, released in November 2020, found that WWF failed to follow up on credible allegations of human rights abuses, and its human rights commitments were particularly weak in Congo Basin countries (report available in WWF 2020a). The WWF management response to the report states:

> we recognize that allegations of human rights abuses have been made against rangers and other third parties not under WWF's direct control.

> These allegations were raised in some of the most conflict-affected and insecure places where we work. The reported atrocities go against all the values for which we stand. Human rights abuses are never acceptable, and we feel great sorrow and sympathy for the people who have suffered.
>
> (WWF 2020b: 1)

The response outlines actions WWF plans to take to better safeguard human rights in its work, yet the language used shifts the blame away from their own organization. Greenpeace (2020) responded: "WWF needs to fully own their responsibility for abuses that are committed by rangers or 'ecoguards' working in the protected areas WWF manages or co-manages".

Rivalries between 'ugly conservationists' and Indigenous rights groups appeared to be rearing their head once more. It is not our intention, however, to replay these well-worn and well-documented debates here (see Chapin 2004; Larsen 2016). Instead, we wish to draw attention to the necropolitical ecology of 'green violence' associated with park rangers and eco-guards, and challenge the narratives that legitimize this violence through the notion of the 'Thin Green Line' and neo-colonial framings of 'poaching'.

The *BuzzFeed* articles and the Independent Report reveal that WWF had prior knowledge of alleged human rights abuses and violations attributed to park rangers and eco-guards that they have funded and/or trained. The absence of prior action on these allegations sits at odds with WWF's concern over park rangers and eco-guards who have lost their lives in the course of their work. Indeed, WWF leaders went on to hand awards to the rangers charged with causing Shikharam's death in Chitwan National Parl, as well as to members of the military accused of sexual assault within the Park (Baker & Warren 2019). In partnership with the Australian Thin Green Line Foundation, WWF have compiled a database of the number of park rangers who have lost their lives in the line of duty since 2009, totalling 871 by 2018.[2] Reporting on the 2018 survey, WWF (2018) notes that "forty-eight rangers of the 107 lost this year were murdered at their place of work whilst protecting wildlife that we all care about". Narratives from friends and colleagues of murdered rangers are included along with the figures, memorializing their service in the protection of wildlife for all. Yet, read alongside the studied silence that has surrounded persistent and endemic allegations of human rights abuses *on the part of* rangers, their celebration by WWF and the Thin Green Line shares much with the discourse surrounding the 'Thin Blue Line' that has proliferated in the USA as part of a backlash to the Black Lives Matter movement.

As Wall (2020) notes, the notion of the Thin Blue Line acts as a 'fiction of legitimate violence' designed to 'render state violence as always defensive in nature while marking unruly populations as not merely transgressors of positive law, but as *hostis humani generis*: "enemies of all mankind"'. Here we see the Thin Green Line – both the organization and the concept as reflected in the memorialization of rangers over those killed by rangers – operating as a fiction of legitimate violence that renders *green* violence as always defensive and frames 'unruly' populations as enemies of wildlife 'that we all care about'. This necropolitical approach to conservation dovetails with a studied silence regarding the atmospheres of violence produced by the actions of extractive industry corporations who maintain partnerships with environmental BINGOs/GONGOs.

The violence of partnership

Over the last two decades, *partnership* has become the dominant rubric through which BINGOs and transnational corporations frame their relationship. As the Internet began to alter NGO engagement with business, risk analysts cautioned that it tended to be overly 'negative' (Bray 1998). Yet, over the last 20 years, environmental BINGOs have steered themselves away from what WWF ambassador and management guru John Elkington (1999) termed the 'Polarizer' role of fighting against businesses, towards engaging with business as 'Integrators'.[3] In the process, critical NGOs have been marginalized. They are hardly welcome at corporate social responsibility and ethical business events, where a clear delineation is made between 'partners of choice' for business actors and "reckless NGOs who destroy brand and reputation with unfounded accusations" (Rajak 2011: 17; Gilbert 2015). Pulp and paper industry representatives, for instance, celebrate the shift from "NGOs attacking our corporate customers about not being 'green enough' to partnering with them to achieve their goals" (Cousin 2014: 27). Representatives of the 'Big Three' environmental BINGOs – The Nature Conservancy (TNC), Conservation International (CI), and WWF – have likewise spoken out to defend partnerships with large corporations representing extractive industries that critics depict as predicated on environmental and social harm (Benson & Kirsch 2010).

Peter Seligmann, founding CEO of CI, epitomizes this antipathy towards 'Polarizers' and embrace of partnership with large transnational corporations. He declared that it is "simply not sufficient to throw stones from the sidelines. . . . We believe that often the biggest improvements to environmental conservation and human well-being can come from effecting change amongst those who have the biggest impact" (Seligmann 2011; see also Foster 2014: 254). A similar inducement towards working with the largest (and perhaps the most harmful) corporations came from Peter Kareiva, Chief Scientist at TNC, following criticism about TNC's partnership with BP in the wake of the Deepwater Horizon oil spill (see also Ottaway & Stephens 2003). Kareiva (2010) declares:

> Look, I know that energy extraction is sometimes environmentally damaging, just as roads, ports, biofuels and even desert solar panels can be. In fact, Conservancy scientists engage with the energy industry *precisely because* that industry often does harm the environment.

Jason Clay and Rob Soutter of WWF likewise insist that working with the largest and most impactful corporations ensures that "when they improve, everyone else in the sector will follow suit", and that "power lies with the corporations. We can only achieve something by working with them" (Huismann 2014: 13–16). Reflecting the widespread legitimation of partnership with harmful industry players, and hostility to critical or oppositional modes of engagement among environmental BINGOs, IUCN Patron of Nature Jon Stryker (2018) has recently written of the need for "teaming up with 'nature's enemies' . . . [and the] need to become more pragmatic about choosing collaborators".

This embrace of collaboration over critique on the part of environmental BINGOs also involves a highly particular approach to corporate personhood. BINGO representatives appear unwilling to attribute specific harms to corporate bodies as a whole, or comment on how operations in one jurisdiction might be more harmful than operations carried out as part of ongoing conservation partnerships (e.g. Huismann 2014: 58). In 1997, WWF Canada nominated Shell for a British Columbia Minister's Environmental Award. This nomination took place two years after the killing of Ken Saro-Wiwa and the Ogoni 9, following their campaigning against Shell in the Niger Delta. Pegi Dover, Director of Communications for WWF Canada, wrote in response to criticism that the nomination "is not an overall endorsement of their environmental practices. . . . WWF has not commented on the overall environmental record of any of the [nominees] and does not anticipate doing so". While environmental BINGOs might be reluctant to cast their partnerships in terms of relations with (or endorsements of) entire corporations, many of the corporations they partner with are happy to present their collaborations in this way – and the ability to identify an accountable corporate person is a vital part of seeking redress for corporate human rights abuses (Grear & Weston 2015).

The partnership of environmental BINGOs with corporations that have questionable human rights and environmental records has been understood by anthropologists studying corporate personhood and CSR to be part of a strategy whereby corporations extract brand value from highly trusted 'superbrands' like WWF (Foster 2014: 251; also Larsen 2016: 23). This extraction of value takes place through provoking ethically charged acts of consumption among those who hold environmental BINGOs in high regard. In theory, what environmental BINGOs get in 'return' (as per the statements from CI, TNC, WWF, and IUCN representatives earlier) is both financial support and improved environmental behaviour and industry or sector leadership from their corporate partners. Critical political ecologists, however, have framed this relation less as a reciprocal exchange of brand value for improved environmental performance and more as a form of 'Faustian Bargain' whereby BINGOs accept market-based approaches to conservation and the notion that capitalism can be fundamentally sustainable (Adams 2017). The result is a shared interest in 'scaling up' conservation and 'offsetting' extractive operations, enabled through the framing of nature as 'natural capital' that is fundamentally substitutable – both for other *units* of natural capital and for other *forms* of (economic, social) capital (Adams 2017: 246; Chapin 2004: 22; Kirsch 2010: 91).

Nonetheless, there is little definitive evidence that corporations who partner with environmental BINGOs reduce their impact or enhance biodiversity conservation and the maintenance of ecosystem services (Robinson 2011). While mining companies are 'leaders' in the setting of 'no net loss' or 'net positive impact' biodiversity conservation goals, most of these goals "have advanced little beyond definition" (Rainey et al. 2015: 6). Against this absence of clear evidence that corporate–BINGO partnerships produce conservation gains, serious questions need to be asked about the consequences of partnership with industries that are often complicit in producing the atmospheres of violence to which defenders are subjected.

In a review of the 'metanarratives' that characterize engagements with conservation BINGOs, Larsen (2016) distinguishes between 'doing good' narratives that emphasize conservation actors as highly moral; 'turning ugly' narratives that focus on blurred boundaries between states, corporations, and NGOs; and the 'Dirty Harry' narrative which foregrounds pragmatism. Partnership may be ugly, but it is necessary. Certainly, the quotations from senior figures in CI, TNC, WWF, and IUCN introduced earlier would suggest that partnership is undeniably seen as necessary, even when it is with industries recognized as 'ugly' (e.g. 'enemies of nature'). Larsen cautions that the 'ugly conservationist' critique is overdone, 'comfortably radical', and in some cases "undermining potential spaces for social change prompted by NGOs" (Larsen 2016: 25). Yet, critique is not merely the crude and unproductive act of denunciation in which critics find solace (Gilbert & Sklair 2018). Rejection of criticism runs the risk of excising from view the atmospheres of violence with which defenders must contend, and the potential complicity of environmental BINGOs and their corporate partners in the harms suffered by defenders.

Indeed, this hostility to critical approaches seems itself to reflect the 'harmony ideology' (Foster 2014) which characterizes partnerships between extractive industry corporations and their 'Integrator' partners among the environmental BINGOs, according to which concern for polite consensus replaces concern for social inequality – or, in some cases, for the rights of environmental defenders. As highlighted earlier, 'necropolitical ecology' operates through 'green violence' that is redolent of the colonial necropolitics which saw 'natives' classified as 'poachers' whose lives are rendered disposable through militarized conservation. In the next section, we examine how environmental BINGOs/GONGOs enter into relations of complicity with extractive industry firms whose operations reproduce atmospheres of violence and expose would-be defenders to risk, threat, and harm.

Royal Dutch Shell: The Nature Conservancy and (formerly) WWF

Shell's long history of entanglement with WWF has been well documented. Shell's General Director, John Loudon, was brought into WWF International by its founder and became its president in 1976 (Huismann 2014: 89–93).

More recently, WWF (particularly WWF-UK) has been more outspoken and confrontational in its relations with Shell. CEO David Nussbaum has recently spoken out against Shell's expansion of drilling into the Arctic, and previous CEO Robert Napier attacked Shell over lax environmental standards on the Sakhalin-2 project in 2005. However, WWF has long been reluctant to speak out against Shell's alleged complicity in human rights abuses and the cultivation of atmospheres of violence with which defenders must contend (see earlier). Shell has been involved in a formal partnership with TNC since 2009 (and has also engaged with IUCN since 2003, albeit not without controversy – see later in this chapter).

TNC present their partnership with Shell as a matter of integrating science and conservation expertise with Shell's business practice. There is a focus on the 'mitigation hierarchy' and the reduction of net carbon footprint through offsetting: Adams' (2017) Faustian Bargain in action. Meanwhile, concerns have been raised by UNEP over Shell's failure to clean up oil spill sites, and its use of discredited and misleading information to attribute oil spills in the Niger Delta to sabotage or theft (Amnesty International & FOE 2014; Amnesty International 2014). At the time of writing, a court case is ongoing against Shell in The Hague, brought by Esther Kiobel and other widows of the Ogoni 9. A previous attempt at holding Shell to account for their part in the trumped-up judicial murder of Ken Saro-Wiwa, Esther's husband Barinem Kiobel, and the rest of the Ogoni 9 has been rejected by American courts on jurisdictional grounds, marking the closure of an important pathway to redress for corporate human rights abuses (Grear & Weston 2015). Nonetheless, proceedings in this earlier case resulted in the release of numerous internal memos. Some of these appear to show Shell requesting support from the military and paying honoraria as a 'show of gratitude' for the 'sustained favourable disposition' shown by military commanders implicated in killings of civilians, both before and after the gratitude was expressed (Dummett 2019; see Zalik 2004).

There are haunting parallels between WWF's displays of gratitude towards park guards accused of abuses and murder in Nepal (see earlier), and Shell's prestations towards abusive military figures in Ogoniland. Divided by decades and continents, these actions seem to be united by a form of necropolitics whereby spaces are governed such that some categories of people are more at risk of death than others. Given TNC's (at least partial) embrace of the rhetoric of 'defenders' in representations of their own conservation work, questions once again need to be raised about the role of environmental BINGOs in policing a 'Thin Green Line' through their partnership with extractive industry corporations. Some 'defenders' of nature *as a whole* are to be valorized; extractive industry partners who engage in mitigation and offsetting are to be lauded; but those corporations' entanglements in human rights abuses and the murder of defenders is met with silence. While there is a tendency for those working in the extractive industries to disconnect 'legacy issues' such as Shell's actions in Ogoniland from contemporary corporate personhood (see Gilbert 2015), events such as the murder of the Ogoni 9 are anything but 'legacy issues' for the family of those involved. If environmental BINGOs continue to support extractive industry corporations that attempt to evade accountability for complicity in creating atmospheres of violence – however long ago – it is difficult to see how decolonial solidarity with defenders can be cultivated.

Chevron: Conservation International and IUCN

CI has been working with Chevron for close to two decades, and Chevron's sponsorship of CI's Integrated Biodiversity Assessment Tool has been the most lauded outcome of their collaboration. Chevron has also engaged in forms of partnership with national branches of IUCN, notably IUCN Bangladesh, on which we focus in this final section. We end our discussion of BINGO/GONGO partnerships with extractive industry corporations and their implications for solidarity with at-risk defenders by focusing on IUCN, because its unique organizational form reflects the possibilities for, as well as the challenges faced by, pursuing decolonial solidarity with defenders. IUCN is, as MacDonald (2003) notes, an ideological actor involved in "producing and circulating a definition of what constitutes conservation". The IUCN's World Conservation Congresses (WCCs) and the resolutions adopted therein have proven fruitful sites of analysis for scholars seeking to understand broad shifts in the legitimation of global environmental actors. Lehmann (2019) identifies three shifts in the IUCN since the 1980s: the first from nature protection to sustainable use and sustainable development; the second to a rights-based focus on conservation that (eventually) made space for Indigenous participation; and finally, a full-on embrace of market-led approaches to conservation. While disaffection towards engagement with actors like Shell (see previous section) was expressed at WCCs through the 1990s and early 2000s, formal expression of discontent towards cooperation with extractive industry corporations trailing controversial human rights records peaked in 2008 (Khan 2013: 116), perhaps reflecting the flowering of Foster's (2014) 'harmony ideology'.

In Bangladesh, IUCN's country office entered into a controversial relationship with Chevron and Chevron's environmental consulting firm, SMEC, to provide monitoring services for a controversial seismic survey that Chevron carried out in the Lawachhara National Park during 2008. Negotiations between IUCNB, SMEC, and Chevron in the run-up to the survey – which violated municipal law – reveal pointedly how far Adam's (2017) Faustian Bargain might take environmental BINGOs (or GONGOs) who come to accept market-based approaches to conservation, and rely on continued corporate partnership. Emails acquired by Tanzim Khan (2013: 160) reveal that IUCN allowed SMEC, hired by Chevron to conduct an Environmental Impact Assessment, to delete suggestions made by IUCN's panel of experts and insert allowances to drill regardless of considerations for particular animals' ranging routes and behaviour patterns. As IUCNB wrote in a February 2008 email to the regional office in Bangkok, "We find this is a very good opportunity to show case business and conservation can go hand in hand" (ibid.: 163).

A willingness to sacrifice 'scientific expertise' to further partnership with business seems to undermine even the core commitments of environmental BINGOs/GONGOs. Where, one might ask, is the willingness to sacrifice partnerships with extractive industry corporations in order to stand in solidarity with defenders? Chevron has been repeatedly subjected to allegations before the UN Human Rights Council that it has perpetrated what might be called slow environmental violence on Indigenous people in Ecuador, polluting 450,000 hectares and digging 880 hydrocarbon waste pits (CETIM 2015). Responsibility has been deflected through creative presentations of corporate personhood (Sawyer 2006), and defenders and their fellow claimants have been depicted as criminal in their pursuit of justice. The lawyer fighting to hold Chevron accountable for the contamination in Ecuador has been placed under house arrest after Chevron came after him with a host of trumped up charges and a campaign to demonize him (Lerner 2020). As we saw above, environmental BINGOs frequently deny corporate personhood in order to distance themselves from an extractive corporation's activities beyond the immediate setting of a specific partnership. Defenders do not often have that privilege: they are subjected to atmospheres of violence that emerge from the actions of large transnational entities, and it should be incumbent on environmental BINGOs/GONGOs, especially those who invoke the discourse of 'defenders', to stand in solidarity with them.

Conclusion: decolonizing solidarity

> I will not waste my time working with [environmental BINGOs]. They are in bed with the very people we are fighting against, with the same people who are killing us and destroying our waters.
>
> – Environmental defender from Ecuador, August 2018

As we argued earlier, environmental BINGOs/GONGOs are frequently complicit in shaping 'necropolitical ecologies' and atmospheres of violence around sites of conservation and/or extraction. It is unsurprising that many environmental defenders do not see BINGOs as allies when they side with extractive corporations and 'enemies of nature' rather than the defenders who attempt to secure land, livelihoods, and environment in the face of extractive harm. The greenwashing of violence resulting from BINGO partnerships with extractive companies, and their complicity in green violence carried out in the name of conservation, reflect a holdover from the colonial past (and present) of conservation. While growing efforts to support environmental defenders are commendable, BINGOs, GONGOs, and other actors need to consider the implications of their wider remit of activities and partnerships, many of which contribute to creating the very spaces and atmospheres of violence that threaten environmental defenders. Silence in the face of necropolitics and in the face of human rights violations, and complicity in greenwashing companies responsible for slow violence and other violences, is inexcusable. Given the increasing number of accusations and reports that point towards their complicity, BINGOs cannot claim ignorance and need to take concrete actions to counteract the human rights violations and violences with which they have been complicit. While defenders attempt to hold transnational corporate persons to account through legal mechanisms, we must ask whether BINGOs are still content to distance themselves from extractive corporations' activities beyond the narrow confines of their partnerships.

We argue for a decolonial approach to conservation, for a "vision of human life that is not dependent upon or structured by the forced imposition of one ideal of society over those that differ" (Mignolo 2007: 459). In essence, a transition towards 'convivial conservation' which Büscher and Fletcher (2019: 1) describe as a "post-capitalist approach to conservation that promotes radical equity, structural transformation and environmental justice and so contributes to an overarching movement to create a more equal and sustainable world". A transition away from the creation of 'deathly spaces' (Margulies 2019) and the atmospheres of violence that put environmental defenders at risk. Many INGOs have begun to engage in discourse around decolonial approaches and 'shifting the power' away from INGOs based in the north towards NGOs and grassroots movements in the south. Such a shift would allow for more effective change but also lead to a drastic change in the structure of these organizations, and has been slow to materialize.

As Doane (2019) noted recently, for

> all the lofty words about 'shifting the power', many INGO staff and board members still seem unable to let go of a model that values technocrats over movement builders, and which places a higher value on their own Northern white role.

Environmental BINGOs are no exception. It is time to move away from a focus on Centers for Environmental Leadership (CI) and One Planet Leader Academies (WWF) that further amplifies the voices and perspectives of personnel from BINGOs' corporate partners. Instead, we need to foreground the voices of defenders, listen to the narratives of those who live in atmospheres of violence, and take care before entering into partnerships with the 'bewilderers' who turn the slow progression of environmental violence into doubt and inaction (Nixon 2011: 40). Decolonizing BINGO solidarity with environmental defenders requires structural change, but perhaps more critically bravery on the part of BINGOs to recognize their complicity and begin to build decolonial, respectful, and equitable relationships with grassroots movements and communities that fight to protect lands, forests, and waters from invasion by extractive industries and thereby protect the wildlife and ecosystems that BINGOs aim to conserve.

Notes

1 These are Shell (TNC, CI, formerly WWF), Chevron (IUCN), and Exxon (CI).

2 These contribute to the problematic Global Witness figures of environmental defenders killed, as noted in the introduction.
3 Conner and Epstein (2007) have worked with a similar dichotomy between 'purity and pragmatism' among 'dark green' and 'light green' NGOs. Hoffman (2009) however, challenges any such rigid distinction, noting that although there are 'dark greens' or 'Isolates' that have no corporate partnerships, social network analysis shows distinct clusters of environmental NGOs in the US with regard to the type and quantity of corporate partnership. In Hoffman's analysis, the 'mediators' with a high diversity of corporate ties and high centrality within the network of environmental NGO-corporate partnerships included CI, TNC, and WWF.

References

Adams, W. (2017). Sleeping with the enemy? Biodiversity conservation, corporations and the green economy. *Journal of Political Ecology*, *24*, 243–357.

Amnesty International. (2014). *Injustice incorporated: Corporate abuses and the human right to remedy*. Amnesty International Ltd.

Amnesty International, & Friends of the Earth. (2014, August 4). No progress: An evaluation of the implementation of UNEP's environmental assessment of Ogoniland, three years on. *ReliefWeb*. https://reliefweb.int/report/nigeria/no-progress-evaluation-implementation-unep-s-environmental-assessment-ogoniland-three

Baker, K., & Warren, T. (2019a, March 8). WWF says indigenous people want this park: An internal report says some fear forest ranger 'repression'. *BuzzFeed News*. www.buzzfeednews.com/article/katiejmbaker/wwf-eu-messok-dja-fears-repression-ecoguards

Baker, K., & Warren, T. (2019b, March 5). A leaked report shows WWF was warned years ago of 'frightening' abuses. *BuzzFeed News*. www.buzzfeednews.com/article/katiejmbaker/wwf-report-human-rights-abuses-rangers

Baker, K., & Warren, T. (2019c, October 17). WWF executives were warned of widespread atrocities by anti-poaching rangers the charity funded. *BuzzFeed News*. www.buzzfeednews.com/article/katiejmbaker/wwf-executives-marco-lambertini-warned-abuses

Banerjee, S. (2008). Necrocapitalism. *Organization Studies*, *29*(12), 1541–1563.

Banerjee, S. (2018). Markets and violence. *Journal of Marketing Management*, *34*, 1023–1031.

Benson, P., & Kirsch, S. (2010). Capitalism and the politics of resignation. *Current Anthropology*, *51*(4), 459–486.

Bernal-Bermudez, L., & Olsen, T. D. (2016). Business, human rights, and sustainable development. In *The SAGE handbook of corporate and public affairs* (pp. 280–299). Thousand Oaks, CA: SAGE Publishing.

Bluwstein, J., & Lund, J. F. (2018). Territoriality by conservation in the Selous-Niassa Corridor in Tanzania. *World Development*, *101*, 453–465.

Bray, J. (1998). Web wars: NGOs, companies and governments in an Internet-connected world. *Greener Management International*, *24*, 115–129.

Büscher, B., & Fletcher, R. (2018). Under pressure: Conceptualising political ecologies of green wars. *Conservation and Society*, *16*, 105–113.

Büscher, B., & Fletcher, R. (2019). Towards convivial conservation. *Conservation and Society*, *17*(3), 283–296.

Büscher, B., & Ramutsindela, M. (2016). Green violence: Rhino poaching and the war to save Southern Africa's peace parks. *African Affairs*, *115*(458), 1–22.

Business & Human Rights Resource Centre. (2015). *Is the UK living up to its business & human rights commitments?* [Briefing]. www.business-humanrights.org/sites/default/files/UK%20Briefing%20-%20FINAL.pdf

Cavanagh, C., & Himmelfarb, D. (2015). 'Much in blood and money': Necropolitical ecology on the margins of the Uganda protectorate. *Antipode, 47*(1), 55–73.

CETIM. (2015). *Chevron denounced before the Human Rights Council for violations of the human rights of indigenous and peasant populations in Ecuador.* www.cetim.ch/chevron-denounced-before-the-human-rights-council-for-violations-of-the-human-rights-of-indigenous-and-peasant-populations-in-ecuador/

Chapin, M. (2004, November/December, 17–31). A challenge to conservationists. *WorldWatch Magazine.*

Climate Accountability Institute. (2019, October 9). *Carbon majors: Update of top twenty companies 1965–2017* [Press release]. https://climateaccountability.org/pdf/CAI%20PressRelease%20Top20%20Oct19.pdf

Conner, A., & Epstein, K. (2007). Harnessing purity and pragmatism. *Stanford Social Innovation Review, 5*(4), 61–65.

Corry, S. (2015, August 9). When conservationists militarise, who is the real poacher? *Truthout.* https://truthout.org/articles/when-conservationists-militarize-who-s-the-real-poacher/

Counsell, S., et al. (2019, April 9). Letter to Marco Lambertini. *Rainforest Foundation UK.* www.rainforestfoundationuk.org/media.ashx/wwfletter.pdf

Cousin, S. (2014). A perfect partnership. *Pulp & Paper International, 56*(6), 26–27.

Doane, D. (2019, December 18). Are INGOs ready to give up power? In *From Poverty to Power.* https://oxfamblogs.org/fp2p/are-ingos-ready-to-give-up-power/

Duffy, R. (2014). Waging a war to save biodiversity: The rise of militarized conservation. *International Affairs, 90*(4), 819–834.

Duffy, R. (2016). War, by conservation. *Geoforum, 69*, 238–248.

Dummett, M. (2019, April 24). Ruling due in Esther Kiobel's epic legal battle against Shell. *Medium: Amnesty Global Insights.* https://medium.com/amnesty-insights/ruling-due-in-esther-kiobels-epic-legal-battle-against-shell-cb76bba37e6d

Elkington, J. (1999). *Cannibals with forks: The triple bottom line of 21st century business.* North Mankato, MN: Capstone.

Fletcher, R. (2018). License to kill: Contesting the legitimacy of green violence. *Conservation and Society, 16*(2), 147–156.

Foster, R. (2014). Corporations as partners: 'Connected capitalism' and the Coca-Cola company. *PoLAR: Political and Legal Anthropology Review, 37*(2), 246–258.

Frontline Defenders. (2014). *Annual report.* www.frontlinedefenders.org/en/resource-publication/2014-annual-report

Ganapin, D., & Osieyo, M. (2019, July 15). Protecting nature and civic space will unlock prosperity. *Medium.* https://medium.com/@WWF/protecting-nature-and-civic-space-will-unlock-prosperity-f3be77277e45

Gilbert, P. R. (2015). Trouble in para-sites: Deference and influence in the ethnography of epistemic elites. *Anthropology in Action, 22*(3), 52–62.

Gilbert, P. R., & Sklair, J. (2018). Introduction: Ethnographic engagements with global elites. *Focaal, 81*, 1–15.

Global Witness. (2018). *At what cost? Irresponsible business and the killing of environmental and land defenders in 2017.* www.globalwitness.org/en/campaigns/environmental-activists/at-what-cost/

Grear, A., & Weston, B. (2015). The betrayal of human rights and the urgency of universal corporate accountability: Reflections on a post-Kiobel lawscape. *Human Rights Law Review, 15*(1), 21–44.

Greenpeace. (2020). *Greenpeace statement on WWF independent review of human rights violations.* www.greenpeace.org/international/press-release/45736/greenpeace-statement-on-wwf-independent-review-of-human-rights-violations/

Hoffman, A. (2009). Shades of green. *Stanford Social Innovation Review*, 7(2), 40–49.

Huismann, W. (2014). *PandaLeaks: The dark side of the WWF*. Bremen: Nordbook.

Kareiva, P. (2010, May 25). *Why we engage with the energy industry: It's for nature*. www.philanthropy.com/article/choosing-wisely-the-nature-conservancy-and-bp/

Khan, M. T. (2013). *The project in Bangladesh: Gas, forests and livelihood!* Doctoral dissertation. University of New England, Australia.

Kirsch, S. (2010). Sustainable mining. *Dialectical Anthropology*, 87–93.

Larsen, P. (2016). The good, the ugly and the Dirty Harry's of conservation: Rethinking the anthropology of conservation NGOs. *Conservation & Society*, *14*(1), 21–33.

Le Billon, P. (2021). Defending territory from the extraction and conservation nexus. In M. Menton & P. Le Billon (Eds.), *Environmental and land defenders: Deadly struggles for life and territory*. Abingdon, UK: Routledge.

Lehmann, I. (2019). From Noah's Ark to Nature+: Legitimating the IUCN. In K. Dingwerth, A. Witt, I. Lehmann, E. Reichel, & T. Weise (Eds.), *International organizations under pressure: Legitimating global governance in challenging times*. Oxford: Oxford University Press.

Lerner, R. (2020, January 29). How the environmental lawyer who won a massive judgement against Chevron lost everything. *The Intercept*. https://theintercept.com/2020/01/29/chevron-ecuador-lawsuit-steven-donziger/

Licker, R., Ekwurzel, B., Doney, S., Cooley, S., Lima, I., Heede, R., & Frumhoff, P. (2019). Attributing ocean acidification to major carbon producers. *Environmental Research Letters*, *14*(12).

Lunstrum, E., & Ybarra, M. (2018). Deploying difference: Security threat narratives and state displacement from protected areas. *Conservation and Society*, *16*(2), 114–124.

MacDonald, K. I. (2003, February). IUCN: A history of constraint. *Address given to the permanent workshop of the centre for the philosophy of law* (Vol. 6). Louvain: Université Catholique de Louvain.

Margulies, J. (2019). Making the 'man-eater': Tiger conservation as necropolitics. *Political Geography*, *68*, 150–161.

Marijnen, E., & Verweijen, J. (2016). Selling green militarization: The discursive (re) production of militarized conservation in the Virunga National Park, Democratic Republic of the Congo. *Geoforum*, *75*, 274–285.

Mbembe, A. (2003). Necropolitics. *Public Culture*, *15*(1), 11–40.

Mignolo, W. D. (2007). Delinking: The rhetoric of modernity, the logic of coloniality and the grammar of de-coloniality. *Cultural Studies*, *21*(2–3), 449–514.

Nature Conservancy. (2016, December 9). *Nature's first defenders*. www.nature.org/en-us/what-we-do/our-insights/perspectives/natures-first-defenders/

Nixon, R. (2011). *Slow Violence and the Environmentalism of the Poor*. Cambridge, MA: Harvard University Press.

Ottaway, D., & Stephens, J. (2003, May 4). Nonprofit land bank amasses billions. *The Washington Post*. www.washingtonpost.com/archive/politics/2003/05/04/nonprofit-land-bank-amasses-billions/10fdb070-d956-40e7-a508-b03483c21899/

Rainey, H., Pollard, E., Dutson, G., Ekstrom, J., Livingstone, S., Temple, H., & Pilgrim, J. (2015). A review of corporate goals of no net loss and net positive impact on biodiversity. *Oryx*, *49*(2), 232–238.

Rajak, D. (2011). Theatres of virtue: Collaboration, consensus, and the social life of corporate social responsibility. *Focaal*, *60*, 9–20.

Robinson, J. (2011). Corporate greening: Is it significant for biodiversity conservation? *Oryx*, *45*(3), 309–310.

Sawyer, S. (2006). Disabling corporate sovereignty in a transnational lawsuit. *Political and Legal Anthropology Review*, *29*(1), 23–43.

Seligmann, P. (2011, June 19). *Partnerships for the planet: Why we must engage corporations.* https://webcache.googleusercontent.com/search?q=cache:niKMCZ5kqccJ:www.huffpost.com/entry/conservation-international-lockheed-martin_b_863876+&cd=1&hl=en&ct=clnk&gl=uk&client=firefox-b-d

Skinner, A., Katebalila, D., Kambi, M., & John, C. (2018, March 21). *An illegal logger in Tanzania becomes a forest defender.* www.worldwildlife.org/stories/an-ill

Stryker, J. (2018, May 18). Unlikely partners: How teaming up with 'nature's enemies' could boost the impact of conservation: Open letter to IUCN members. *Crossroads Blog.* www.iucn.org/crossroads-blog/201805/unlikely-partners-how-teaming-nature-s-enemies-could-boost-impact-conservation

Survival International. (n.d.). *Hunters or poachers? Survival, the Baka and WWF.* www.survivalinternational.org/campaigns/wwf

Wall, T. (2020). The police invention of humanity: Notes on the 'thin blue line'. *Crime, Media, Culture, 16*(3), 319–336.

WWF. (2018, July 31). *New survey finds, one in seven wildlife rangers have been seriously injured over the past year in the line of duty.* https://wwf.panda.org/wwf_news/press_releases/?332051/New-survey-finds-one-in-seven-wildlife-rangers-have-been-seriously-injured-over-the-past-year-in-the-line-of-duty

WWF. (2019, March 22). *Statement of objectives: Independent panel of experts: WWF Independent review.* https://wwf.panda.org/?344901

WWF. (2020a, November 17). *Embedding human rights in nature conservation: From intent to action.* https://wwfint.awsassets.panda.org/downloads/independent_review___independent_panel_of_experts__final_report_24_nov_2020.pdf

WWF. (2020b, November 24). *WWF management response to recommendations from independent panel report: Embedding human rights in nature conservation.* https://wwfint.awsassets.panda.org/downloads/4_ir_wwf_management_response.pdf

Ybarra, M. (2018). *Green wars: Conservation and decolonization in the Maya forest.* Berkeley, CA: University of California Press.

Zalik, A. (2004). The Niger delta: 'Petro violence' and 'partnership development'. *Review of African Political Economy, 31*(101), 401–424.

21 Defending territories of life through Indigenous and Community Conserved Areas (ICCAs)

Cora Shaw

Protection of the environment does not equate to support or protection for environmental defenders. Histories of dispossession and violence accompany the creation of protected areas under the colonial pretense of exclusion and restriction, while marginalizing and disenfranchising Indigenous peoples and local communities (IPLC). Conservation practice, in this sense, seeks to protect designated areas of land, waters, biodiversity, and resources through formalized protected areas (PAs) – even though the ecological integrity of many of these landscapes results from the deep, historical relationships and traditional management practices of the evicted communities. The establishment of a PA makes a territorial claim over accessibility and managerial rights. When such a claim overlaps with IPLC territories, this external imposition is frequently met with resistance. Increased acknowledgement of the faulty assumption that human presence inherently harms biodiversity and the devastating impacts to the local or evicted communities have given rise to alternative forms of governance for PAs, opting for decentralized management and more 'inclusive' processes. Indigenous and Community Conserved Areas (ICCAs) constitute an important instrument to better support Indigenous peoples and local communities in defense of their ancestrally managed territories and promote healthy ecosystems without fueling violence through imposed PAs. This chapter will argue that the appropriate recognition of ICCAs can support the objectives of environmental and land defenders, but that the risks of perpetuating institutionalized colonial conservation practices remains in need of redress.

Indigenous and Community Conserved Areas

Indigenous peoples and local communities have managed and protected sacred spaces and resources for millennia. The motivations for such protection stem not from the flaws depicted in the Western imaginary, but rather from deep relationality and understanding of the interconnectedness of all humans and non-humans. The custodian communities considered as ICCAs have deep, historical connections to their territory, are responsible for making and upholding decision regarding the territory's land and resources, and make decisions through their governing institutions to promote the conservation of nature and support the community's well-being (ICCA Consortium n.d.).[1] The valuation of human and non-human relationality between ancestral communities and their territories determines the positive ecological outcomes, whether intentional or not (Artelle et al. 2020; Tran et al. 2020).

Indigenous and Community Conserved Areas as a term is intended to represent the characteristics of territories managed by their ancestral custodians around the world, and as such are known under many names, including ICCAs, Indigenous Protected Areas (IPAs), Indigenous Protected and Conserved Areas (IPCAs), and Community Conserved Areas (CCAs) (see Smyth 2015; Kothari et al. 2012; and ICCA Consortium n.d.). As expressed by the ICCA Consortium, the title "ICCA" seeks to encompass the qualities of "territories of life". A singular image or term cannot encompass the diversity of histories, practices, geographies, lives, and livelihoods associated, though considering territories of life as ICCAs, both legally and otherwise, provides a means to support Indigenous and local governance and self-determination (Artelle et al. 2020; Dominguez & Benessaiah 2017).

The recognition of IPLC in global conservation efforts has significantly increased since recognition of Community Conserved Areas at the 2003 World Parks Congress in Durban, with the evolution toward recognition of ICCAs emphasizing a "greater degree of identity, territoriality, and heritage" (Dominguez & Benessaiah 2017: 166; see also Forest Peoples Program 2020; Kothari et al. 2012; Knox 2018; Tauli-Corpuz et al. 2018.). ICCAs are now a formalized category of protected area within international conservation policy (Smyth 2015; Kothari et al. 2012), though to appropriately recognize and support the custodian IPLC to govern traditionally managed lands, waters, and resources, it is crucial that status of ICCA be self-declared rather than imposed (Artelle et al. 2020).

Considering the pressures on and vulnerabilities of IPLC

Indigenous peoples and local communities seeking to establish ICCAs often do so as a result of pressures within their socio-environmental and political contexts, including the violence that they face in seeking to defend their land and environment (Butt et al. 2019; Le Billon & Lujala 2020; Middeldorp & Le Billon 2019), as demonstrated by the disproportionate number of Indigenous peoples killed for defending their lands and territories. Global Witness (2020) reported that of the 212 confirmed murders of environmental defenders in 2019, 40% of these were Indigenous people. Noting that Indigenous peoples comprise 5% of the total global population illuminates the disparity of threats faced specifically by Indigenous land and environmental defenders and amplifies the need for increased protections (Bille Larsen & Lador, this volume; Scheidel et al. 2020).

Defense of ancestral territories by Indigenous and community defenders occurs within a nexus of multiple vulnerabilities. Disproportionate impacts of both localized environmental degradation and global climate change threaten the lives and livelihoods of Indigenous and community defenders reliant upon their environment for physical, cultural, and ecological well-being (Knox 2018). A lack of free, prior, and informed consent (FPIC) by Indigenous and community defenders for 'development' projects like extractive activities are commonly associated with elevated levels of environmental degradation, increased conflict between workers and local community members, or displacement (Conde & Le Billon 2017), exacerbating colonial disenfranchisement and marginalization. The deep connection of IPLC to their lands, waters, and territories elicits resistance to such impositions, and as Le Billon and Lujala (2020) find, Indigenous peoples are the most likely to mobilize in defense of their lands, waters, and resources. However, conservation practice through protected areas poses what some consider "the number one threat to Indigenous territories" and local communities (Dowie 2009: xviii, as quoted in Domínguez & Luoma 2018).

Western tactics of conservation seek to preserve lands, resources, and biodiversity through the designation creation of protected areas (PAs), inherently claiming and operationalizing some forms of 'sovereign authority' over the management of a territory (Robbins 2011; see also Massé 2020). Conservation practices and policies undermine the rights of Indigenous and community defenders (Colchester 2004), who bear the burdens of exclusive, strictly enforced protected areas, deemed 'fortress conservation' (Brockington 2002). Conflicts frequently result from the restriction of access to or displacement away from natural resources or sacred sites important for the lives and livelihoods of (former) IPLC (Stevens & Dean 1997; West et al. 2006; see also Ndoinyo, this volume), especially when PAs and rules are established on customary lands without the free, prior, and informed consent of communities.

International policy recognizes the importance of biodiversity and healthy ecosystems for the realization of all human rights (Knox 2018), but the mainstream approach of imposing externally designated protected areas to reduce rates of biodiversity loss often undermines human rights.[2] Lands and territories traditionally managed by IPLC are of particular interest in resolving this tension. Indigenous lands and territories comprise 40% of ecologically intact regions (Garnett et al. 2018) and over 80% of the planet's remaining biodiversity (Forest Peoples Programme 2020), in many cases with higher rates of biodiversity than that of adjacent PAs (Schuster et al. 2019). The lands and territories managed under customary tenure by Indigenous peoples and local communities comprise an estimated 50% of the planet's land area, though only 10% hold legal title (Forest Peoples Program 2020; Rights & Resource Initiative 2015). This disparity poses a challenge to environmental and land defenders as the inappropriate designation of Indigenous or communally managed lands as PAs threaten to undermine the right to self-determination, among others.

Protected areas, criminalized defenders

Histories of forced displacement, harsh disciplining of dissent, restricted access to essential resources and sacred sites, and erasure of cultural identity of IPLC for the creation of PAs have been well documented (Agrawal & Redford 2009; Brockington & Igoe 2006; West et al. 2006). This exclusionary model imposes a singular understanding of the relationship between humans and nature (Cronon 1996; Büscher et al. 2012), while criminalizing traditional means of subsistence (Duffy et al. 2019). Acknowledgement of the flawed assumption that sought to preserve an idealized 'pristine wilderness' and its devastating impacts on millions of lives and livelihoods (Agrawal & Redford 2009) has proven instrumental in the introduction of alternative categorizations of PAs and their governance types, such as ICCAs (Berkes 2009; Stevens & Dean 1997).

The shifting dynamics of PA management with the transition to Community Conserved Areas and ICCAs introduce new forms of governance, often transferring management responsibilities to NGOs and including more participatory models (Holmes & Cavanagh 2016). This can exacerbate the reliance upon international funding to maintain the current status of the PA, and frequently fails to offer full support for Indigenous and community defenders in realizing their full rights to self-determination (Holmes & Cavanagh 2016; Stevens & Dean 1997). The costs of protected areas are borne by local communities, the greatest benefits are experienced on a global scale, and local benefits are unevenly distributed (see Pelletier et al. 2019: 297).

Despite these flaws, the practice of creating strict, exclusionist PAs persists (Domínguez & Luoma 2020) with the institutionalized logics of colonial regimes in post- or settler-colonial contexts (Eichler & Baumeister 2018). PAs frequently create zones of "double exception" (Le Billon, this volume) by allowing extractive industrial activities within its borders (MacKenzie et al. 2017), failing to address root causes of global biodiversity loss and climate change (see Büscher et al. 2012, 2017). IPLC resistance to imposed PAs seeks to protect traditional knowledge and management practices undermined by imposed industry and conservation alike.

Scheidel et al. (2020) found that one in eight cases of environmental conflict involved murders associated with establishing a protected area. Alongside the increasing rates of militarization within conservation practice (Duffy et al. 2019), this illuminates the urgency of attending to the faulty assumptions and power grabs underpinning conservation practice, particularly in communities with disproportionate vulnerability (Le Billon & Lujala 2020) or those already experiencing unrest (Marijnen et al. 2020). Militarized conservation depicts a continuation of the colonial exclusionary logics pervasive in externally imposed PAs, weaponizing its ideology through those who will defend and uphold the same 'pristine wilderness' they purport and disciplining those who do not.

Defending territories of life

Indigenous and local community leaders called for the respect and support of Indigenous and local governance in conservation practice long before the current categorizations of PAs or the scholarly literature detailing the negative effects of conservation on Indigenous and local communities (Stevens & Dean 1997; Jonas et al. 2017). Legal tenure of ancestral lands and territories and FPIC are constituted in the full and appropriate recognition of the rights of Indigenous peoples under the UN Declaration on the Rights of Indigenous Peoples (Artelle et al. 2020; Le Billon & Lujala 2020; Scheidel et al. 2020), while the right to self-determination for all IPLC is found in other international legal and human rights instruments (See Kothari et al. 2012: 40).

The high rates of biodiversity of territories ancestrally managed by IPLC, as noted by Artelle et al. (2020), present an opportunity to support Indigenous and local governance institutions through conservation and beyond. ICCAs diverge from externally imposed PAs, including through 'participatory' approaches, by supporting self-determination to defend and manage their territory, and as an act of communal defense. However, national-level support and recognition significantly lags behind that of international, and what national recognition does occur frequently fails to fully support ICCAs, with limited transfer of power and rights (Kothari et al. 2012).

Successes of ICCAs

Formal declaration of an ICCA acts to legitimize territorial claims as a practice of self-determination (Artelle et al. 2020; Kothari et al. 2012). Tran et al.'s (2020) analysis of 86 ICCA case studies finds that common themes motivating the declaration of an ICCA are to affirm Indigenous Rights and Title, to enforce FPIC requirements, and to support the intergenerational transfer of traditional knowledge.[3] This act of territorial defense can be a response to a present threat by an extractive project, like the 1984 establishment of the *Tla-o-qui-aht* Tribal Parks to halt industrial logging on the *Tla-o-qui-aht* people's ancestral lands in Western Canada (Murray & Burrows 2017). ICCAs also enable the reaffirmation of rights to continue traditional management practices eliminated by exclusionary and externally managed PAs, which often categorize traditional subsistence hunting as 'poaching', wrongly equating it with large-scale commercial poaching operations (Domínguez & Luoma 2020). ICCAs have demonstrated 'successes' in supporting Indigenous and community defenders by increasing financial and capacity building support for IPLC to practice traditional management, affirm tenure, conserve traditional knowledge, and offer improved access to traditional food sources, among others (see Tran et al. 2020 for specific case studies).

Challenges and limitations of ICCAs

Challenges to ICCAs and their custodian communities stem from both lack of recognition and inappropriate recognition (Kothari et al. 2012). Inappropriate recognition is a significant barrier to self-determination, especially when recognition is exclusively granted through colonial understandings and in frameworks that have historically undermined IPLC rights (Coulthard 2014; Tran et al. 2020). Interpolation into national PA systems poses the threat of 'colonial entanglements' for Indigenous and community defenders (Dennison 2012), and resistance reflects justifiable fears that recognition in conservation policy will mirror prior experiences of loss of access to territory and resources (Berkes 2009; Tran et al. 2020). Recognition as a co-managed PA can still come with strings attached. As Davies et al. (2018) demonstrate, the Walpiri people of central Australia receive financial and institutional support from external entities and NGOs in managing the Northern Tanami Indigenous Protected Area, though their management priorities do not always align, causing community tensions.

National and sub-national policies which recognize the rights of IPLC to govern their territory in a singular context, but are contradicted by superseding policies undermine the ability for ICCAs to practice traditional management. Granting management rights but failing to support secure land tenure limits an ICCA's ability to enforce their management decisions and practices (Kothari et al. 2012). For example, Indigenous territories in Bolivia experience what is considered "the most advanced form of recognition in the Americas", where all Indigenous peoples and territories are legally recognized as *Tierras Comunitarias de Origen* (ibid.: 124). However, this recognition fails to support these communities in practice, because State-determined extractive projects undermine Indigenous and community defenders' rights by proceeding without respecting free, prior, and informed consent (Gambon & Rist 2018).

The Ogiek peoples of the Kenyan Mau Forest's fight for appropriate recognition provides an example of both the successes and challenges of IPLC environmental defenders. After years of mobilizing and legal actions, in May of 2017, the Ogiek were granted territorial rights over their forest by the African Court of Human and Peoples' Rights, after years of unjust evictions and undue blame for forest degradation resulting from decades of State and external extractive operations (ICCA Consortium 2020). However, even as the legal territorial rightsholders, the Ogiek peoples were targeted, criminalized, and displaced in July 2020, again accused of causing forest degradation that had been carried out by external communities and entities - both allowed and supported by the State. The Kenyan government justified the eviction, claiming that "preserving this ecosystem takes priority over land claims of the Ogiek" (ibid.) though this "flies in the face of a growing body of evidence that honoring land rights is the foundation for equitable and sustainable conservation of forests" (ibid.). The Ogiek communities were again left without their homes or territory in the middle of the COVID-19 global pandemic, to which Indigenous peoples face elevated risks.[4] Marginalization and criminalization through 'environmental protection' undermines land and environmental defenders and their ability to practice self-determination and promote healthy ecosystems, including the custodian communities of ICCAs.

Conclusion

Indigenous people and local communities face higher risk as environmental defenders and are among the most vulnerable to the impacts of extractive activities, climate change, and the negative impacts of conservation associated with PAs. Protectionist conservation practices have excluded and evicted local communities, by privileging Western ontologies, erasing Indigenous peoples, and monopolizing both how the environment is to be protected and who is able to do so. In the words of Kenyan ecologist and author Mordecai Ogada (2019), "We must dismantle and discard the model of conservation developed in the West, and funded by the West to conserve for the West". Increasing calls for PAs must support IPLC as the defenders of territories of life. To eliminate the colonial pretense and threads extending through conservation practice, support for ICCAs must encompass support for the full realization of rights in international, national, and sub-national legal contexts.

While ICCAs can be seen as the ideal solution to the flaws – and crimes – of conservation practices by supporting Indigenous and local community defenders, ICCAs are not a 'perfect' solution to protect all who identify as environmental defenders or to end all violence against IPLC. Recognition within conservation practice must be a component of broader sociopolitical support and appropriate recognition for IPLCs. Legal recognition of an ICCA as a PA, within land use or conservation policies, provides one step toward greater recognition of the rights of Indigenous and community defenders, but the full socio-legal context must also be supportive of these goals. As noted by Tran et al. (2020), it is doubtful that ICCAs alone can shift colonial practices on larger scales (see also Carroll 2014; Muller 2003; Ross et al. 2009), though they prove a significant ability to secure tenure and support practices of traditional management.

Support for IPLC to determine and carry out their management goals for their ancestral territory may well aid in global conservation goals, though without appropriate recognition, State and conservation organizations continue to undermine the fundamental human rights of these communities. Büscher and Fletcher (2019) suggest a discursive and conceptual shift, moving from *protected* areas to *promoted* areas. This call beckons consideration of the realities of *from* whom and *for* whom PAs are protected. In an effort both to respond to prior violence enacted upon Indigenous peoples and local communities in the name of conservation and to prohibit a continuation of colonial violence, there is a need also to emphasize *by whom promoted* areas should be established and managed. With such priority in mind, Indigenous and local communities will be better supported in asserting local governance, reclaiming ancestral rights, and practicing self-determination as defenders of territories of life.

Notes

1 See ICCA Consortium for greater explanation and resources on defining and identifying, as well as case studies. www.iccaconsortium.org/index.php/discover/.
2 The Post-2020 Global Biodiversity Framework, in its current draft, calls for 60% of 'sites of particular importance for biodiversity' to be protected areas. Globally, the draft sets a goal of protecting 30% of global land and marine areas, with 10% of that as strictly protected. (CBD/WG2020/2/4) www.cbd.int/doc/c/b14d/6af5/a97c4f2c9d58203f5e2e059c/wg2020-02-04-en.pdf.
3 See Tran et al. (2020) for a meta-analysis of 86 case studies on ICCAs, detailing the common themes of motives, successes, and challenges found in the literature.
4 See UN Department of Economic and Social Affairs: COVID-19 and Indigenous peoples. www.un.org/development/desa/indigenouspeoples/covid-19.html. For more on the Ogiek case, see www.forestpeoples.org/en/press-release-kenyan-authorities-burn-down-28-Sengwer-homes-in-Embobut-Forest and www.iccaconsortium.org/wp-content/uploads/2020/08/CLAN-Press-Statement-on-Ogiek-and-Sengwer.pdf.

References

Agrawal, A., & Redford, K. (2009). Conservation and displacement: An overview. *Conservation and Society*, 7(1), 1.

Artelle, K. A., Zurba, M., Bhattacharyya, J., Chan, D. E., Brown, K., Housty, J., & Moola, F. (2020). Supporting resurgent Indigenous-led governance: A nascent mechanism for just and effective conservation. *Biological Conservation*, *240*, 108284.

Berkes, F. (2009). Community conserved areas: Policy issues in historic and contemporary context. *Conservation Letters*, *2*(1).

Brockington, D. (2002). *Fortress conservation: The preservation of the Mkomazi game reserve, Tanzania*. Bloomington, IN: Indiana University Press.

Brockington, D., & Igoe, J. (2006). Eviction for conservation: A global overview. *Conservation and Society*, *4*(3), 424–470.

Büscher, B., & Fletcher, R. (2019). Towards convivial conservation. *Conservation and Society*, *17*(3), 283.

Büscher, B., Fletcher, R., Brockington, D., Sandbrook, C., Adams, W. M., Campbell, L., Corson, C., Dressler, W., Duffy, R., Gray, N., Holmes, G., Kelly, A., Lunstrum, E., Ramutsindela, M., & Shanker, K. (2017). Half-Earth or whole Earth? Radical ideas for conservation, and their implications. *Oryx*, *51*(3), 407–410.

Büscher, B., Sullivan, S., Neves, K., Igoe, J., & Brockington, D. (2012). Towards a synthesized critique of neoliberal biodiversity conservation. *Capitalism Nature Socialism*, *23*(2), 4–30.

Butt, N., Lambrick, F., Menton, M., & Renwick, A. (2019). The supply chain of violence. *Nature Sustainability*, *2*(8), 742–747.

Carroll, C. (2014). Native enclosures: Tribal national parks and the progressive politics of environmental stewardship in Indian Country. *Geoforum*, *53*, 31–40.

Colchester, M. (2004). Conservation policy and indigenous peoples. *Environmental Science & Policy*, 7(3), 145–153.

Conde, M., & Le Billon, P. (2017). Why do some communities resist mining projects while others do not? *The Extractive Industries and Society*, *4*(3), 681–697.

Coulthard, G. S. (2014). *Red skin, white masks: Rejecting the colonial politics of recognition*. Minneapolis: University of Minnesota Press.

Cronon, W. (Ed.). (1996). *Uncommon ground: Rethinking the human place in nature*. New York: Norton.

Davies, J., Walker, J., & Maru, Y. T. (2018). Warlpiri experiences highlight challenges and opportunities for gender equity in Indigenous conservation management in arid Australia. *Journal of Arid Environments, 149*, 40–52.

Dennison, J. (2012). *Colonial entanglement: Constituting a Twenty-First-Century Osage Nation*. Chapel Hill, NC: University of North Carolina Press.

Domínguez, L., & Luoma, C. (2020). Decolonising conservation policy: How colonial land and conservation ideologies persist and perpetuate Indigenous injustices at the expense of the environment. *Land, 9*(3), 65.

Dominguez, P., & Benessaiah, N. (2017). Multi-agentive transformations of rural livelihoods in mountain ICCAs: The case of the decline of community-based management of natural resources in the Mesioui agdals (Morocco). *Quaternary International, 437*, 165–175.

Dowie, M. (2009). *Conservation refugees: The hundred-year conflict between global conservation and Native peoples*. Cambridge, MA: MIT Press.

Duffy, R., Massé, F., Smidt, E., Marijnen, E., Büscher, B., Verweijen, J., Ramutsindela, M., Simlai, T., Joanny, L., & Lunstrum, E. (2019). Why we must question the militarisation of conservation. *Biological Conservation, 232*, 66–73.

Eichler, L., & Baumeister, D. (2018). Hunting for justice: An Indigenous critique of the North American model of wildlife conservation. *Environment and Society, 9*(1), 75–90.

Forest Peoples Program (2020). *The contributions of indigenous peoples and local communities to the implementation of the strategic plan for biodiversity 2011–2020 and to renewing nature and cultures*. A complement to the fifth edition of the Global Biodiversity Outlook. Moreton-in-Marsh.

Gambon, H., & Rist, S. (2018). Moving territories: Strategic selection of boundary concepts by Indigenous people in the Bolivian Amazon-an element of constitutionality?. *Human Ecology, 46*(1), 27–40.

Garnett, S. T., Burgess, N. D., Fa, J. E., Fernández-Llamazares, Á., Molnár, Z., Robinson, C. J., Watson, J. E. M., Zander, K. K., Austin, B., Brondizio, E. S., Collier, N. F., Duncan, T., Ellis, E., Geyle, H., Jackson, M. V., Jonas, H., Malmer, P., McGowan, B., Sivongxay, A., & Leiper, I. (2018). A spatial overview of the global importance of Indigenous lands for conservation. *Nature Sustainability, 1*(7), 369–374.

Global Witness. (2020). *Defending tomorrow: The climate crisis and threats against land and environmental defenders*. London.

Holmes, G., & Cavanagh, C. J. (2016). A review of the social impacts of neoliberal conservation: Formations, inequalities, contestations. *Geoforum, 75*, 199–209.

ICCA Consortium (n.d.). www.iccaconsortium.org/index.php/discover/

ICCA Consortium (2020). *Kenya: Indigenous and forest-dwelling communities report illegal evictions from ancestral lands during COVID-19 pandemic*. www.iccaconsortium.org/index.php/2020/07/23/kenya-indigenous-ogiek-sengwer/

Jonas, H., Lee, E., Jonas, H., Matallana-Tobon, C., Wright, K., Nelson, F., & Ens, E. (2017). Will 'other effective area-based conservation measures' increase recognition and support for ICCAs? *Parks, 23*(2), 63–78.

Knox, J. (2018). *Framework principles on human rights and the environment*. Report of the Special Rapporteur on Human Rights and the Environment. HRC/37/59.

Kothari, A., Corrigan, C., Jonas, H., Neumann, A., & Shrumm, H. (2012). *Recognising and supporting territories and areas conserved by indigenous peoples and local communities: Global overview and national case studies*. Montreal: Secretariat of the Convention on Biological Diversity.

Le Billon, P., & Lujala, P. (2020). Environmental and land defenders: Global patterns and determinants of repression. *Global Environmental Change, 65*, 102163.

MacKenzie, C. A., Fuda, R. K., Ryan, S. J., & Hartter, J. (2017). Drilling through conservation policy: Oil exploration in Murchison falls protected area, Uganda. *Conservation and Society, 15*(3), 12.

Marijnen, E., de Vries, L., & Duffy, R. (2020). Conservation in violent environments: Introduction to a special issue on the political ecology of conservation amidst violent conflict. *Political Geography*, 102253.

Massé, F. (2020). Conservation law enforcement: Policing protected areas. *Annals of the American Association of Geographers, 110*(3), 758–773.

Middeldorp, N., & Le Billon, P. (2019). Deadly environmental governance: Authoritarianism, eco-populism, and the repression of environmental and land defenders. *Annals of the American Association of Geographers, 109*(2), 324–337.

Muller, S. (2003). Towards decolonisation of Australia's protected area management: The Nantawarrina Indigenous protected area experience. *Australian Geographical Studies, 41*(1), 29–43.

Murray, G., & Burrows, D. (2017). Understanding power in Indigenous protected areas: The case of the Tla-o-qui-aht Tribal Parks. *Human Ecology, 45*(6), 763–772.

Ogada, M. (2019, June 27). Decolonising conservation: It is about the land, stupid! *The Elephant*. www.theelephant.info/culture/2019/06/27/decolonising-conservation-it-is-about-the-land-stupid/

Pelletier, J., Gélinas, N., & Potvin, C. (2019). Indigenous perspective to inform rights-based conservation in a protected area of Panama. *Land Use Policy, 83*, 297–307.

Rights and Resource Initiative. (2015). *Who owns the world's land? A global baseline of formally recognized Indigenous and community land rights*. Washington, DC: Rights and Resource Initiative.

Robbins, P. (2011). *Political ecology: A critical introduction*. Hoboken, NJ: John Wiley & Sons.

Ross, H., Grant, C., Robinson, C. J., Izurieta, A., Smyth, D., & Rist, P. (2009). Co-management and Indigenous protected areas in Australia: Achievements and ways forward. *Australasian Journal of Environmental Management, 16*(4), 242–252.

Scheidel, A., Del Bene, D., Liu, J., Navas, G., Mingorría, S., Demaria, F., Avila, S., Roy, B., Ertör, I., Temper, L., & Martínez-Alier, J. (2020). Environmental conflicts and defenders: A global overview. *Global Environmental Change, 63*, 102104.

Schuster, R., Germain, R. R., Bennett, J. R., Reo, N. J., & Arcese, P. (2019). Vertebrate biodiversity on indigenous-managed lands in Australia, Brazil, and Canada equals that in protected areas. *Environmental Science & Policy, 101*, 1–6.

Smyth, D. (2015). Indigenous protected areas and ICCAs: Commonalities, contrasts and confusions. *PARKS, 21*(2). https://doi.org/10.2305/IUCN.CH.2014.PARKS-21-2DS.en

Stevens, S. F., & Dean, T. D. (1997). *Conservation through cultural survival: Indigenous peoples and protected areas*. Washington, DC: Island Press.

Tauli-Corpuz, V., Alcorn, J., & Molnar, A. (2018). *Cornered by protected areas: Replacing 'fortress' conservation with rights-based approaches helped bring justice for Indigenous peoples and local communities, reduces conflict, and enables cost-effective conservation and climate access*. Washington, DC: Rights and Resource Initiative.

Tran, T. C., Ban, N. C., & Bhattacharyya, J. (2020). A review of successes, challenges, and lessons from Indigenous protected and conserved areas. *Biological Conservation, 241*, 108271.

West, P., Igoe, J., & Brockington, D. (2006). Parks and peoples: The social impact of protected areas. *Annual Review of Anthropology, 35*(1), 251–277.

22 Interrogating international cooperation in support of environmental human rights defenders

The Geneva Roadmap 40/11 and the power of connecting solutions

Peter Bille Larsen and Yves Lador

This chapter explores the role and potential of international cooperation for environmental human rights defenders through four different lenses. Firstly, we discuss the emergence of the very idea of environmental *human rights* defenders. Secondly, we offer an overview of the background, content, strengths and limitations of the Human Rights Council resolution 40/11 dedicated to their cause. Thirdly, we present the main content and modalities of the Geneva Roadmap. Finally, we seek to draw out some more general lessons for international cooperation to support environmental human rights defenders through the challenges and opportunities involved in connecting solutions beyond sectoral and organizational boundaries.

While the deepening environmental human rights defender crisis is increasingly documented by both national organizations and international efforts such as Global Witness reports (e.g. Global Witness 2020), the continued high levels of killings and other forms of violence reveal the complexity of building and effectively connecting national and international responses (Bille Larsen et al. 2020). Such complexity, we argue, reflects a number of key challenges, such as: (i) the gap between international commitments and realities on the ground, (ii) the difficulty of shifting from a fire-fighting to a preventive approach, (iii) silo tendencies in international cooperation and (iv) high levels of fragmentation.

On the one hand, a growing number of NGO activities, reports and multilateral efforts reveal unprecedented internationalized attention to the rights of environmental human rights defenders. The levels of documentation, the presence of emergency support and social media have rendered both attacks against and protection of environmental defenders a public concern subject to multiple types of responses.

On the other hand, persistent levels of killings and deepening patterns of systemic violence reveal the limitations of both domestic and international mechanisms in place. A Human Rights Council resolution may have been adopted, but it remains good intentions with few implications for many States concerned. Regional instruments such as the Escazú Agreement offer new language and measures, yet ratification and actual implementation take time. Special procedure mandate holders may raise their voices and have a certain level of political voice and legitimacy, yet procedures and dialogue with specific countries are also fraught with challenges. While much is being done, it is rarely enough and often happens too late. Despite the proliferation of different organizational initiatives, major divides between, but also within, the environmental conservation and human rights communities prompt the need for new thinking around how to connect and build bridges on environmental human rights defender issues.

Growing attention to environmental human rights defenders

Why do we use the term "environmental human rights defenders" and how is it defined?

The term originates from the Declaration on human rights defenders adopted by the UN General Assembly in 1998, on the occasion of the fiftieth anniversary of the Universal Declaration of Human Rights (Resolution *A/RES/53/144*). The Declaration's full name is "Declaration on the Right and Responsibility of Individuals, Groups and Organs of Society to Promote and Protect Universally Recognized Human Rights and Fundamental Freedoms". This longer title is frequently abbreviated to "Declaration on human rights defenders".

A "human rights defender" is a person who, individually or together with others, acts to promote or protect human rights. Experience has shown that when individuals engage in promoting or protecting human rights, they become more exposed than other individuals just because of this engagement. This raised concerns and led to the drafting and eventual adoption of the Declaration in 1998, aimed at addressing this vulnerability from speaking out and acting for human rights.

Human rights defenders are therefore identified above all by what they do, and it is through a description of their actions and of some of the contexts in which they work that the term can best be explained. This means, clearly, that being a "defender" is not a status. It concerns a situation in which anybody can find him or herself when protecting or promoting human rights in general or the rights of others more specifically. Being a human rights defender is to act to address any human right(s) on behalf of individuals or groups, whether civil and political rights or economic, social and cultural rights.

Until the Declaration of 1998, terms such as human rights "activist", "professional", "worker", "advocate", or "monitor" were the most commonly used terms. Since the Declaration, the more general term "human rights defender" has been increasingly used and is seen as a more relevant and useful term.

The development of the Declaration on human rights defenders began in 1984. A collective effort by a number of non-governmental human rights organizations and some States helped to ensure that the final result was a strong and useful text. Perhaps most importantly, the Declaration is addressed not just to States and human rights defenders, but to everyone. It outlines that all have a role to play and that the global human rights movement involves everyone.

After the Declaration, in 2000, a mandate of the UN Special Rapporteur (SR) on the situation of human rights defenders was established by the Commission on Human Rights (*E/CN/2000/61*) and later extended by the Human Rights Council. The Special Rapporteur on several occasions focused on selected groups of defenders at risk, like women human rights defenders in 2011 (*A/HRC/16/44*) and 2019 (*A/HRC/40/60*), human rights defenders on the rights of people on the move in 2018 (*A/HRC/37/51*), and now the emerging notion of children as rights defenders, which is relevant to the children and youth movement for the climate. In 2012, the SR report focused on journalists and media workers, defenders working on land and environmental issues and youth and student defenders. In 2016, the SR report was entirely devoted to environmental human rights defenders (*A/71/281*), spelling out clearly for the first time this designation.

What is an "environmental" human rights defender?

Resolution 40/11, after recalling a number of fundamental principles, expresses concerns "that human rights defenders working in environmental matters, **referred to as environmental human rights defenders**, are among the human rights defenders most exposed and at risk" (emphasis added). An environmental human rights defender is a person who, individually or together with others, acts for environmental protection grounded in human rights, even if it is not explicitly done. This is exactly what the resolution acknowledges when it recognizes "the positive, important and legitimate role played by human rights defenders in the promotion and protection of human rights as they relate to the enjoyment of a safe, clean, healthy and sustainable environment".

During the negotiation of the resolution in 2019, some States opposed this term, but all the alternatives formulated appeared complicated, vague and as weakening any attempt to secure protection. Actually, the term seemed more and more appropriate as the discussion proceeded. It refers clearly to the obligations that States have under ratified international binding treaties and conventions.

While the term is not very familiar to everyone engaged in land and environment protection, it is foreseeable that over time a similar evolution will occur, like the one that took place among human rights activists who progressively adopted and used the term "defenders".

Human Rights Council resolution 40/11, its negotiation and its significance

On 21 March 2019, the Human Rights Council adopted by consensus a resolution recognizing the role of environmental human rights defenders in protecting ecosystems, addressing climate change and attaining the sustainable development goals (SDGs). The resolution was presented by Norway, on behalf of 60 States from all parts of the world. In particular, many Latin American States strongly supported the resolution, which is significant given the dangerous situation defenders face in that region.

The coalition of NGOs led by the International Service for Human Rights (ISHR), which permanently follows the work of the Human Rights Council on human rights defenders and on the resolutions on defenders, was joined by environmental NGOs, whose voice was carried by Earthjustice, to actively contribute to the discussion on the drafting of this resolution.

Reaching a consensus on protecting environmental human rights defenders was quite an achievement. Securing consensus remained uncertain until the very end, yet its achievement sends a much-needed signal from the international community. It formally acknowledges the legitimacy of defenders and their crucial contribution to safeguarding a healthy and sustainable environment on this planet. This also helps to counter stigmatization and contribute to maintaining and expanding the civic space they need.

The resolution was drafted first as a continuation of the Council's work on human rights defenders, recognizing that the protection of these defenders can only be achieved through an approach which promotes and celebrates their work. It expresses concerns about reports that Indigenous human rights defenders also face attacks and criminalization, while the specific conditions of Indigenous peoples and local communities can aggravate their vulnerability.

Recognizing the importance of gender equality and women's empowerment in the management of natural resources and safeguarding of the environment, the resolution pays particular attention to women human rights defenders, stressing the intersectional nature of violations and abuses against them.

It also stresses the intersectionality of violations against Indigenous peoples, children, persons belonging to minorities and rural and marginalized communities, and calls for root causes of violations to be addressed by strengthening democratic institutions, combating impunity and reducing economic inequalities.

The resolution reminds States that they have to respect, protect and fulfill all their human rights obligations when addressing environmental challenges. For example, information held by public authorities relating to land, natural resources and development must be publicly available. Therefore, it welcomes regional instruments such as the Aarhus Convention and the Escazú Agreement.

Having acknowledged the importance of environmental human rights defenders' work, the resolution expresses alarm at the increasing rate of violations against them, including killings, gender-based violence, threats, harassment, intimidation, smear campaigns, criminalization, judicial harassment, forced eviction and displacement. It acknowledges that violations are also committed against defenders' families, communities, associates and lawyers.

States are urged to adopt laws guaranteeing the protection of defenders, to put in place holistic protection measures for and in consultation with defenders and to ensure investigation and accountability for threats and attacks against environmental human rights defenders. Businesses are also reminded of their responsibilities as defined in the United Nations Guiding Principles on Business and Human Rights, in particular, to carry out human rights due diligence and to hold meaningful and inclusive consultations with defenders, potentially affected groups and other relevant stakeholders.

To have the resolution adopted by consensus came at the price of a lack of specificity in certain areas. The push from civil society to list the main root causes of the insecurity facing environmental human rights defenders was not followed, even though they have been documented by UN experts. The human rights obligations of finance institutions were not clearly spelled out, nor were the obligations to consult, respect and protect the work of environmental human rights defenders included in the end.

Even with these shortcomings, it is the first time that the UN Human Rights Council, the main universal human rights political body, explicitly and with one voice has called for the protection of environmental human rights defenders. It constitutes a real step towards better recognition and protection of their rights and echoes a host of national, regional and global initiatives to harness measures for the protection of people acting for the environment. These include research and organizational activities in both the human rights and the environmental communities. For example, UNEP has developed its "Environmental Rights Initiative" and IUCN is considering this issue for its coming World Conservation Congress.

Like all resolutions, however, this one needs to be translated "from Geneva" (i.e. on paper) into action on the ground through implementation at all levels by States, international organizations, businesses, development agencies and financial institutions. Civil society organizations also play a critical role in ensuring concrete follow-up and much-needed results. Roadmap 40/11 was created precisely with this purpose of supporting, facilitating and monitoring this translation from standards to practice.

While the resolution is not a legally binding instrument, it calls on States to respect existing legal obligations as a basis for preventing and responding to abuses identified. It condemns the failure of their implementation and serves as a reminder that if such events occur, it is an obligation under human rights law to provide access to effective remedies. A key strength of the resolution, indeed, is its articulation of how human rights principles and obligations apply in the context of persons and groups engaged in land and environmental protection.

As pointed out by the former UN Special Rapporteur on the rights of Indigenous peoples, Vicky Corpuz-Tauli: "How can we protect the environment if we don't protect those who are protecting it?"

The Geneva Roadmap 40/11: connecting solutions

How did the Geneva Roadmap emerge?

In late 2018, researchers from the Universities of Geneva, Oxford and Sussex started a conversation about organizing a joint event in Geneva to exchange information on environmental defender dynamics. The conversation in 2019 quickly broadened to include NGOs, States and UN partners active in the Geneva arena, including members of the support group behind Resolution 40/11. Key interlocutors included Earthjustice, OHCHR and UNEP as well as cooperation with the Geneva Science Policy Interface. This also connected to long-standing conversations with the United Nations Special Rapporteur on the right to a healthy environment (Knox 2017; Boyd 2018) and the United Nations Special Rapporteur on Human Rights Defenders (Forst 2020).

Simultaneously, discussions were also taking place in the context of preparing for the IUCN World Conservation Congress, then scheduled to take place in June 2020, including an effort to formulate a motion on environmental defenders as well as a number of targeted workshops (Bille Larsen & Balsiger forthcoming). This conversation included an IUCN Netherlands initiative on environmental defenders, the IUCN Council and the Commission on Environmental, Economic and Social Policy, numerous IUCN members and experts.

In February 2020, the small group of researchers, defenders, conservation and human rights professionals, including the two UN Special Rapporteurs (human rights defenders and the environment), came together in the Palais Eynard, made available to the organizers by the City of Geneva (Figure 22.1).

Figure 22.1 Participants of the Geneva Roadmap dialogue

Just a few weeks before the meeting, two environmental defenders in the Monarch Butterfly Biosphere Reserve in Mexico were found dead after challenging illegal logging and land grabbing. The biosphere reserve, inscribed as a World Heritage site in 2008, is the site of the spectacular return to the area every autumn of millions of butterflies. The site was now also another crime scene in the globally occurring assault against those who speak up and act for the environment.

A minute of silence at the beginning of the meeting was little consolation for this or any other killing, yet it sparked two days of intense discussions to strengthen initiatives to come up with solutions. The event aimed to take stock of analyses about challenges, review existing support initiatives and debate possible responses. From early on, organizers sought to frame follow-up action through the idea of a Geneva Roadmap. The aim was to inform both human rights and conservation policy processes and kick off more systematic collaboration in the long term.

A panel of defenders from Russia, Philippines, Brazil, Kenya, Turkey and Mozambique brought participants back to the serious challenges faced on the ground. Aside from the importance of their personal and organizational commitments, support from academic, civil society and multilateral sources enabled their travel, underlining the importance of making defenders' voices heard, but also of the potential of connecting support responses.

Testimonies of personal threats, loss of family members, criminalization and lack of voice in decision-making left little doubt as to the urgency of the crisis. Academic analysis in turn drew attention to the underlying drivers, patterns and trends of violence identified in emerging research. Reviews from both human rights and conservation practitioners synthesized lessons learned from growing civil society, State-led and multilateral protection efforts. A particularly strong emphasis was put on the growing networking and dialogues organized with or by defenders themselves. Ideas explored included how to provide a collective platform through which initiatives and commitments of States, civil society, research and academia or private actors could be supportive of each other.

As a follow-up to Human Rights Council Resolution 40/11, the Geneva Roadmap aims to ensure the effective implementation of the right to act for the protection of the environment and to promote free and safe spaces for information and discussion on environmental matters. It aims to inform both human rights and conservation policy processes and to kick off more systematic collaboration in the long term. Deliberately scheduled during the first week of the Human Rights Council, an official side-event co-sponsored by Switzerland and Fiji was organized at the United Nations to make sure that conclusions and recommendations were shared with the wider international community.

The Geneva Roadmap is a living document currently made up of four action goals and a commitment to follow up, build synergies and pool efforts among both civil society and multilateral efforts. Each action goal includes both general recommendations for action as well as a specific set of commitments of existing, planned actions. The Geneva Roadmap is conceived as a collective platform and forum for debate in which initiatives and commitments of States, civil society, research and academia or private actors can support each other through a process of continuous engagement and yearly follow-up currently projected to take place in conjunction with the Human Rights Council. The following preliminary set of four action goals were identified to bridge efforts, guide reflection and stimulate synergies.

1 **Reverse the tide** of marginalization and attacks against environmental actors
2 **Reinforce environmental rights**, enabling civic spaces and accountability
3 **Bridge initiatives** and enhance cooperation
4 **Break isolation** and ensure effective access to protection

None of the four goals are set in stone, yet they seek to capture key challenges identified and flesh out avenues for action and coordination. For each action goal, the group brainstormed about **possible fields of action** as well as laying out already planned next steps (Table 22.1).

Aside from the articulation of goals and contents, the group also provided input for other on-going policy efforts, such as the IUCN motion on environmental defenders and whistleblowers being prepared for the World Conservation Congress. Indeed, by targeting events in both the human rights sphere (Human Rights Council and Resolution 40/11 follow-up) and the environmental conservation sphere, the Roadmap dynamic aimed to underline important collective action potential, whether in the realm of policy making and coordination or in terms of connecting hands-on training or emergency support. A more extensive description of the Roadmap results can be found in the Annex of this chapter.

Concluding remarks: Roadmap during the COVID pandemic and beyond

Environmental and conservation organizations widely launched 2020 as a "super" year for nature in an orchestrated attempt to build global policy attention around a string of high-level decision-making processes in the biodiversity arena; the Conference of Parties of the Convention on Biological Diversity, the IUCN World Conservation Congress and the United Nations Summit on Biodiversity. The COVID pandemic not only transformed or prevented these events from taking place, it also revealed the fragile state of protection measures for environmental defenders targeted. Rather than a super year, global reports reveal record levels of killings, shrinking civic spaces and the further targeting of defenders. Not only have Indigenous people made up a large proportion of defenders and been disproportionately affected by COVID, reports also indicate lockdowns have been turning into deepening resource pressures and crackdowns on defenders in countries such as the Philippines.

Given such dynamics, it is all the more urgent that international cooperation shifts from a fire-fighting approach to one of strengthening prevention,

Table 22.1 The Geneva Roadmap 40/11 action goals

Action goals	*Action goal 1: reverse marginalization and attacks on environmental defenders*	*Action goal 2: reinforce rights, civil society spaces and accountability*	*Action goal 3: bridge initiatives and enhance cooperation*	*Action goal 4: break isolation and ensure effective access to protection*
Types of action	Inclusive narratives Address root causes Enhance accountability Strengthen data management Consolidate group efforts Promote security	Engage businesses Promote ratification and implementation of Aarhus and Escazú Agreements Strengthen UN processes Enhance civil society spaces Promote the right to a healthy environment Implement international rights at the national level Reinforce rights, enabling civic society spaces and accountability – planned Roadmap activities	Promote new connections Strengthen regional and international connections Support new forms of engagement National human rights institutions Environmental organizations and networks Build effective platforms	Strengthen international to local connections Enhance action on the ground Toolkits that make sense to defenders Representative networks

(See Annex 1 for a more detailed overview.)

notably by putting in place rights-based building blocks of safe and enabling civic spaces. Recognizing the constraints of international mechanisms, it is not enough to simply couch plans in the language of Human Rights Council Resolution 40/11, nor for States or organizations to reiterate normative commitments to such standards. Indeed, the emerging lessons from the Roadmap process suggest the persistent gap between international commitments and realities on the ground, even if a growing body of domestic measures are paving the way for more effective responses.

It is one thing to wish to protect environmental defenders; it is quite something else to achieve it in highly varying country contexts. On the one hand, as the protagonists and problems vary from State-driven red-tagging and complicit public support to illegal economies, weak domestic measures and generalized contexts of violence, the roles of international cooperation efforts will differ considerably. On the other hand, the ability of human rights and environmental organizations to cooperate and act consistently is also at times constrained by the tendency to work in silos and with high levels of fragmentation.

While the constraints of both national and international cooperation are all too clear, the Geneva Roadmap process demonstrates the need and potential of connecting solutions across sectoral divides and individual organizations and of connecting the wider community of defender networks and support organizations. There are real, concrete steps to be taken to break the isolation and to ensure effective protection measures, just as both the human rights and environmental communities can enhance efforts to create safe and enabling environmental civic spaces. The simple action goal framework of the Roadmap has proven to offer straightforward language and goals that organizations could consider in their respective efforts.

Throughout 2020, dialogue processes have taken place with national and global defender networks as well as within wider environmental, peace and human rights networks. While the COVID pandemic largely disrupted the initial calendar of joint activities, it also demonstrated the urgency of responding with collective solution building that challenges the deep-running divides between environment and human rights communities. In the case of IUCN networks, for example, a virtual dialogue was held entitled "Environmental human rights defenders in the pandemic: the Geneva Roadmap & strengthening IUCN action".

As defenders' testimonies remind us, the problems have not gone away. The core goal of bridging initiatives and enhancing cooperation remains as important as ever. Yet, at the same time, the volunteer-based model of cooperation across sectors, networks and individual organizations is also stretched to its limits. Rome was not built in a day, and profound challenges remain in terms of consolidating a connected solution space that genuinely unites international and national efforts for the effective protection of environmental human rights defenders.

References

Bille Larsen, P., & Balsiger, J. (forthcoming). Environmental defenders, human rights and the growing role of the IUCN policy: Retired, red-tagged or red listed? *Policy Matters.*

Bille Larsen, P., Le Billon, P., Menton, M., Aylwin, J., Balsiger, J., Boyd, D., et al. (2020). Understanding and responding to the environmental human rights defenders crisis: The case for conservation action. *Conservation Letters*. https://doi.org/10.1111/conl.12777

Boyd, D. R. (2018). Evaluating forty years of experience in implementing the right to a healthy environment. *The Human Right to a Healthy Environment, 99*(5), 17.

Forst, M. (2020). *Defending and protecting the defenders*. Paris.

Knox, J. (2017). *Environmental human rights defenders*. Geneva: Universal Rights Group.

For further information

Further information about the Resolution 40/11 Geneva Roadmap on can be found on the following websites:

1 https://environment-rights.org/geneva-roadmap/
2 www.unige.ch/gedt/recherches/projets/public-roundtable-supporting-environmental-defenders/

An English and Spanish version of a film with testimonies from defenders and other roadmap participants can be accessed through the following:

English version: https://vimeo.com/475116723
Spanish version: https://vimeo.com/488742529

Annex – Geneva Roadmap action goals

The Roadmap is made up of four action goals to bridge efforts, guide reflection and stimulate synergies. These are the following:

1. **Reverse the tide** of marginalization and attacks against environmental actors
2. **Reinforce environmental rights**, enabling civic spaces and accountability
3. **Bridge initiatives** and enhance cooperation
4. **Break isolation** and ensure effective access to protection

None are set in stone, yet they seek to capture key challenges that have been identified and to flesh out avenues for action and coordination. For each action goal the group brainstormed about **possible fields of action** as well as laying out already **planned next steps.** Given that the calendar of planned activities has been affected by the COVID pandemic, already planned next steps are not included here. The Roadmap comes with a call to international organizations, governments and civil society organizations to join forces and **support Roadmap activities**. Suggested approaches and activities are listed here in their "raw" form with only slight editing to facilitate comprehension.

Action goal 1: Reverse the tide of marginalization and attacks against environmental defenders

How is the negative spiral of marginalization and attacks against defenders broken? To reverse the tide of violence, the dialogue identified the following series of issues and activities ranging from inclusive narratives and addressing root causes to the enhancement accountability.

Inclusive narratives

- Invest in reshaping a value-based narrative
- Show value of EHRDs for achieving the SDGs
- Use Stockholm +50 in 2022 to put environment as a key geopolitical issue
- Use the momentum of the Zero Tolerance Initiative to highlight the collective nature of EHRDs
- Display positive images of the actors across media
- Control and fight hate speech
- Promote human rights in the society
- Challenge the "development paradigm"
- Work with governments and other actors to celebrate the effort of defenders through public recognition awards and prizes and connect all of these champions in a global network

Address the root causes

- Weak rule of law
- Lack of civic spaces

- Connect to trade and investment
- Lack of consultation and FPIC
- Insecure land tenure

Enhance accountability

- Monitor cases of impunity
- Expose attacks by companies, industries, investors and embassies
- Counter and denounce criminalization
- Hold accountable the crimes through effective justice systems
- Help spread awareness of crimes against human rights defenders
- Use communication functions (Special Rapporteurs and United Nations Working Group business and human rights directed at companies)
- Engage in strategic litigation at UN treaty bodies
- Use the HRC EHRD resolution as a tool to rate States' efforts and capacity
- Effective mechanisms like a certification for organizations and individuals supporting environmental defenders
- Promote accountability of businesses and investors, including the promotion of full and effective human rights due diligence processes
- Engage with companies and investors to commit to zero tolerance of attacks against EHRDs

Strengthen data management

- Collect data on the threats and attacks and get these out in the open to increase public pressure
- Define how a defender can get protection and escape social isolation/criminalization
- Strengthen the voice of science

Consolidate group efforts

- Organize groups of support for defenders
- Support from government officials
- Build across actors and sectors
- Fund the frontline and local organizations

Promote security

- Train EHRD on security management and the monitoring and documentation of human rights violations
- Offer advocacy support
- Train in digital security and evidence collection
- Ask defenders what they need and do it
- Develop a knowledge and learning network on the topic of collective community-led protection

Action goal 2: Reinforce environmental rights, enabling civic spaces and accountability

How are enabling and safe conditions created for environmental human rights defenders? Protecting environmental human rights defenders requires respecting human rights, enabling civic spaces and reinforcing mechanisms of accountability.

Engage businesses

- Mobilize and work with responsible businesses and investors to take steps towards a Zero Tolerance approach to violence in the supply chain
- Push embassies to encourage companies to respect EHRD
- Create strategy to make environmental defenders part of an international focus concerning trade and investment
- Engage the UN Working Group on business and HRs on the topic of EHRDs
- National mandatory environmental and human rights due diligence obligations for companies
- Advocacy work on HRD and business engagement (e.g. EU)

Promote ratification and implementation of Aarhus and Escazú Agreements

- Raise awareness on the Agreements
- Promote Escazú ratification and implementation
- Encourage EHRD efforts in Aarhus Convention
- Adopt environmental rights at constitutional level
- Support lawyers defending defenders, track progress on accountability and case follow-up
- Support NGOs to reach operational stage of Escazú
- Develop similar regional instruments where absent (Pacific and Africa)

Strengthen UN processes

- Promote Roadmap action goals in UN processes
- Reinforce environmental rights
- Support defender-led litigation against transnational companies
- Promote events for large-scale advocacy and increase publicity
- Make use of HR mechanisms (e.g. Treaty Bodies and Universal Periodic Reviews)
- Create a UN code of conduct for the protection of environmental defenders
- Work with UNEP and Special Rapporteurs to promote implementation of environmental rights and advocate for universal recognition of this right

Enhance civil society spaces

- Support jurisdictional solution and support
- Extend funding to civil society projects and spaces
- Engage with new spaces and build understanding of EHRD issues within these
- Network with EHRD and grassroots-level CSOs to understand what is happening on the ground
- Allow space for defenders to create alternatives, not simply defend
- Encourage IUCN to set up standard for enabling civil society spaces

Promote the right to a healthy environment

- Promote the right to a healthy environment with everyone's dignity and wellbeing
- Stand for clean air, safe and non-toxic environments, clean water, healthy sustainability, health in biodiversity
- Strengthen tools like access to information, participation, access to justice and effective remedies

Implement international rights at the national level

- Include Indigenous rights, right to a healthy environment, law as a priority
- Protection of EHRD should be according to the reality of each country

Action goal 3: Bridge initiatives and enhance cooperation

How to counter the risk of multiple and fragmented efforts to support environmental human rights defenders?

Promote new connections

- Reinforce cooperation between Special Rapporteurs thematically and by country
- Promote widely the UN declaration on HRD and its definition of HRD
- Engage UN organizations to coordinate efforts on EHRD
- Consolidate, coordinate and create focal points across organizations with existing initiatives
- Reinforce cooperation between mechanisms

Strengthen regional and international connections

- Reinforce cooperation with regional organizations IACHR, ACPHR, COE, OSCE

- Connect across silos, seek to connect beyond the usual suspects
- International and regional conferences assembled by environmental defenders
- Promote and reinforce national or regional national networks of HRD
- Establish regional defenders and generate a globally shared platform
- Regional networks of HRD
- Create an international alliance of defenders

Support new forms of engagement

- Promote dialogue spaces
- Support involvement of more scientists and activists
- Decentralized working groups and dialogue spaces
- Engage with organizations from countries with similar cases and promote inclusiveness of communities
- Aim for diversity when working on human rights for diverse perspectives

National human rights institutions

- Include NHRIs in all discussions
- Have inclusive systems that involved defenders from a grassroots level

Environmental organizations and networks

- HR and environmental organizations engage with each other more
- HR community share lessons learned with environmental organizations
- Reach out to diverse environmental communities
- Support in joining Zero Tolerance and be a part of the coalition
- Create a comprehensive coalition initiative of all emerging and pre-existing groups to create an action plan with collective calendar
- Connect existing groups: URG, N1M, UNEP, Frontline, Global Witness, . . .
- Link national human rights institutions between local and national levels. Assist connecting between HRD and these institutions

Build effective platforms

- Have an open platform as a clearing house
- Roadmap action goals, bridge initiative and enhance cooperation- join this group with EHRD and find support to energize an open platform to share work and news
- Online services to better deal with existing problems

Action goal 4: Break isolation and ensure effective access to protection

How can the common isolation of defenders be broken and better access to protection be enhanced? The Geneva dialogues stressed strengthening local–international connections, deepening action on the ground, adapting toolkits and improving networks through more direct engagement with defenders themselves.

Strengthen international to local connections

- Involve environmental defenders in decision-making processes and have emergency funds and safe places for protection of EHRDs that face threats in their line of work
- Identify blind spots and defenders in isolation
- Identify training for key actors/NGOs/associations in different regions as focal points/bridges between defenders and support networks (access often depends on trust and relationships)
- UN special bodies to review situation of environmental defenders to highlight their concerns
- Bring UN agencies and embassies on board
- Push embassies and diplomatic missions to be more proactive and reach out to remote EHRDs

Enhance action on the ground

- Enhance actual protection measures, not only perspectives and networks
- Look at and support collective protection mechanisms and increase investments in protection
- Share successful strategies by conflict type
- Provide support funds for families of affected defenders
- Strengthen information, communication and awareness-raising among target communities
- Special education to police force about the rights of defenders
- Implement tools of communication with isolated groups of defenders

Create toolkits that make sense to defenders

- Make access easier to support and opportunities for defender networks
- Develop practical action toolkits, using existing kits and systems and translate into local languages and culturally appropriate
- Improve resources for community-led protection.
- Information sharing and training systems for EHRD
- All NGOs and UN mechanisms to translate, create audio format for their info and access
- Support paralegals and train up security specialists within communities
- Make toolkits that can self-disseminate/implement

Representative networks

- Build coalitions of HRDs including EHRD and connect to existing ones

- Work with global alliances of national human rights institutions to publicize and disseminate good practices at the national level to protect defenders
- Build capacity of environmental networks to recognize and respond to threats
- Reach across intersectional divides
- Activists who are working class are still less likely to report violations or see their complaints being responded to
- Create space for defenders to network with each other
- Facilitate peer-to-peer exchanges among local communities on risk and protection methodologies
- Support defender solidarity meetings
- Propose any international meeting is at least 50% defender

Index

Note: Numbers in **bold** indicate a table. Numbers in *italics* indicate a figure.

For Product Safety Concerns and Information please contact our EU
representative GPSR@taylorandfrancis.com
Taylor & Francis Verlag GmbH, Kaufingerstraße 24, 80331 München, Germany

www.ingramcontent.com/pod-product-compliance
Ingram Content Group UK Ltd.
Pitfield, Milton Keynes, MK11 3LW, UK
UKHW022317250926
475702UK00046B/907
* 9 7 8 0 3 6 7 6 4 9 7 0 8 *